Advances in Wastewater Treatment 2024

Advances in Wastewater Treatment 2024

Guest Editors

Marin Ugrina
Jelena Milojković

MDPI

Basel • Beijing • Wuhan • Barcelona • Belgrade • Novi Sad • Cluj • Manchester

Guest Editors
Marin Ugrina
Department of
Environmental Engineering
University of Split
Split
Croatia

Jelena Milojković
Institute for Technology of
Nuclear and Other Mineral
Raw Materials
Belgrade
Serbia

Editorial Office
MDPI AG
Grosspeteranlage 5
4052 Basel, Switzerland

This is a reprint of the Special Issue, published open access by the journal *Energies* (ISSN 1996-1073), freely accessible at: https://www.mdpi.com/journal/energies/special_issues/5T3IH28ODT.

For citation purposes, cite each article independently as indicated on the article page online and as indicated below:

Lastname, A.A.; Lastname, B.B. Article Title. *Journal Name* **Year**, *Volume Number*, Page Range.

ISBN 978-3-7258-6456-0 (Hbk)
ISBN 978-3-7258-6457-7 (PDF)
https://doi.org/10.3390/books978-3-7258-6457-7

Contents

About the Editors

Marin Ugrina

Marin Ugrina, Ph.D. in chemical engineering, has been a scientist and associate professor at the Department of Environmental Engineering, Faculty of Chemistry and Technology, University of Split, Croatia. His primary research interests are related to the tertiary treatment of wastewater loaded with heavy metals using natural and chemically modified zeolites and other low-cost sorbents. His recent research has focused on developing procedures for the chemical modification of natural zeolite for targeted applications, specifically mercury sorption. For this purpose, he is developing environmentally friendly modified sulfur-based natural zeolites for in situ and ex situ remediation of mercury-contaminated soil and water from the area of the Idrija mine in Slovenia. His latest research focused on examining the possibilities of solidification/stabilization of mercury-saturated zeolites in cement composites. He was a collaborator on several projects and the leader of two bilateral Croatian-Slovenian projects: "Natural modified sorbents as materials for remediation of mercury contaminated environment, (2020-2023)", and currently, "Application of modified natural sulfur-based zeolites for remediation of mercury-contaminated soil from the Idrija mine area in Slovenia, (2025–2027)." He has been a member of the scientific and organizational committees of several international conferences. He has attended several training courses at international institutions (Germany, Italy, Poland, Serbia) and actively participates in reviewing scientific papers and mentoring students. In addition, he is actively involved in musical performance, and since 2012 has held the position of titular organist and regens chori of the Cathedral of St. Lawrence in Trogir.

Jelena Milojković

Jelena Milojković, Ph.D., is a Senior Research Associate specializing in environmental pollution, biosorption, water and wastewater treatment, biochar, and the phytoremediation of uranium. She holds a Ph.D. in Environmental Engineering and a B.Sc. in Biochemical Engineering and Biotechnology from the Faculty of Technology and Metallurgy, University of Belgrade. Since 2008, she has been affiliated with the Institute for Technology of Nuclear and other Mineral Raw Materials (ITNMS) Belgrade, Serbia. Jelena Milojković is actively engaged in several research projects addressing environmental sustainability and wastewater treatment. She is the coordinator of Work Package 7 in the project (2023–2026), "From Waste to Food and Soil Enrichment - minimizing waste by applying circular economy in fruits/vegetables processing industry -WasteBridge" (Project No. 7439). She was also the project leader of two projects financed by Innovation Fund of Republic of Serbia: AWARENESS (Project No. 1138) (2022–2024) and PoC (Project No. 5099) (2020–2021).

Preface

Clean water is one of the most important, and challenging to implement, United Nations Sustainable Development Goals, SDG 6, which emphasizes access to water and sanitation for all. Therefore, systematic monitoring and preservation of natural water resources is necessary, as well as appropriate wastewater treatment and management strategies. Namely, the primary goal of any wastewater treatment strategy is to meet national effluent standards in order to protect ecosystems and human health. However, the conventional linear model of wastewater treatment (use–treat–discharge) must be reconsidered, since wastewater represents a valuable source of renewable chemical and thermal energy. On the other hand, during above-ground wastewater treatment, gaseous decomposition products are often emitted into the atmosphere. Furthermore, biological aerobic treatment results in the permanent loss of chemically stored energy, thereby eliminating potential energy recovery opportunities. Therefore, the concept of controlled anaerobic digestion in accordance with the circular economy is desirable in order to use gaseous decomposition products as an energy source. The wastewater–energy nexus has become increasingly important in recent times, highlighting the potential for simultaneous treatment and energy production. This approach represents a clear pathway to clean energy production, which is in line with SDG 7. In this context, biogas production through anaerobic digestion and the installation of a hydropower plant is one of the possible solutions for reducing energy needs from an external source. Through this model, wastewater treatment plants can achieve self-sustainability. Moreover, the integration of the aforementioned technologies for energy generation in wastewater treatment plants represents a logical and environmentally friendly strategy that contributes to increasing energy efficiency, sustainability and reducing dependence on electricity from the external grid. This Special Issue Reprint, "Advances in Wastewater Treatment 2024" investigates innovative wastewater management technologies and strategies aimed at exploiting renewable energy sources contained in wastewater, for the purpose of energy-efficient wastewater treatment plant management practices. Ultimately, in addition to the dedicated authors, we express our sincere gratitude to the MDPI staff, the editorial team of *Energies*, the assistant editors and Section Managing Editor of this Special Issue, and the expert reviewers for their valuable contributions and support.

Marin Ugrina and Jelena Milojković
Guest Editors

energies

Editorial

Advances in Wastewater Treatment, 2024

Marin Ugrina [1,*] and Jelena Milojković [2]

1 Department of Environmental Engineering, Faculty of Chemistry and Technology, University of Split, Ruđera Boškovića 35, 21000 Split, Croatia
2 Institute for Technology of Nuclear and Other Mineral Raw Materials, 86 Franchet d'Esperey St., 11000 Belgrade, Serbia; j.milojkovic@itnms.ac.rs
* Correspondence: mugrin@ktf-split.hr; Tel.: +385-21-329-451

Water is undoubtedly the most important and invaluable natural resource that humans utilize. The main source of water for human consumption and various industrial uses on a global scale is groundwater. Different types of organic and inorganic contaminants from natural and, more often, anthropogenic sources, leach into all water systems, especially groundwater. A lack of water for human consumption is also a result of inadequate wastewater treatment, rising freshwater usage, and climate change. Therefore, it is necessary to adequately and conscientiously manage wastewater to ensure safe drinking water and sanitation for all, which is in line with Sustainable Development Goal 6 on water and sanitation (SDG 6) [1].

The systematic treatment of wastewater is carried out through three successive stages: primary (physical–chemical), secondary (biological), and tertiary (advanced physical–chemical) treatment. The primary goal of any wastewater treatment plant is to meet the legal regulations for the discharge of treated wastewater into water bodies. Currently, it is very challenging to achieve the required quality of purified wastewater, especially due to the appearance of emerging environmental pollutants. For this reason, advanced knowledge is necessary and desirable in developing new advanced technologies or improving existing technologies at all levels of wastewater treatment. However, even in the 21st century, almost all existing treatment methods for managing wastewater occur linearly (water is used, treated, and then discharged). This approach has several disadvantages that might affect sustainability and the environment. Some of the deficiencies include a loss of nutrients, pollutant discharge, non-renewable energy use, unsustainable use of finite water resources, waste generation, valuable resources not being fully recovered, limited climate resilience, high operational costs, public health concerns, difficulties in meeting increasingly stringent environmental regulations and standards, etc. Hence, the goal of any sustainable management, including wastewater management, should be circular sustainability. In this context, wastewater should not be viewed exclusively as a waste stream, but as a valuable resource since it contains a significant amount of stored chemical and heat energy, as well as nutrients [2]. Therefore, it is substantial to manage wastewater holistically, i.e., wastewater treatment, obtaining energy and resources at the same time. The reuse of purified wastewater with energy recovery is imperative for sustainable development, which is in agreement with two of the seventeen Sustainable Development Goals: SDG 6 (clean water and sanitation) and SDG 7 (affordable and clean energy) [1]. Thus, purified wastewater can be used for agricultural purposes, industry, etc., as well as for the supply of hydroelectric power plants to obtain electricity [3]. Recent findings indicate that chemical energy contained in wastewater is destroyed by aerobic biological oxidation [4]. Accordingly, the chemical energy contained in wastewater can be recovered in the form of biogas by anaerobic digestion; hydrogen can be obtained by electrohydrolysis, as can nitrogen and phosphorus. The heat contained in wastewater can be recovered using heat pumps and heat exchangers; alternatively, it can be obtained through sludge incineration [2–4]. The mentioned benefits as a result of wastewater treatment point to the water–energy nexus.

Therefore, the future perspective of wastewater treatment plants is ecological sustainability, i.e., the adaptation and redesign of existing plants and the design of new integrated plants for the simultaneous treatment and recovery of energy and resources.

Finally, it should be pointed out that water is the most essential resource contained in wastewater because it cannot be obtained from alternative sources, unlike energy. Ultimately, the production of cleaner water and energy through wastewater recycling is an important pathway to achieving the circular economy and sustainable development. Therefore, in dealing with environmental, economic, and social concerns, it demonstrates a comprehensive and responsible approach to water resource management.

"Advances in Wastewater Treatment, 2023" describes the progress, innovations, and improvements in the technology and processes utilized in wastewater treatment. All knowledge about the inter-relationship and synergy between wastewater and energy is welcome in this Special Issue, exploring the development of solutions for exploiting the hidden potential of wastewater.

Funding: This study received no external funding.

Conflicts of Interest: The authors declare no conflicts of interest.

References

1. United Nations. *The UN Sustainable Development Goals*; United Nations: New York, NY, USA, 2015; Available online: http://www.un.org/sustainabledevelopment/summit/ (accessed on 18 January 2024).
2. Kehrein, P.; van Loosdrecht, M.C.; Osseweijer, P.; Garfi, M.; Dewulf, J.; Posada, J. A critical review of resource recovery from municipal wastewater treatment plants—Market supply potentials, technologies and bottlenecks. *Environ. Sci. Water Res. Technol.* **2020**, *6*, 877. [CrossRef]
3. Zarei, M. Wastewater resources management for energy recovery from circular economy perspective. *Water-Energy Nexus* **2020**, *3*, 170–185. [CrossRef]
4. Hao, X.; Li, J.; van Loosdrecht, M.C.M.; Jiang, H.; Liu, R. Energy recovery from wastewater: Heat over organics. *Water Res.* **2019**, *161*, 74–77. [CrossRef] [PubMed]

energies

MDPI

Article

Treatment of Wastewater from Soil Washing with Soluble Humic Substances Using Biochars and Activated Carbon

Mariusz Z. Gusiatin [1,*], Sylwia Pasieczna-Patkowska [2], Magdaléna Bálintová [3] and Marcin Kuśmierz [4]

1 Department of Environmental Biotechnology, Faculty of Geoengineering, University of Warmia and Mazury in Olsztyn, Słoneczna St. 45G, 10-709 Olsztyn, Poland
2 Department of Chemical Technology, Institute of Chemical Sciences, Faculty of Chemistry, Maria Curie-Skłodowska University, Maria Curie-Skłodowska 3, 20-031 Lublin, Poland; sylwia.pasieczna-patkowska@mail.umcs.pl
3 Institute of Environmental Engineering, Faculty of Civil Engineering, Technical University of Kosice, 04200 Kosice, Slovakia; magdalena.balintova@tuke.sk
4 Analytical Laboratory, Institute of Chemical Sciences, Faculty of Chemistry, Maria Curie-Skłodowska University, Maria Curie-Skłodowska 3, 20-031 Lublin, Poland; marcin.kusmierz@mail.umcs.pl
* Correspondence: mariusz.gusiatin@uwm.edu.pl

Abstract: Energy can be obtained by pyrolysis of organic wastes, and the solid residue of pyrolysis (biochar) can be used as an adsorbent for the treatment of various types of wastewater. Although soil washing can effectively remediate metal-contaminated soils, it can generate significant amounts of soil washing wastewater (SWW). This study investigated the effectiveness of using activated carbon and various biochars to treat SWW from the remediation of soil heavily contaminated with cadmium (Cd), copper (Cu), nickel (Ni), lead (Pb), and zinc (Zn) with soluble humic substances (SHS) from municipal sewage sludge. Willow biochar (BW), plant biomass biochar (BPB), coconut shell biochar (BCH), and Norit SX2 activated carbon (ACN) were tested at different dosages (12.5–100 g/L) and adsorption times (30–1440 min) for the treatment of SWW. At 100 g/L dosage, biochar removed Cd, Cu, Ni, Pb, and Zn with 56–83%, 32–41%, 18–42%, 75–83%, and 44–83% efficiency, respectively, while ACN removed them with 87–95% efficiency. Only BW and ACN removed soluble organics with efficiencies of 49% and 94%, respectively, at the highest dosage. Adsorption of metals and soluble organics was mainly controlled by physisorption and chemisorption. Diffusion of metals and soluble organics into the different pore sizes was not the most important rate-limiting step. ACN and BW had better structural properties and treated SWW most effectively. BPB and BCH removed metals but not soluble organics, which could be beneficial for SHS recycling.

Keywords: soil washing; treatment; adsorption; pyrolysis; biochar; heavy metals; humic substances; soluble organics

1. Introduction

Soil washing is one of the most widely used and studied off-site techniques for remediating metal-contaminated soil [1]. Some conventional washing agents used in soil washing (e.g., acids, EDTA) can diminish the quality of the treated soil. Nowadays, there is a trend toward using new-generation washing agents in the form of naturally soluble organic substances to minimize harmful effects on the physicochemical properties of the washed soil and to make remediation more sustainable [2,3]. The main sources of these novel agents can be geochemical resources, unprocessed wastes (e.g., municipal sewage sludge, agricultural wastes), and processed wastes (e.g., residue after anaerobic digestion, compost) [4]. Utilizing waste as a source of washing agents for soil washing is consistent with the principles of a circular economy. In the previous studies, it was demonstrated that soluble humic substances (SHS) derived from municipal sewage sludge could serve as an alternative washing agent to synthetic EDTA. SHS from sewage sludge contains fulvic acids (FAs) and humic acids (HAs), and they possess great potential for remediating soils highly polluted with Cu, Pb, and Zn (removal efficiency of 60–90%), promote recovery of soil function, and have limited soil toxicity [5,6].

Treating soil with a given washing agent generates large amounts of soil washing wastewater (SWW) [7,8], also known as spent soil washing solution [1]. Real SWW have complex compositions. They contain soluble metals at different concentrations, residual spent washing solution, competing cations washed from the soil, and other coexisting soil contaminants [7,9–11]. The resulting physicochemical properties of SWW (e.g., pH, conductivity, contaminant concentrations) can vary substantially depending on the type of soil remediated, the type and concentration of contaminants, the type and properties of the washing solution, and the operating conditions used for soil washing.

For example, the SWW from soil treatment with $FeCl_3$ contained high concentrations of Zn (212.5 mg/L) and Cd (84.4 mg/L) [12]. For comparison, SWW from soil treatment with EDTA had an initial pH of 6.64 and contained metals at various concentrations: 258 mg/L of Pb, 290.1 mg/L of Zn, 5.9 mg/L of Cu, and 1.8 mg/L of Cd [8]. The SWW produced by soil washing with NaOH mixed with an amphoteric surfactant had an alkaline pH (13.1) and high concentrations of inorganic and organic pollutants: 14.5 mg As/L, 2.41 mg Cr/L, 8.22 mg Cu/L, 6.42 mg/L of pentachlorophenol, and 127 ng/L of polychlorinated dioxins and furans [13]. Due to its complex composition, a complete treatment of SWW can be challenging. Nevertheless, SWW should be treated before final disposal, or it can be regenerated and reused for subsequent soil treatments. Proper disposal of SWW is an essential part of a complete remediation process.

So far, experiments on the treatment of SWW have been conducted mainly with conventional washing agents such as chelating agents, i.e., EDTA, EDDS [8,14], HCl solutions [7,10], $FeCl_3$ solutions [12,15]; and plant biosurfactants, e.g., saponin, with the aim of recovering these agents and making the soil remediation process more cost-effective [16]. Adsorption, precipitation, advanced oxidation processes, and membrane technology can be used to treat SWW [16]. A combination of soil washing with selective adsorption of pollutants from SWW and final disposal of the adsorbent [17] may be an important alternative due to its high efficiency, lower cost, and ease of operation compared to other methods [18]. Recently, adsorbents tested for treating SWW from the washing of metal-contaminated soils have mainly included agricultural wastes, e.g., garlic peels [19] and coffee grounds [7], metal oxides in modified nanoparticles [20], or graphene oxide [10].

The growing demand for sustainable and renewable energy sources has led to increased interest in converting biomass into valuable bioproducts through pyrolysis. Biochar, as a solid carbon bioproduct, is a well-known adsorbent for removing complex inorganic and organic contaminants. Current research aims to replace activated carbon with biochar to reduce the cost of treatment processes and take advantage of the specific properties of biochar. The properties of biochar depend on the type of feedstock, the technology used to pre-treat the feedstock, the thermal process, and the post-treatment of the biochar. Due to its porous properties, high surface area, and functional groups, such as phenol, hydroxyl, and carboxyl groups, biochar is highly adsorbent [21]. Biochar is an attractive adsorbent also due to its low cost of production, economic advantages, possibility of conversion to biochar, and wide availability of raw materials [22,23].

The biomass for producing biochar comes from a wide range of organic materials, such as plant residues, sewage sludge, manure, agro-industrial biomass, etc. Thus, biochar production enables improvements in waste management and sustainable resource utilization. Biochar is becoming increasingly important as a solution for the removal of pollutants in industry and agriculture to improve environmental quality. It has great potential for treating various types of wastewater, such as industrial, municipal, and agricultural waters, as well as stormwater [24]. Its ability to simultaneously remove inorganic and organic pollutants makes it a potentially suitable adsorbent for treating SWW generated by washing soils contaminated with heavy metals. Biochar is a popular additive for immobilizing metals in contaminated soils, but its application for treating SWW is a new approach. There is only one paper in the literature on applying pristine and modified biochar for treatment of $FeCl_3$ SWW after washing soil contaminated with Cd, Cu, Pb, and Zn [12]. In general, there is a lack of information on the treatment of wastewater generated by soil washing with "green" washing agents [1], such as soluble humic substances (SHS) recovered from waste. Given the advantages of adsorption processes for treating wastewater and post-treatment waters and the possibility of using biochar to remove complex pollutants, the present study aimed to compare the efficiency of biochar produced from different feedstocks with that of commercial activated carbon (as a reference adsorbent) for treating SWW from soil remediation with SHS from municipal sewage sludge at various adsorbent dosages, and adsorption times. We hypothesized that willow biochar, mixed plant biomass biochar, and coconut shell biochar would have different effects when used to treat SWW from soil washing with SHS.

2. Materials and Methods

2.1. Soil Characterization and SHS Preparation

For remediation, soil spiked with high levels of Cd, Cu, Ni, and Pb and aged for two years (at room temperature and a moisture content of 65% of its water holding capacity) was used. The soil had the following characteristics: loamy texture, pH in KCl of 6.3, organic matter of 10.6%, 47 mg/kg (Cd), 1020 mg/kg (Cu), 498 mg/kg (Ni), 4094 mg/kg (Pb), and 2110 mg/kg (Zn).

Soluble humic substances (SHS) extracted from dried and ground municipal sewage sludge from the municipal wastewater treatment plant in Olsztyn (Poland) were used as washing agents for soil treatment. Previous research has shown that municipal sewage sludge is an important source of SHS [6]. The full procedure for SHS extraction has been described in [25]. Briefly, prior to SHS extraction, the sewage sludge was pretreated with distilled water to remove non-humus substances (e.g., sugars and proteins) and defatted with a mixture of chloroform:methanol to remove fats, waxes, and bitumen in a MarsXpress microwave oven (CEM, Corporation, Matthews, NC, USA) at 65 °C [16]. The extraction of SHS was conducted with 0.1 M NaOH for 24 h. Then, the suspension was centrifuged and filtered with a 0.45-µm filter. The concentration of the obtained SHS was 3654 mg TOC/L. For soil washing, the SHS was diluted with water, and the pH was adjusted with 1 M HNO_3.

2.2. Soil Washing with SHS

Batch soil washing was performed at a ratio of soil to SHS of 1:40 (*w*/*v*) in polyethylene tubes for 24 h. Samples were shaken using an Intelli-Mixer RM-2L. After washing, the supernatants were separated from the soil by centrifugation and filtration through a 0.45 µm filter. The solution from soil washing is referred to as soil washing wastewater (SWW). The characteristics of the SHS and SWW are compared in Table 1.

Table 1. Characteristics of SHS and SWW solutions (mean ± SD, *n* = 3).

Property	Unit	SHS	SWW
pH	-	4.0 ± 0.1	4.9 ± 0.1
Electrical conductivity	mS/cm	36.6 ± 4.5	64.9 ± 1.5
Total organic carbon	mg/L	2200.0 ± 20.6	1934.0 ± 11.6
Total inorganic carbon		6.5 ± 0.6	6.4 ± 0.2
Calcium (Ca)		15.6 ± 2.6	127.6 ± 6.5
Magnesium (Mg)		15.1 ± 1.9	29.9 ± 3.2
Sodium (Na)		1025.5 ± 8.9	1022.9 ± 10.6
Cadmium (Cd)		0.1 ± 0.02	0.5 ± 0.08
Copper (Cu)		6.1 ± 0.4	10.3 ± 1.3
Nickel (Ni)		0.1 ± 0.03	3.1 ± 0.6
Lead (Pb)		0.05 ± 0.1	7.8 ± 0.02
Zinc (Zn)		3.0 ± 0.3	14.2 ± 0.9

2.3. Types of Adsorbents and Their Characterization

Three types of biochar were made from pyrolysis at 650 °C (pyrolysis time 15 min, heating rate 3 °C/s) of willow biomass (BW); a mixture of plant biomass, including willow, alder, and oak (BPB); and coconut husks (BCH) (Fluid S.A., Sędziszów, Poland) and then used to treat SWW via batch adsorption. As a reference adsorbent, commercially available Norit SX2 activated carbon (ACN) was purchased from Chempur Company (Piekary Śląskie, Poland) (Figure 1).

Figure 1. Adsorbents used for treatment of SWW: (**a**) BW, (**b**) BPB, (**c**) BCH, and (**d**) ACN.

For the experiment, the biochars were crushed and sieved through a 1 mm sieve. The ACN was used in the original form provided by the manufacturer. The adsorbents were characterized in terms of their textural properties (surface area, pore volume, and pore size); content of moisture, volatile matter, ash, fixed carbon, and elemental composition; pH; electrical conductivity (EC); total heavy metals (Cd, Cu, Fe, Ni, Pb, and Zn); and macroelement concentrations (Ca, Mg, and Na). In addition, a qualitative analysis of surface oxygen-containing functional groups and a mineralogical analysis were performed.

2.4. Treatment of SWW

The SWW was treated with the biochars and ACN under batch laboratory conditions. The effectiveness of SWW treatment with the tested sorbents was determined depending on:

- The adsorbent dosage (12.5; 25; 50; and 100 g/L): in this experiment, 20 mL of SWW was shaken at different adsorbent dosages for 24 h at 90 rpm; then samples were centrifuged at 8000 rpm for 10 min and filtered through a 0.45 μm filter;
- Adsorption time (30, 60, 120, 180, and 1440 min): this experiment was performed with an adsorbent dosage of 100 g/L; the samples were prepared in the same way as in the previous experiments.

In the treated SWW, the following characteristics were determined: pH, conductivity, and concentrations of total organic carbon (TOC), heavy metals (Cd, Cu, Ni, Pb, and Zn), and macroelements (Ca, Mg, and Na).

2.5. Analytical Methods

A pH meter (HI 221, Hanna Instruments, Woonsocket, RI, USA) and a conductivity meter (HI 8733, Hanna Instruments, Woonsocket, RI, USA) were used to measure both pH and electrical conductivity (EC) of distilled water extracts (1:10 w/v). To determine moisture content, volatile matter (VM) content, ash content, and fixed carbon, proximate analyses were performed on the sample by drying it at 105 °C for 18 h, 950 °C for 10 min, and 750 °C for 6 h. Using 1–3 mg of the finely ground and sieved sample, the elemental composition of the adsorbents (C, H, and N) was analyzed using a FLASH 2000 elemental analyzer (Thermo Scientific, MA, USA). The oxygen content was calculated from the difference, including C, H, N, and ash. The H/C and O/C molar ratios of the adsorbents were calculated on the basis of their elemental compositions. The pore size distribution was calculated using the Barret–Joyner–Halenda method (Micrometrics ASAP

2420 M, Norcross, GA, USA) based on the desorption plot of the N2 adsorption-desorption isotherm. The Brunauer-Emmet-Teller (BET) specific surface area was calculated by fitting the BET equation to the linear part of the BET plot. Fourier transform infrared spectra with attenuated total reflectance (FT-IR/ATR) were recorded in the 3600–600 cm^{-1} range with 4 cm^{-1} resolution at room temperature (Nicolet 6700 spectrometer and Meridian Diamond ATR accessory; Harrick, NY, USA). Samples were applied directly to the diamond crystal prism and brought into intimate contact with the diamond surface by a pressure tower. The interferograms of each spectrum from 512 scans were averaged. ATR spectra were recorded using dry potassium bromide (48 h, 105 °C) as reference material. All ATR spectra were corrected for carbon dioxide and water vapor. No smoothing algorithms were applied. At least three replicates were performed for each spectral measurement. Mineralogical XRD analysis of the adsorbents was performed using an X'Pert Pro diffractometer with an X'Celerator Scientific detector (Malvern Panalytical, Malvern, UK) at an angular range of 7–120° (2Theta). X-ray photoelectron spectroscopy (XPS) was performed using the multi-chamber UHV system (PREVAC, Rogów, Poland). The hemispherical Scienta R4000 electron analyzer was used to obtain the spectra. A complementary piece of equipment was the Scienta SAX-100 X-ray source (Al K, 1486.6 eV, 0.8 eV band) coupled with the XM 650 X-ray Monochromator (0.2 eV band). The pass energy of the analyzer was set to 50 eV for regions (high resolution spectra): C1s, O1s, and N1s with a 50 meV step and 200 eV for survey spectra (with a 750 meV step). The analytical chamber had a base pressure of 2×10^{-9} mbar. It did not rise above 3×10^{-8} mbar during the spectrum collection. All the spectra processing and fitting were performed using CasaXPS software v. 2.3.25. For quantitative analyses, theoretical cross-sections were used. Envelopes of C1s and O1s spectra were fitted using asymmetric peaks for aromatic carbon components (line shape: LF(1.32,1.92,20,30,1)) and Gaussian–Lorenzian profiles (70% Gaussian, 30% Lorenzian) after applying Shirley backgrounds.

The concentration of SHS (as TOC) in the samples before and after the adsorption experiments was measured using a Shimadzu Liquid TOC-VCSN analyzer. To measure the total metal (Cd, Cu, Fe, Ni, Pb, and Zn) and macroelement concentrations (Ca, Mg, and Na), the dried adsorbent samples were digested in a MarsXpress microwave oven (CEM, Corporation, Matthews, NC, USA) with 6 mL of HNO_3 acid, 2 mL of H_2O_2, and 0.4 mL of HF acid. The samples were heated to 190 °C in 15 min, then digested at 190 °C for 20 min [26]. After cooling, the samples were filtered through Whatman 42 filters into volumetric flasks. The concentrations of heavy metals in the SHS, the SWW both before and after adsorption, and the digested adsorbents and soil were measured using a flame atomic absorption spectrometer (FAAS) (Varian, AA28OFS, Mulgrave, Australia). Metals in soils were fractionated into exchangeable and acid-soluble (F1), reducible (F2), oxidizable (F3), and residual (F4) fractions [27]. The accuracy of metal analysis by FAAS was validated by the analysis of CRM 142 R reference material.

2.6. Calculation and Statistical Methods

The adsorption capacities (q) and removal efficiencies (E) for heavy metals and soluble organics at different adsorption conditions were calculated using Equations (1) and (2):

$$q = \frac{(C_0 - C_e)}{m} \times V \quad (1)$$

$$E = \frac{(C_0 - C_e)}{C_0} \times 100\% \quad (2)$$

where C_0 (mg/L) and C_e (mg/L) are the concentrations of heavy metals or soluble organics (as TOC) in SWW before and after adsorption, respectively; m (g) is the mass of the adsorbent; and V (L) is the volume of SWW.

Pseudo-first-order, pseudo-second-order, and intraparticle diffusion models were employed to describe the adsorption kinetics of metals and soluble organics as given in Equations (3)–(5), respectively:

$$q_t = q_e\left(1 - e^{-k_1 t}\right) \tag{3}$$

$$q_t = \frac{k_2 q_e^2 t}{(1 + k_2 q_e t)} \tag{4}$$

$$q_t = k_i t^{0.5} + C_i \tag{5}$$

where q_e (mg/g) is the capacity of the adsorbent at a given time and equilibrium; k_1 (1/min) and k_2 (g/(mg·min)) are the rate constants of pseudo-first-order and pseudo-second-order models', respectively; k_i (mg/g·min$^{0.5}$) is the rate constant of the intraparticle diffusion model; and C_i (mg/g) is the adsorption constant of the intraparticle diffusion model, which describes the effect of boundary layer thickness on adsorption.

The validity of the kinetic models was evaluated by the correlation coefficient (R^2) (Statistica 13.3). Adsorption experiments were performed in triplicate, and the mean and standard deviation are reported. A one-way analysis of variance (ANOVA) was performed to examine differences between treatments of SWW with BW, BPB, BCH, and ACN. Tukey's HSD test was performed to further analyze the significant differences identified by ANOVA (Statistica 13.3).

3. Results and Discussion

3.1. Metal Removal from Soil

Figure A1 (Appendix A) shows the distribution of heavy metals in the contaminated soil and the removal of heavy metals from the soil with the SHS in the single-stage soil washing process. In the spiked soil, aged for two years, the heavy metals occurred in varying total concentrations, which is typical for real contaminated soils. The removal of individual metals from heavy contaminated soils with SHS was differentiated, which was related to their total content, their distribution in the soil, and the affinity of SHS for individual metals. The heavy metal removal efficiency decreased in this order: Cd (44.3%) > Cu (40.4%) > Zn (27.0%) > Ni (25.2%) > Pb (7.6%). The highest removal was obtained for Cd, whose total concentration was the lowest and mobility (the proportion in the F1 fraction) the highest. Cd, such as Zn, is characterized by high mobility in contaminated soils, which increases under acidic conditions [25].

Although Zn was also mobile in soil, its removal efficiency was lower than that of Cd since the total concentration of Zn was about 45 times higher than that of Cd. Although Ni was more mobile than Cu (152.7 mg/kg of Ni and 126.3 mg/kg of Cu in the F1 fraction), the removal efficiency was higher for Cu than for Ni. Many studies confirm that SHS from various sources is effective in removing Cu due to the affinity of the carboxyl and phenolic groups of SHS to Cu [5,28,29]. In addition, the efficiency of metal removal through soil washing with SHS strongly depends on soil washing conditions [28]. The SHS can also be effective for Pb removal [29]. Under the soil washing conditions used in the present study (SHS concentration of 2 g/L, washing time of 24 h, single-stage washing), the removal efficiency of Pb was below 10%. The previous study, with the use of SHS extracted from compost, revealed that the efficiency of metal removal with SHS at a concentration of 2.1 g/L and pH 4 was comparable to those values obtained for SHS from sewage sludge: 44.1% (0.52 mg/L) for Cd, 22.1% (5.64 mg/L) for Cu, 17% (2.12 mg/L) for Ni, 12.9% (13.2 mg/L) for Pb, and 11% (5.80 mg/L) for Zn [25]. A considerable improvement in metal removal was observed after triple soil washing with SHS. Because the aim of this study was not the optimization of soil washing with SHS but the treatment of wastewater generated by soil washing, wastewater generated in single-stage soil washing was used.

3.2. Adsorbent Characterization

The selected physicochemical properties of the tested adsorbents are listed in Table 2. The adsorbents differed in their structural properties, including surface area, pore volume, and pore size. Total pore volume and surface area were highest for ACN and BW, providing more active sites for pollutant adsorption. Based on the micropore volume and total pore volume, the proportion of micropores in the total pore volume decreased in this order: BW (76%) < ACN (24%) < BPB (9%) < BCH (4%). The average pore diameter of BW and ACN was characteristic of mesoporous materials (2–50 nm) [30], while the average pore diameter of BPB and BCH was typical for macroporous materials (>50 nm). Surface area and porosity are the most important physical properties affecting the metal sorption capacity of biochar [21]. Additionally, pore size is important for the sorption of metals. Adsorption of large sorbates, regardless of their charge or polarity, onto biochar with small pore sizes may be limited [31], because mesopores are needed for their adsorption [32]. The structural properties of biochar vary significantly depending on the type of feedstock and the temperature of pyrolysis. Lignin-rich biomass (e.g., bamboo and coconut shells) produces microporous-structured biochar, while cellulose-rich biomass (e.g., shells) yields predominantly macroporous-structured biochar [33]. Jindo et al. [34] showed that, with an increase in pyrolysis temperature from 400 to 800 °C, the surface area of oak tree biomass increased from 5.6 to 398.1 m^2/g, while that of risk husk biomass increased from 193.7 to 295.6 m^2/g. In general, the surface area of biochar can vary over a wide range, from less than 1.0 to about 500 m^2/g [35]. Some commercial activated carbons have a high surface area of nearly 1000 m^2/g, a total pore volume of 0.5 cm^3/g, and an average pore diameter of 2–3 nm [30,32].

Table 2. Physicochemical properties of adsorbents (n = 3, mean ± standard deviation).

Characteristic	Adsorbent			
	BW	BPB	BCH	ACN
Average pore diameter (nm)	5.3	127.0	128.6	7.2
BET surface area (m^2/g)	313.7	2.36	2.88	659.4
Micropore surface area (m^2/g)	272.9	0.52	0.29	347.3
Total pore volume (cm^3/g)	0.1132	0.0089	0.0012	0.6116
Micropore volume (cm^3/g) (%) *	0.0859 76	0.00077 9	0.00052 4	0.1456 24
Moisture (%)	5.2	4.1	4.6	4.7
Volatile matter (%)	26.3	27.2	35.1	13.7
Ash content (%)	8.8	7.0	7.8	8.2
Fixed carbon (%)	59.7	61.7	52.5	73.4
pH	10.35 ± 0.29	7.90 ± 0.1	7.65 ± 0.09	9.58 ± 0.18
EC (mS/cm)	2.9 ± 0.0	2.0 ± 0.0	0.06 ± 0.0	1.2 ± 0.0
Total C (%)	75.27 ± 0.45	77.21 ± 2.32	67.64 ± 3.42	78.33 ± 0.64
Total H (%)	1.94 ± 0.13	2.14 ± 0.18	3.00 ± 0.05	0.71 ± 0.15
Total N (%)	1.78 ± 0.24	0.50 ± 0.09	0.90 ± 0.14	1.63 ± 0.09
Total O (%)	12.2	13.1	20.7	11.1
H/C mole ratio	0.31	0.30	0.53	0.11
O/C mole ratio	0.12	0.13	0.23	0.11

* The % value is the ratio of the micropore volume to the total pore volume × 100.

The content of volatiles in the tested adsorbents ranged from 13.7% (ACN) to 35.1% (BCH). Such values are typical for biochar produced at high pyrolysis temperatures, as the loss of volatiles is greater. All adsorbents contained a small amount of ash (7.0–8.8%), which is consistent with the data in the available references. Wood-derived biochar has an ash content of less than 10%, while non-wood-derived biochar has a high ash content (>50%) [36]. Ash contains potassium (K), calcium (Ca), magnesium (Mg), carbonates, and heavy metal compounds that prevent the formation of aromatic structures, which contribute significantly to the fixed carbon content [37,38]. In the present study, the fixed carbon content in the adsorbents was high and ranged from 52.5% (BCH) to 73.4% (ACN), indicating that the adsorbents were highly stable.

All the adsorbents had an alkaline pH: BPB and BCH were slightly alkaline, while BW and ACN were more alkaline (pH of 10.35 and 9.58, respectively). The alkaline character of biochar is usually due to the presence of basic cations (mainly Ca, Mg, K, and Na), which are converted to oxides, hydroxides, and carbonates during pyrolysis, and to high electrical conductivity (EC), which indicates the presence of soluble salts. In the present study, BW and ACN had higher ash contents (8.8% and 8.2%, respectively) compared to the other biochars. BW had the highest EC, while ACN had the highest Na content (9.3 g/kg) (full data not shown). When biochar is produced at high temperatures, its content of basic cations (e.g., Na, Ca, and Mg) and carbonates increases while that of acidic functional groups decreases, contributing to a higher pH [21]. For example, when the temperature was increased from 200 to 800 °C, the basic functional groups on the surface of biochar prepared from Conocarpus waste increased from 0.15 to 3.55 mmol/g, while the acidic functional groups decreased from 4.17 to 0.22 mmol/g, which was reflected by the increase in the pH of the biochar from 7.37 to 12.4 [39]. Among the adsorbents tested in the present study, BW had the highest salinity (as EC), while BCH had the lowest salinity.

All the adsorbents had relatively high C contents (67.6–78.3%), which are typical of biochars produced at high temperatures and fulfilled the European Biochar Certificate (EBC) requirements for biochar quality (>50% d.m.). The contents of O, H, and N were progressively lower than those of C. The adsorbents had different H/C ratios, but all exhibited a high degree of carbonization and stability (H/C ratio < 0.7, according to [26]). As shown by the H/C ratio, ACN was the most condensed of the adsorbents, while BCH was the least condensed. These results were consistent with those regarding the volatile matter and fixed carbon contents (Table 2). During pyrolysis, the increase in C content due to polymerization, dehydration, and volatilization is inversely related to the change in H content. The H content is lost through dehydration, while the C content increases through condensation and graphitization, which affect the H/C ratio [38].

The adsorbents had relatively low molar ratios of O/C (0.11–0.23), which could be due to the fact that the biochars were prepared at high temperatures (650 °C), favoring a greater loss of H and O than of C [34]. The H/C and O/C ratios indicate that BW and ACN were more hydrophobic than BPB and BCH. These properties can be important for the treatment of SWW, which contains SHS, heavy metals, and some macroelements. Aromatic and aliphatic moieties in biochar have been found to be important for the sorption of organic compounds [40].

The XPS spectra of BW, BCH, BPB, and ACN are shown in Figure 2a, and their surface elemental compositions are listed in Table 3. The XPS confirmed that C and O were the main components of the adsorbents. BCH contained the least amount of surface C, 84.0%at. (atomic percentage), while the content in ACN exceeded 91.4%at. The amount of O was lowest in ACN (7.7%at.). In the other adsorbents, the content ranged from 11.5 to 14.5%at. Small amounts of N and Ca or Si were present in all adsorbents, which can be attributed to the composition of the raw biomass used as substrates. The individual spectra of the materials studied are shown in Figures A2–A5 (Appendix A). The elemental surface composition alone is not sufficient to explain the differences in the overall removal efficiency of the heavy metals. As mentioned earlier, aromatic/graphitic structures play an important role in the sorption of heavy metals due to the interactions of their π-electrons with the metals. To a lesser extent, carboxyl groups may also play a role. The results of fitting C1s spectra and O1s spectra are shown in Figures A6 and A7 and in Tables A1 and A2 (Appendix A). The concentration of surface aromatic structures was clearly highest in ACN, reaching 63.4%at. The percentage of such structures in BCH, BW, and BPB was 33.6, 29.1, and 40.3%at, respectively. These results are in agreement with the FTIR measurements (Figure 2c).

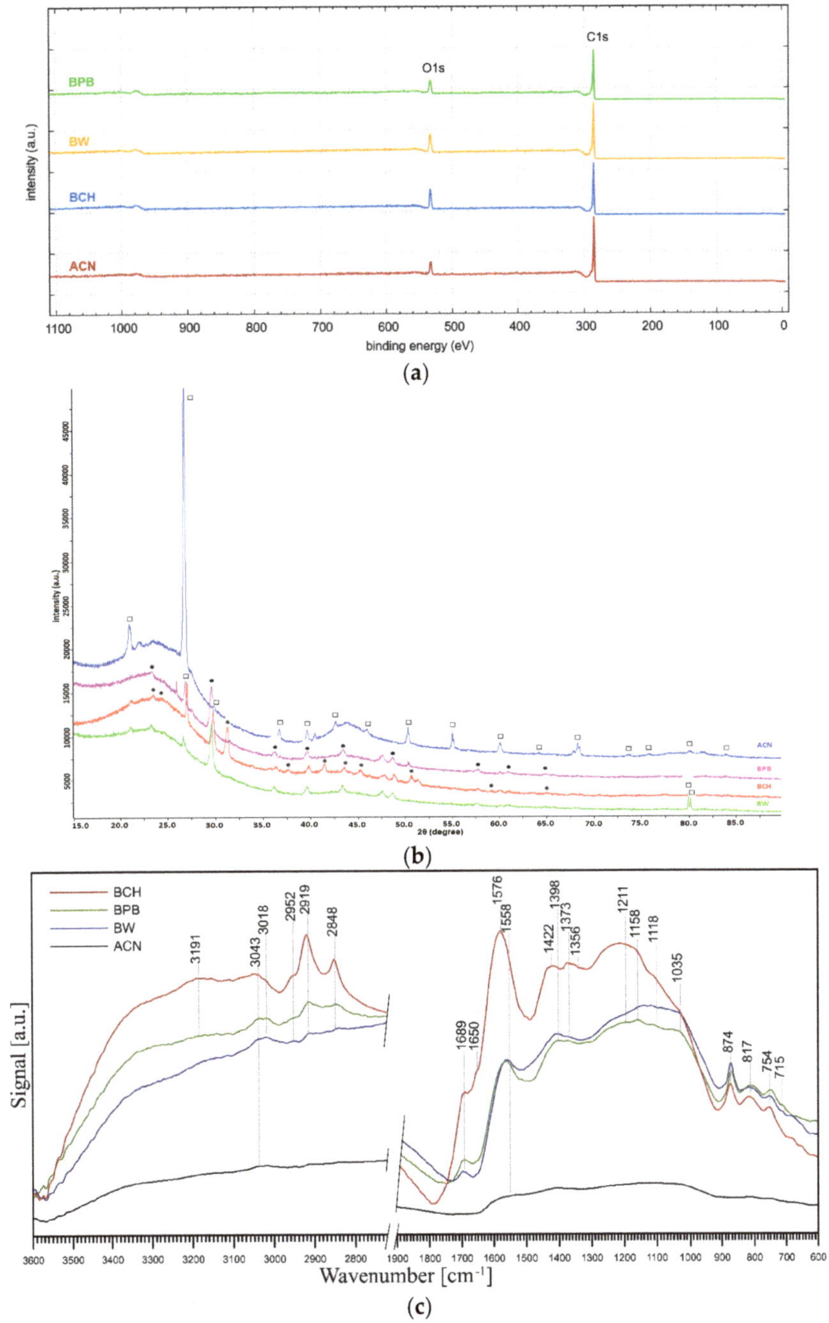

Figure 2. XPS spectra (**a**), XRD spectra (**b**), and FT-IR/ATR spectra (**c**) of studied adsorbents. The XRD spectra were matched with different colors: green for BW, pink for BPB, red for BCH, and blue for ACN (□ SiO_2 quartz; • $CaCO_3$ calcite; ✱ $CaMg(CO_3)_2$ dolomite).

Table 3. Surface elemental composition of adsorbents (atomic percents) determined by XPS.

Adsorbent	Element	Binding Energy (eV)	Conc. (%at.)
BW	C	285.0	85.2
	N	399.8	0.8
	O	533.3	13.5
	Si	103.5	0.4
BPB	C	285.0	87.2
	N	399.8	0.6
	O	532.5	11.5
	Ca	347.3	0.7
BCH	C	285.0	84.0
	N	399.0	1.2
	O	533.3	14.4
	Ca	348.0	0.5
ACN	C	285.0	91.4
	N	402.0	0.5
	O	533.3	7.7
	Si	102.8	0.4

Several principal bands can be observed in the FT-IR/ATR spectra of adsorbent samples (Figure 2a). The band at ~3200 cm^{-1} is responsible for the presence of O–H groups in carboxylic acids [41,42].

The band at about 3043 cm^{-1}, which is the most intensive in the FT-IR/ATR spectrum of the BCH sample, is attributed to =C–H stretching in aromatic structures. The presence of these structures can be confirmed by analyzing the spectral range below 1600 cm^{-1}, where relatively intense bands at 1576 cm^{-1} (BCH spectrum) and 1558 cm^{-1} (BW and BPB spectra) as well as multiple bands within the 880–700 cm^{-1} range (aromatic structures out-of-plane bending modes) [42,43] are present. The lowest intensity of the bands indicating the presence of aromatic structures is visible in the spectrum of the ACN sample. However, the band at ~1576 cm^{-1} may be attributed to both C=C stretching in aromatic structures and C=O stretching in ketones, quinones, and/or carboxylates [44]. Similarly, the band at ~3043 cm^{-1} may be the result of the overlapping of both stretching modes of aromatic and/or unsaturated aliphatic hydrocarbons [45]. The bands of aliphatic C–H stretching, mainly $-CH_3$ substituents of aromatic rings (2952 and 2848 cm^{-1}) and methylene bonds connecting aromatic rings ($-CH_2-$ at 2919 cm^{-1}), are present in all biochar spectra. These bands are the most intensive in the BCH spectrum.

The presence of C=O groups is confirmed by the shoulders at 1689 and 1650 cm^{-1} (aldehyde/ketone, carboxyl) [42]. Peaks between 1422 and 1035 cm^{-1} indicate the presence of C–O and O–H in alcoholic, phenolic and carboxylic groups (1356 cm^{-1}) or C–O–C stretching in ethers. Bands indicating the C–H deformation in aromatic structures can also appear within this range (1200–1000 cm^{-1}). Additionally, the presence of bands of inorganic compounds in biochars (i.e., carbonates, sulfates, phosphates, and silica) cannot be excluded. The wide band in the range of 1300–1000 cm^{-1} may have the contribution of both oxygen functional groups (e.g., C–O) and stretching vibrations of Si–O–Si bridges in quartz. The vibrations of the carbonate ions may also have a contribution in ~1409, ~1100, and 874 cm^{-1} bands [41]. These data can be confirmed by XRD analysis (Figure 2b). The intensity of oxygen functional group bands is the highest for BCH biochar, and the O/C mole ratio value seems to confirm it (Table 2). Those bands have a lower but similar intensity for BW and BPB samples, and the O/C molar ratio for these biochars is also similar. Spectroscopic data are therefore consistent with the physicochemical characteristics of the examined biochars.

3.3. The Effect of Adsorbent Dosage on SWW Treatment

3.3.1. Changes in pH and EC

The application of the adsorbents at different dosages changed the pH and EC in the treated SWW (Figure 3). Initially, SWW was strongly acidic (pH 4.88). The addition of the lowest dosage of BW (12.5 g/L) considerably increased the pH to an alkaline value (7.49); the same dosage of ACN increased it to a nearly neutral value (6.97); and the same dosages of BPB and BCH increased the pH by lesser amounts (to pH 5.88–5.95). As the dosage of each of the adsorbents was increased, the pH of the treated SWW also increased, with the values being highest at a dosage of 100 g/L (Figure 3a). At this dosage, the pH in the SWW treated with BW, ACN, and BPB was alkaline (pH 8.26, 7.94, and 7.26, respectively), while the pH in the SWW treated with BCH remained somewhat acidic (6.26). BW biochar increased the pH to the greatest extent, indicating that it has the highest alkalinity of the four adsorbents. BCH increased the pH by the smallest amount, which may indicate that it has a lower ion-exchange capacity compared to other adsorbents.

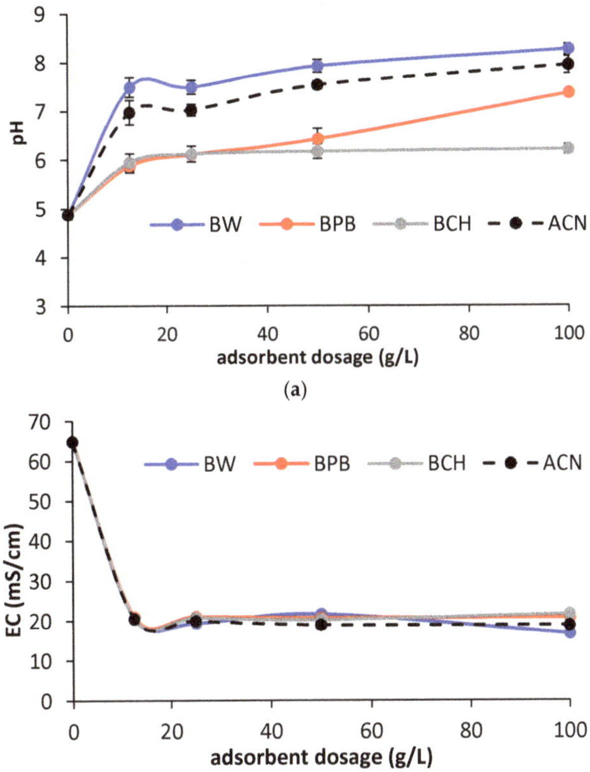

(a)

(b)

Figure 3. Values of pH (**a**) and EC (**b**) in SWW after addition of adsorbents at different dosages (sorption time 24 h, SWW concentration 1.9 g TOC/L) (n = 3, for EC standard deviation ±(0.1–0.5)).

All of the adsorbents lowered the EC in the SWW. At the lowest dosage of the adsorbents, the EC in the SWW was more than 3-fold lower than it was before adsorption (Figure 3b). Although the use of BW biochar at the highest dosage (100 g/L) lowered the EC by the largest amount, from 65 to 11.7 mS/cm, increasing the dosages of the adsorbents beyond 12.5 g/L did not result in substantial further changes to the EC. The decreases in the EC of the SWW can be connected to the sorption of different dissolved ions affecting salinity.

In the present study, the concentrations of three representative constituents of salinity (Na, Ca, and Mg) were monitored. The concentrations of the investigated alkali cations in the treated SWW are shown in Table 4. The removal of Na, Ca, and Mg from SWW with the different adsorbents varied depending on the type of adsorbent and its dosage. A concentration of these alkali cations that is lower than their initial concentration indicates that they were sorbed onto the adsorbents (positive values of RE), whereas a higher concentration indicates that they were released from the adsorbent during the adsorption process (negative values of RE). Such releases of alkali metals were particularly evident after treatment with BW biochar.

Table 4. Concentrations of Na, Ca, and Mg in SWW and their removal efficiency (RE) with different adsorbent dosages (n = 3).

Adsorbent	Dosage (g/L)	Na (mg/L)	RE (%)	Ca (mg/L)	RE (%)	Mg (mg/L)	RE (%)
BW	0	1022.9 ± 22.1		127.6 ± 5.9		29.9 ± 2.1	
	12.5	1167.9 ± 15.6	−14.2	182.6 ± 9.4	−43.1	57.9 ± 4.7	−93.9
	25	1167.7 ± 23.1	−14.2	179.8 ± 3.9	−41.0	58.6 ± 3.1	−96.3
	50	1041.8 ± 10.9	−18.0	129.0 ± 6.4	−1.2	49.6 ± 4.4	−66.2
	100	977.4 ± 8.9	4.4	88.0 ± 3.3	31.1	53.5 ± 4.9	−79.1
BPB	0	1022.9 ± 16.7		127.6 ± 2.7		29.9 ± 1.3	
	12.5	514.2 ± 9.1	49.7	145.3 ± 6.2	−13.9	40.1 ± 2.3	−34.3
	25	512.4 ± 5.9	49.9	162.0 ± 3.1	−27.0	25.7 ± 1.2	14.0
	50	495.4 ± 7.1	51.6	155.2 ± 2.9	−21.7	25.0 ± 0.9	16.2
	100	497.0 ± 7.3	51.4	158.1 ± 2.8	−23.9	42.7 ± 2.8	−43.0
BCH	0	1022.9 ± 19.1		127.6 ± 3.1		29.9 ± 1.0	
	12.5	520.3 ± 10.1	49.1	121.3 ± 1.2	4.9	51.5 ± 2.1	−72.3
	25	530.9 ± 8.2	48.1	121.9 ± 3.9	4.4	47.5 ± 3.7	−59.1
	50	501.5 ± 9.4	51.0	123.9 ± 4.1	2.9	47.6 ± 1.7	−59.5
	100	484.7 ± 7.2	52.6	146.6 ± 2.4	−14.9	74.9 ± 4.2	150.8
ACN	0	1022.9 ± 27.0		127.6 ± 4.1		29.9 ± 1.2	
	12.5	497.0 ± 18.1	51.4	98.9 ± 3.1	22.5	37.2 ± 0.9	−24.7
	25	489.3 ± 13.7	52.2	78.1 ± 3.8	38.7	32.7 ± 1.3	−9.5
	50	489.3 ± 10.6	52.2	61.1 ± 2.8	52.1	30.3 ± 1.0	−1.3
	100	488.8 ± 9.8	52.2	46.9 ± 1.9	63.3	21.4 ± 2.1	28.4

For example, Ca^{2+} and Mg^{2+} cations can be released from complexed functional groups (such as R–O–Me or R–COO–Me) as well as from Ca or Mg precipitates on the adsorbent surface. During the process, these complexes and precipitates may be dissociated or dissolved due to the acidic environment and/or the formation of new precipitates or complexes containing heavy metal cations [46]. Despite the poor removal of the analyzed alkali cations with BW biochar, the decrease in the EC in the treated SWW was similar to that observed with the other adsorbents. This suggests that other dissolved substances causing salinity (e.g., K, Fe, and SO_4) could be removed with the adsorbents, which was reflected by the decrease in the EC.

3.3.2. Removal of Heavy Metals

The efficiency of heavy metal removal at different dosages of adsorbents is shown in Figure 4. Although the heavy metal concentrations in the SWW were not high, the complex composition of the SWW meant that its treatment required higher dosages of adsorbents. The higher the dosage of the adsorbent, the greater the number of active sites available to bind metal ions, which leads to an increase in the efficiency of treatment. Removal of Cu, Ni, Cd, Pb, and Zn was highest with ACN at a dosage of 50 or 100 g/L. At these dosages, ACN treated SWW with the highest RE (≥88%), which made it the most efficient among those tested. Increasing the dosage tended to increase the adsorption efficiency of each biochar, and the highest efficiency of metal removal with the biochars was obtained at the highest dosage (100 g/L). BW biochar at dosages of 50 and 100 g/L can serve as an alternative to ACN for treating SWW contaminated with Cd (77–83% RE), Pb (77–83% RE), and Zn (78–83% RE). The BPB and BCH biochars removed Zn with lower efficiency (maximum RE 44–66%). With the use of the three biochars at a dosage of 100 g/L, the efficiency of Cu removal ranged from 32% (BCH) to 41% (BPB), while that of Ni removal ranged from 18% (BCH) to 42% (BW).

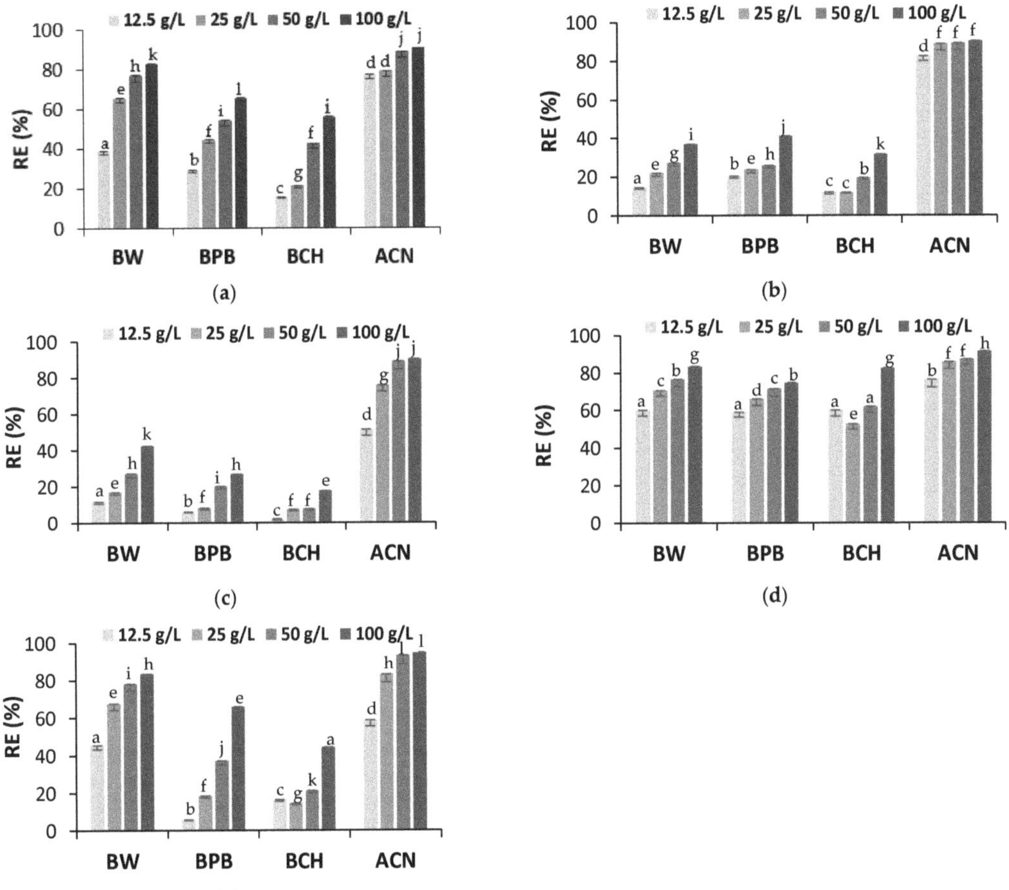

Figure 4. Removal efficiency (RE) of heavy metals from SWW at different adsorbent dosages: (**a**) Cd, (**b**) Cu, (**c**) Ni, (**d**) Pb, and (**e**) Zn (sorption time 24 h). For a given metal and specific dosage of adsorbent, values of RE followed by different letters are significantly different (ANOVA followed by Tukey's HSD test, $p < 0.05$).

When the dosage of biochar reaches a certain value, the adsorption capacity ceases to increase. Therefore, maximum removal can be achieved with the optimal dosage [47]. According to the obtained results, the optimal dosage for metal removal from SWW with ACN is 50 g/L. Such an optimal dosage cannot be given for the biochars because the metal removal efficiency increased almost linearly with the increase in biochar dosage.

Usually, the optimal biochar dosages for metal removal from aqueous solutions are much lower than the dosages that were used for metal removal from the SWW after soil washing. For example, Shan et al. [48] found that the optimum dosage of peanut shell biochar for Cd, Cu, Ni, Pb, and Zn removal from an aqueous solution was only 2 g/L. Chen et al. [49] found that, at a dosage of 10 g/L, ≥90% of Cu and Zn were removed by corn straw biochar produced at 600 °C, and 40–70% of metals were removed with biochar made of a mix of hardwoods at 450 °C.

The major mechanisms controlling the sorption of metals by biochar from aqueous solutions include complexation, cation exchange between protons or alkali metals, precipitation, and electrostatic interactions [50]. The role that each of these mechanisms plays can vary greatly depending on the target metal [21]. The complexation of metals mainly involves interactions with functional groups. In addition, complexation can be facilitated via π-cation interactions with more aromatic and condensed adsorbent structures (linking metal d-electrons and C=C (π-electrons) through coordination bonds) [48,51]. The last mechanism may be particularly relevant for ACN, which had the lowest H/C ratio, the highest content of fixed carbon (Table 2), and the most aromatic structure according to XPS analysis.

Metal precipitation plays an important role in biochars containing carbonates, phosphates, sulfides, hydroxides, and silicates [52]. In the present study, the XRD patterns showed the presence of SiO_2 only (in ACN), $CaCO_3$ and SiO_2 (in BW, BPB, and BCH), and $CaMg(CO_3)_2$ (only in BCH) (Figure 2b), which could have been responsible for the formation of insoluble compounds with the metals. Fei et al. [52] demonstrated the participation of Si–O–Si bonds in the Cd sorption onto rice straw-derived biochar. According to Li et al. [53], silicates in biochar can provide sorption sites for Pb.

The role of electrostatic interactions between the negatively charged adsorbent and the positively charged metal could have differed depending on the type of adsorbent. All the adsorbents were prepared at high temperatures, which reduced the amounts of negatively charged functional groups (e.g., –COO–, –COH, and –OH). Based on the H/C and O/C molar ratios (Table 2) and FTIR spectra, the greatest role of functional groups in metal removal was for BCH, which had the highest abundance of hydroxyl and carboxylic groups, while ACN had the lowest abundance. Taking into consideration the higher pH of BW and ACN, these adsorbents could possibly have more deprotonated functional groups than BPB and BCH biochars.

Cation exchange could be a co-existing adsorption mechanism, especially with BW biochar (highest release of Na, Ca, and Mg, Table 4). Zhang et al. [54] showed that the amount of cations released from water hyacinth biochar (sum of K, Ca, Na, and Mg) was almost equal to the amount of Cd sorbed, suggesting that cation exchange plays a dominant role in Cd sorption by biochar and exchangeable cations may play a more important role in the sorption of metals than the surface area [52]. In the present study, the release of Na, Ca, and Mg was largest during SWW treatment with BW biochar, which may indicate that ion exchange plays a role in metal adsorption by this adsorbent (Table 4). Some alkali metals in biochar (such as Ca, Mg, K, and Na), which are retained by electrostatic effects, inner sphere complexes with carboxyl and hydroxyl groups, and precipitation, can be replaced by heavy metal cations in solution during the sorption process [46]. In situations where functional groups are not fully deprotonated, metals can also be adsorbed by exchange with protons from carboxyl and hydroxyl groups [55]. Such a mechanism is also possible in the present study. Lu et al. [46] showed that sludge-derived biochar at pH 5.0 adsorbed most of the Pb that was present by releasing Ca^{2+} and Mg^{2+} and by Pb complexation with COOH groups.

In this study, all the biochars effectively removed Pb. This could be because it is more easily hydrolyzed and has a larger ionic radius than the other metals. In addition, Pb has a greater affinity for surfaces than many other metals, which provides a competitive advantage in a coexisting system [51]. Taking into account the release of Ca, Mg, and Na from the biochars and the efficiency of Pb removal, cation exchange could be the dominant mechanism of Pb removal by these adsorbents. With ACN, Pb complexation and precipitation might play more important roles [21] due to the reduced exchange of alkali cations (Table 4).

Considering the total pore volume, the average pore diameter of the adsorbents, and the ionic radii of the metal cations (0.097 nm for Cd, 0.073 nm for Cu, 0.069 nm for Ni, 0.119 nm for Pb, and 0.074 nm for Zn), adsorption through filling of the adsorbent micropores by metal cations may also have improved metal removal, especially in the cases of BC biochar and ACN [12].

The removal efficiency of Cu and Ni with BW, BPB, and BCH biochars was much lower than with can, which might be due to the greater competition of these metals with other coexisting components of the SWW for adsorption sites. Biochar has good removal efficiencies in single-metal systems but lower capacities in multiple-metal systems due to the competition between the heavy metals present in wastewater [55].

3.3.3. Removal of Soluble Organics

In this study, the concentration of soluble humic substances in the SWW after soil washing was assumed to be equal to the concentration of total organic carbon (TOC). To determine the changes in the concentration of organic compounds after treating the SWW with the adsorbents, TOC was also used. The changes in TOC in the treated SWW are shown in Figure 5.

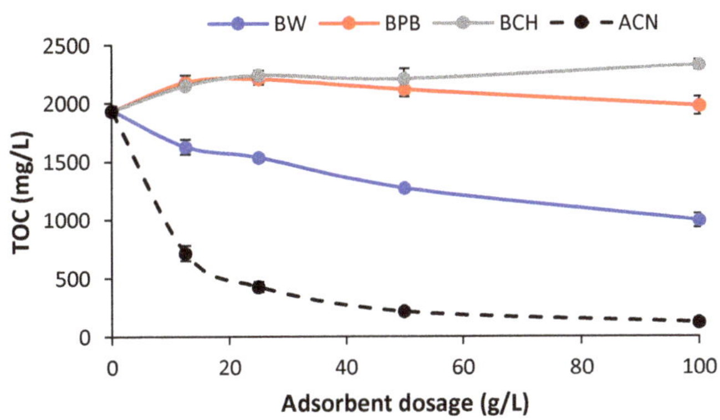

Figure 5. Removal of soluble organics (as TOC) from SWW at different adsorbent dosages ($n = 3$).

The initial TOC concentration in the SWW was 1934 mg/L, and this concentration was lower than that in the SHS used for soil washing (Table 1), indicating that some portion of the SHS (12%) was lost from solution during soil washing. On this basis, it can be assumed that the remaining fraction of soluble organics in the SWW consisted of soluble humic substances. Based on the methodology of SHS extraction, SHS contains macromolecular compounds such as humic and fulvic acids [16].

The TOC in the SWW decreased significantly when BW biochar and ACN were used, even at the lowest dosage (12.5 g/L). At a dosage of 100 g/L, the efficiencies of soluble organics removal with ACN and BW were 94% and 49%, respectively. The application of BCH increased the TOC concentration from 2148 mg/L (at 12.5 g/L) to 2326 mg/L (at 100 g/L). Similar trends were observed for BPB. This indicates that BCH and BPB biochars are not suitable for removing residual dissolved organics from the SWW. Although biochar is refractory, it can release dissolved organics, the amount and composition of which can vary greatly depending on the feedstock. The TOC for BW and ACN was 27 and 23 mg/L, respectively, while for BPB and BCH it was 120 and 132 mg/L, respectively. This could be an explanation for the increased TOC concentration in SWW treated with BPB and BCH.

The removal of dissolved organics from SWW may have been higher with ACN and BW than with the other adsorbents because of the mechanisms of organic removal with biochars. The first mechanism to consider is a pore-filling mechanism, in which the organics are at the surface of the biochar, which possesses mesopores (2–50 nm) and micropores (<2 nm) [50]. The presence of mesopores was shown to be crucial for the adsorption of humic acids onto biochar and activated carbon [56]. In the present study, ACN and BW had the best textural characteristics in terms of total pore volume and volumes of meso- and micropores (Table 2). In experiments with activated carbons and organics of different molecular weights and molecular sizes (phenol: 94 g/mol and 0.62 nm in diameter; iodine: 254 g/mol and 0.56 nm in diameter; tannic acid: 1701 g/mol and 1.6 nm in diameter), Hsieh and Teng [32] found that mesopores had a stronger effect on the capacity to adsorb larger adsorbates because mesopores reduced the length of the diffusion path of the micropores, and this length generally has a stronger effect on the diffusion of larger molecules than that of smaller molecules. According to Liu et al. [57], biochar prepared at 900 °C after modification with hydrochloric acid had a large specific surface area (459.3 m^2/g) and a high porosity, and it removed over 80% of humic acids in aqueous solution at pH 5.0–9.0. Yan et al. [56] demonstrated that humic acids are better adsorbed on bamboo biochar than on helmpalm biochar and activated carbon. This is because bamboo biochar has a larger pore volume and wider pore size distribution and contains more polar functional groups on the surface, which makes it easier for it to bind humic acid than the other adsorbents.

It was shown that organics can also be adsorbed on the biochar surface through hydrophobic interaction. The removal took place via hydrophobic interaction due to the hydrophobic nature of the organics and the adsorbent [50]. In this study, ACN was the most condensed of the adsorbents (H:C ratio = 0.11, aromaticity of 63.4%at), which could have facilitated hydrophobic interactions with the hydrophobic parts of the humic substances from the SWW.

3.4. *The Effect of Contact Time on the SWW Treatment*

3.4.1. Heavy Metals

By investigating various contact times, it can be determined how long it takes an adsorbent to adsorb pollutants until adsorption equilibrium is reached. Figure 6 shows the effect of contact time between the adsorbents and the SWW on metal adsorption capacities. In general, the capacities of the different adsorbents to adsorb the metals decreased in this order: Zn > Pb > Cu >Ni > Cd. Only BCH had a higher capacity for adsorbing Pb than Zn during 180 min of adsorption (Figure 6c).

The adsorption process was fast, but its efficiency depended on the type of metal adsorbed on a given adsorbent. During the first 30 min of adsorption, the efficiency of metal adsorption was 72–99% with ACN, 29–85% with BW, 15–82% with BPB, and 6–76% with BCH. Pb was adsorbed most rapidly by all the adsorbents, with 76–99% of this metal adsorbed during the first 30 min of adsorption. During rapid adsorption, there are usually a large number of active sites on the adsorbent surface. Over time, more sites are occupied by heavy metal ions, and the rate of adsorption slows until equilibrium is reached [48].

The capacities for adsorbing the metals differed during the adsorption time. This is because metal removal depends not only on the adsorbent properties but also on the competition between pollutants, which is typical of mixed-pollutant solutions. In the present study, the SWW from soil washing contained heavy metals, alkaline ions, and soluble humic substances. Usually, the capacity for removal of a heavy metal is lower in a mixed solution than it is in a solution with a single solute, which might be due to faster saturation of the adsorption sites on biochar by mixed pollutants [48].

In the present study, adsorption equilibrium for all adsorbents and metals was reached at 180 min (Figure 6). Only in the case of Zn adsorption onto BPB was the adsorption capacity after 180 min about 30% lower than it was after 1440 min (Figure 6b). The literature indicates that the adsorption time required to achieve equilibrium adsorption of metals onto biochar varies widely, ranging from a few minutes to a few hours up to 24 h. Zhao et al. [55] found that the time to reach equilibrium adsorption of multi-metals (Cr, Cd, Cu, and Pb) depended on the feedstock used for biochar preparation: 5 min with biochar from agricultural waste, 5 min to 24 h (depending on the metal) with wood biochar, and 8–24 h with sewage sludge biochar. Chen et al. [49] reported that Cu and Zn adsorption onto hardwood and corn straw was fast, with 77–83% of adsorption occurring during the first 120 min, followed by a very slow approach to equilibrium. Liu and Zhang [58] used biochars prepared from hydrothermal liquefaction of pinewood or rice husk and found that the contact time needed to reach 95% of the Pb removal equilibrium was less than 5 h with concentrations of 10–20 mg/L. Mohan et al. [59] used biochars made from oak wood, pine wood, oak bark, and pine bark and found that 40–70% of total Pb adsorption occurred within the first hour.

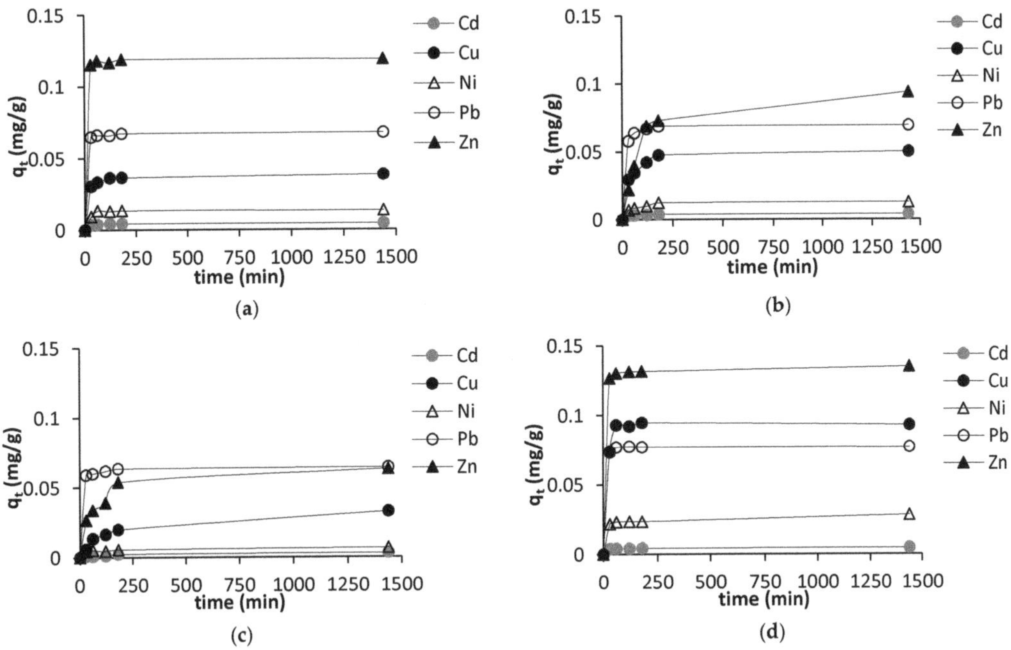

Figure 6. Effect of contact time on the metal adsorption capacities at different times (q_t) of the tested adsorbents: (**a**) BW, (**b**) BPB, (**c**) BCH, and (**d**) ACN.

For insight into the mechanisms of heavy metal adsorption onto the adsorbents, the data were analyzed with a pseudo-first-order kinetics model, a pseudo-second-order kinetics model, and an intraparticle diffusion model (Tables 5 and A3).

Pseudo-first-order kinetics are controlled by physical processes (e.g., weak van der Waals forces), while pseudo-second-order kinetics are controlled by chemical processes [55]. The high values of R^2 (from 0.96 to over 0.99) for these two kinetic models indicate that metal removal from SWW can be explained by these two processes. Other studies on the application of carbon-based adsorbents have also found that different kinetic models may describe the rate of adsorption of heavy metals equally well [60,61]. A pseudo-first-order model is appropriate for the initial contact time but not for the entire time [55].

Adsorption occurs rapidly due to the binding of metal ions to the biochar surface and is predominantly physical adsorption. Then the adsorption rate slows down, and chemisorption controls the adsorption process until equilibrium is reached. Additionally, at this stage, metal ions can diffuse into the macropores and micropores of the biochar and bind to the inner surface of the biochar to reach the adsorption endpoint [62,63]. The rate-limiting steps in adsorption are external mass transfer (interfacial diffusion) and/or intraparticle diffusion [61].

Table 5. Parameters of pseudo-first-order and pseudo-second-order kinetics models of metal adsorption.

Adsorbent/ Metal		Pseudo-First Order			Pseudo-Second Order		
		k_1 1/min	q_e mg/g	R^2 -	k_2 g/(mg·min)	q_e mg/g	R^2 -
BW	Cd	0.0621	0.0042	0.9965	35.1408	0.0043	0.9992
	Cu	0.0553	0.0364	0.9962	3.2075	0.0382	0.9996
	Ni	0.0423	0.0133	0.9954	6.0823	0.0140	0.9874
	Pb	0.1189	0.0667	0.9997	12.4141	0.0673	0.9998
	Zn	0.1252	0.1179	0.9998	9.7425	0.1187	0.9998
BPB	Cd	0.0543	0.0036	0.9821	26.6563	0.0039	0.9920
	Cu	0.0257	0.0475	0.9877	0.7811	0.0516	0.9964
	Ni	0.0220	0.0120	0.9821	2.5908	0.0131	0.9912
	Pb	0.0621	0.0679	0.9987	2.2940	0.0704	0.9998
	Zn	0.0096	0.0934	0.9963	0.1217	0.1022	0.9897
BCH	Cd	0.0066	0.0029	0.9656	2.6020	0.0032	0.9631
	Cu	0.0063	0.0321	0.9879	0.2224	0.0352	0.9949
	Ni	0.0463	0.0055	0.9559	10.5117	0.0061	0.9755
	Pb	0.0943	0.0623	0.9984	5.5857	0.0637	0.9996
	Zn	0.0125	0.0603	0.9725	0.2860	0.0650	0.9878
ACN	Cd	0.0747	0.0043	0.9926	40.6366	0.0045	0.9966
	Cu	0.0536	0.0939	0.9990	1.4020	0.0972	0.9948
	Ni	0.0668	0.0248	0.9830	5.0726	0.0262	0.9903
	Pb	0.1028	0.0771	0.9999	8.6864	0.0778	0.9998
	Zn	0.1049	0.1321	0.9995	4.1552	0.1339	0.9999

The adsorbents had different structural properties that could have affected the diffusion process. Because the pseudo-first-order and pseudo-second-order kinetic models could not clearly identify the diffusion mechanism [64], the possibility of intra-particle diffusion was explored by using the intra-particle diffusion model. Figure A8 presents the plots of metal adsorption capacities versus $t^{0.5}$ for all the adsorbents. Two lines were fitted to the data, indicating that two steps influenced the adsorption process, and the kinetic parameters determined from the intraparticle diffusion model are shown in Table A3. The first straight line may reflect macropore and mesopore diffusion, and the second line may reflect micropore diffusion [65]. The increases in metal adsorption capacities were higher in the first phase than in the second one. Thus, the first stage had the strongest effect on the rate of adsorption. In the second phase, the diffusion of most metals remained limited due to the low metal concentrations [66,67]. The straight lines in the first phase showed deviation from the origin, which may be due to differences in the rate of mass transfer of a specific metal onto a given adsorbent. The values of C_i were higher than 0, indicating that intraparticle diffusion was not the main rate-limiting step of metal adsorption in this experiment, although it could have served to some extent as a co-mechanism.

Some film diffusion effects could have affected the adsorption of Cd and Cu on BCH biochar and the adsorption of Zn on BPB biochar, for which negative C_i values were obtained [68]. The BPB and BCH biochars had the smallest pore volumes, but their average pore diameter (128 nm) was in the macropore range. For adsorbents with a pore structure that consists of macropores, it is believed that the adsorbate undergoes film diffusion resistance not only on the outer surface of the adsorbent but also inside the large macropores (i.e., a film exists between the bulk of the pore and the inner pore surface) [68].

In general, the values of the intra-particle diffusion rate constants ($k_{i,1}$) were not high due to the relatively low metal concentrations in the SWW. The highest values were obtained for Zn and Cu, and the lowest for Cd. Usually, intraparticle diffusion has a faster adsorption rate at higher pollutant concentrations [69].

3.4.2. Soluble Organics (as TOC)

The effect of adsorption time was also shown in the removal of soluble organics from SWW (Figure 7). As shown, the removal of soluble organics (as TOC) during adsorption differed between the adsorbents. BCH biochar did not remove soluble organics. Their concentration during adsorption was higher than the concentration of organics in the SWW, increasing from 2104 mg/L (after 30 min) to 2326 mg/L (after 1440 min). These results suggest that BCH biochar is a source of soluble organics released under adsorption conditions. Similarly, with BPB biochar, the TOC concentration increased to 2142 mg/L within 180 min after adsorption. However, when the adsorption time was extended up to 1440 min, the organics released from the biochar were readsorbed, and the organics from the SWW were partially adsorbed, as the TOC concentration decreased to 1973 mg/L (Figure 7a). In contrast, BW and ACN were able to remove soluble organics from the SWW during the entire time of the experiment. The soluble organics were rapidly adsorbed by ACN and gradually by BW biochar. After 30 min, the TOC concentration in the SWW treated with ACN had decreased by 87%, and equilibrium was reached after 120 min (TOC removal efficiency of 91%). The use of BW biochar decreased the TOC concentration from 1874 mg/L (3% removal after 30 min) to 992 mg/L (49% removal after 1440 min), indicating that adsorption required more time than when using ACN. These results indicate that ACN was more effective than BW biochar for removing soluble organics from the SWW (Figure 7b).

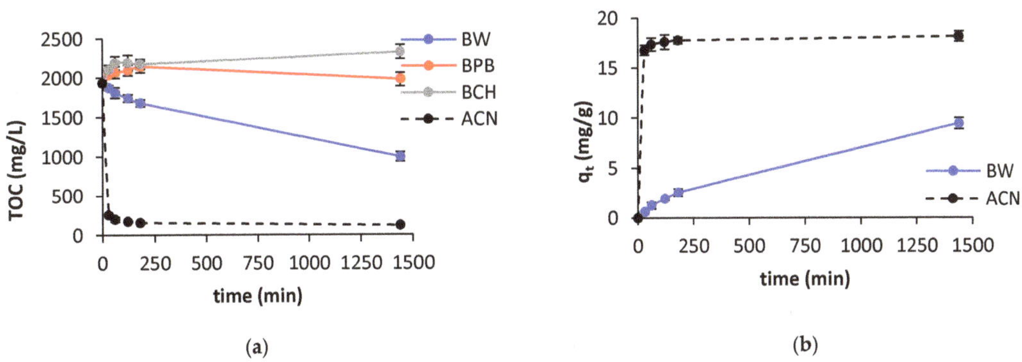

Figure 7. Effect of adsorption time on the concentration of soluble organics (as TOC) (**a**) and capacities for adsorption of soluble organics (**b**) (pH 4.88 with 100 g/L adsorbent) (n = 3).

Omri et al. [70] found that adsorption of humic acid onto almond shell activated carbon from aqueous solutions at different concentrations (10–100 mg/L) increased sharply in the first 180 min of contact time and that equilibrium adsorption was reached within 300 min. Yan et al. [56] observed that the capacities of bamboo biochar, helm palm biochar, and activated carbon to adsorb humic acid were highest during the first 4 h. The fast rate of adsorption at the beginning of the process was caused by humic acid adsorption on the exterior surface of activated carbon. After saturation of the exterior adsorption surface, humic acids can enter pores and are adsorbed by the interior surface [70].

Since only BW and ACN adsorbed soluble organics from SWW within the adsorption time, the kinetic models were fitted to the experimental data for only these two adsorbents to explore the mechanisms of soluble organic adsorption. The kinetic parameters of adsorption are listed in Table 6. All three kinetic models fit the data on the adsorption of soluble organics on BW and ACN well.

Table 6. Kinetic constants for soluble organics (as TOC) adsorption onto BW and ACN with different kinetic models.

Kinetic Model			BW	ACN
Pseudo-first order		k_1 (1/min)	0.0015	35.2196
		q_e (mg/g)	10.9698	17.7280
		R^2	0.9942	0.9987
Pseudo-second order		k_2 (g/mg·min)	0.0001	0.021
		q_e (mg/g)	14.0056	18.1490
		R^2	0.9855	1.0000
Intraparticle diffusion	Stage $1_{30–180min}$	k_i (mg/g·min$^{0.5}$)	0.2731 *	0.1155
		C_i (mg/g)	−0.9825 *	16.291
		R^2	0.9982 *	0.9007
	Stage $2_{180–1440min}$	k_i (mg/g·min$^{0.5}$)	–	0.0152
		C_i (mg/g)	–	17.556
		R^2	–	1.000

* Values determined for adsorption time of 30–1440 min.

This indicates that the adsorption of soluble organics from the SWW, which had a very high initial concentration (1934 mg/L), was not determined by only one type of mechanism; instead, physical adsorption, chemisorption, and intraparticle diffusion were responsible for the removal of soluble organics (R^2 values for all three models were between 0.90 and 1.00). Many studies that have used carbon-based adsorbents to adsorb humic acids from aqueous solutions with concentrations up to 100 mg/L have indicated that the adsorption of humic acids appears to be controlled mainly by chemisorption processes [56,69,70].

In the present study, the adsorption rate constants (k_1 and k_2) from the pseudo-first and pseudo-second order models indicated that physical adsorption and chemisorption of soluble organics proceeded at a higher rate with ACN than with BW biochar. The faster and greater adsorption onto ACN is connected with the presence of a higher number of active sites on the surface of the adsorbent and the high probability of adhesion. This is because ACN has more pore structure and a greater specific surface area than BW biochar. These characteristics provide channels for adsorbate dispersion and improve mass transfer, which provide more active sites for the adsorption of soluble organics [57].

Adsorption data modeled with the intraparticle diffusion model usually indicate that there are two or three steps in the sorption process of different soluble organics [65]. According to the intraparticle model, the plot q_t against $t^{0.5}$ for BW biochar was linear over the entire adsorption time (R^2 = 0.9982, Figure A9a), revealing that only one stage was involved in the adsorption of soluble organics with this type of biochar. In this study, the plot of q_t against $t^{0.5}$ for BW did not pass through the origin, and the intercept (C_i) value was negative. This suggests that the adsorption of soluble organics could have been the result of the combined effects of film diffusion and adsorbent-surface interaction (chemical or physical interactions) [71]. The soluble organics were gradually adsorbed onto BW, but the final equilibrium adsorption stage was not reached. Thus, longer adsorption of soluble organics is needed to obtain equilibrium. Similarly, Zhu et al. [68] observed a time lag (hence a negative intercept, C_i) in the initial stage of the adsorption of pyridine onto synthetic polymers, which indicates that pyridine adsorption onto adsorbents experienced external film diffusion resistance.

In contrast, the plot qt against $t^{0.5}$ for ACN was not linear over the whole range of time but was divided into two linear sections, implying that more than one process affected the adsorption of soluble organics, which is consistent with the higher efficiency of adsorption of these organic compounds with ACN. The pore diffusion rate constant decreased from 0.1555 (1st stage) to 0.0152 mg/g·min$^{0.5}$ (2nd stage) (Table 6), indicating the diminishing role of intraparticle diffusion. Unlike adsorption with BW, the C_i was positive, indicating a lack of external film diffusion resistance during organic adsorption. Zhang [69] reported that the use of an intra-particle model indicated that the sorption of humic acid onto modified biochar from forestry waste was also a multi-stage process. Three stages were indicated, each of which differed in the rate constants of intraparticle diffusion. The first stage was a fast process involving the transfer of the humic acid molecules from the main solution to the surface of the biochar. Due to the strong electrostatic attraction, it was interpreted as external surface adsorption or instantaneous adsorption. The second phase was interpreted as diffusion within the particles, as humic acid diffused into the pores or capillaries of the internal structure, which was referred to as mesopore diffusion and adsorption. The last phase was considered to be the final adsorption-desorption equilibrium, which is reached only after a long time. In the study by Zhang et al., equilibrium humic acid adsorption was reached between 60 and 750 min. In this study, equilibrium was reached after 180 min. of adsorption. Based on the C_i values, the rate-controlling mechanism of HA adsorption was mainly chemisorption but simultaneously affected by multiple mechanisms [69]. Other authors have confirmed that the sorption of soluble organics in the form of humic acids is not dominated by intraparticle diffusion [56]. Omri et al. [70] concluded that humic acid molecules are slowly transported via intraparticle diffusion into the particles of activated carbon and finally retained in their pores.

4. Conclusions

This study investigated soil washing with soluble humic substances combined with the treatment of generated wastewater (SWW) with biochar as an innovative and promising wastewater treatment technology. The SWW was characterized by a complex composition due to the presence of soluble organic matter (as humic substances), alkali elements (Na, Ca, Mg), and heavy metals (Cd, Cu, Ni, Pb, Zn). The differences in the physicochemical properties of the adsorbents affected their usability. All adsorbents had a positive effect on increasing the pH of SWW from acidic to slightly acidic and alkaline, as well as reducing salinity. Treatment of SWW required the use of high dosages of adsorbents in the range of 50–100 g/L. Activated carbon and willow biochar, which had the highest surface area, porosity, and pH and were more hydrophobic than other biochars, provided the best results in removing heavy metals and soluble organics from SWW.

The treatment of SWW was not determined by only one type of mechanism; instead, several mechanisms were involved. Analysis of the adsorption behavior of heavy metals and soluble organics with different kinetics models revealed that physical and chemical processes dominated, and diffusion into mesopores and micropores made lesser contributions. Depending on the type of adsorbent, the chemisorption of heavy metals was mainly determined by cation exchange, precipitation, complexation with functional groups and aromatic structures, and hydrophobic interactions with soluble organics.

Among the tested biochars, willow biochar was the most effective for the simultaneous removal of heavy metals and soluble organics from SWW. It is believed that willow biochar, after modification of its properties, has the potential to improve the treatment of SWW and be as effective as activated carbon. Using biochars made from a mixture of plant biomass or coconut husks in soil remediation could be considered for the recovery and reuse of soluble humic substances for further soil treatment by soil washing. Thus, the application of pyrolyzed biomass in soil washing to remove pollutants and/or recover washing agents can make this technology more sustainable.

Author Contributions: Conceptualization, M.Z.G.; Methodology, Investigation, M.Z.G., S.P.-P., M.B. and M.K.; Data curation, M.Z.G.; Writing—Original Draft, M.Z.G., S.P.-P. and M.B.; writing—review and editing, M.Z.G., S.P.-P. and M.K.; Visualization, M.Z.G., S.P.-P. and M.K.; project administration, M.Z.G.; funding acquisition, M.Z.G. All authors have read and agreed to the published version of the manuscript.

Funding: Linguistic verification of the paper was financed by the project supported by the Minister of Education and Science under the program entitled "Regional Initiative of Excellence" for the years 2019–2023, Project No. 010/RID/2018/19, the amount of funding was 12.000.000 PLN.

Data Availability Statement: Not applicable.

Acknowledgments: The authors gratefully thank Mark Leonard, a native speaker of English, for English language support.

Conflicts of Interest: The authors declare no conflict of interest.

Appendix A

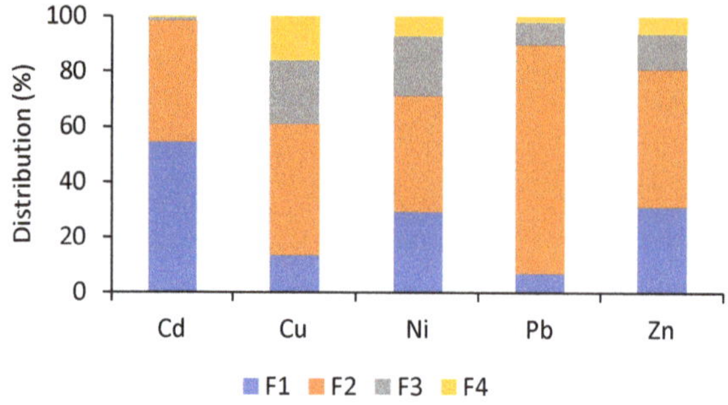

(**a**)

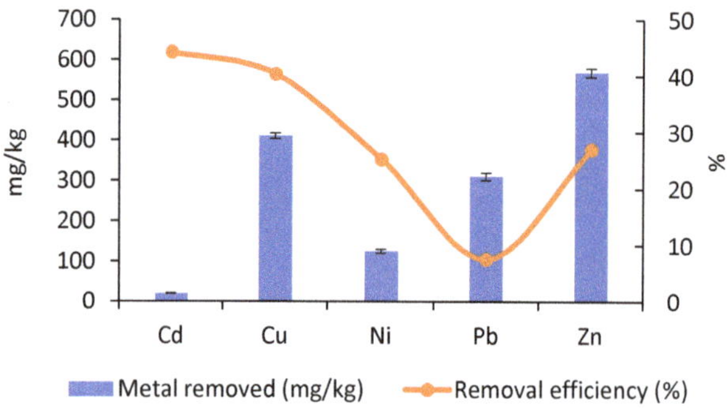

(**b**)

Figure A1. Metal distribution (**a**) and metal removal (**b**) from soil with SHS solution (n = 3). F1 is an exchangeable and acid-soluble fraction; F2 is a reducible fraction; F3 is an oxidizable fraction; and F4 is a residual fraction.

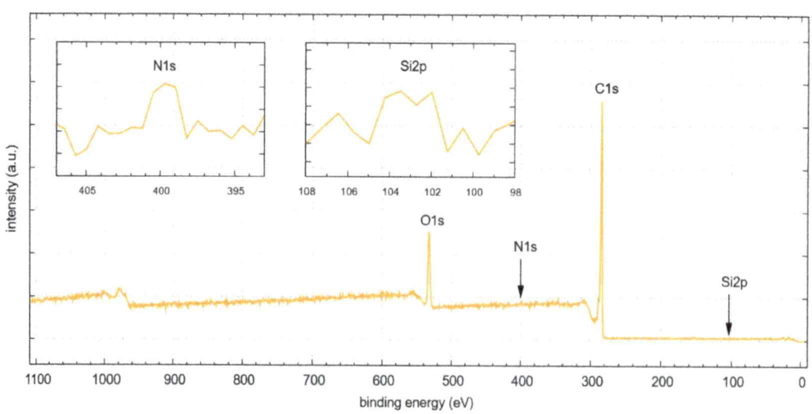

Figure A2. XPS survey spectrum of BW.

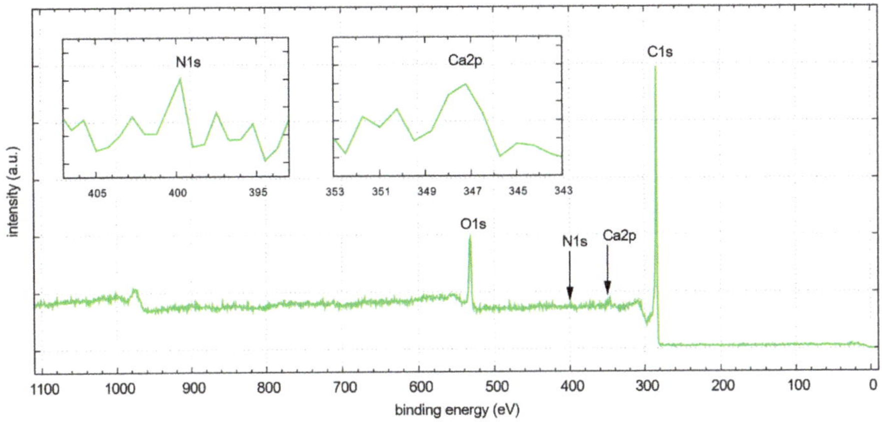

Figure A3. XPS survey spectra of BPB.

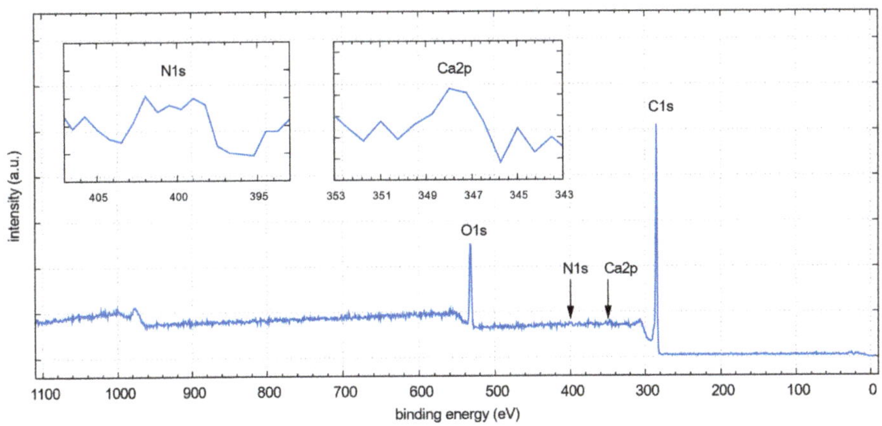

Figure A4. XPS survey spectrum of BCH.

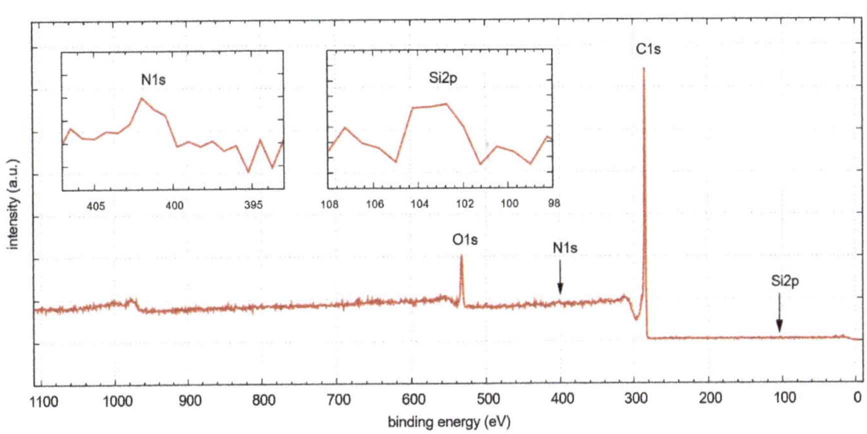

Figure A5. XPS survey spectrum of ACN.

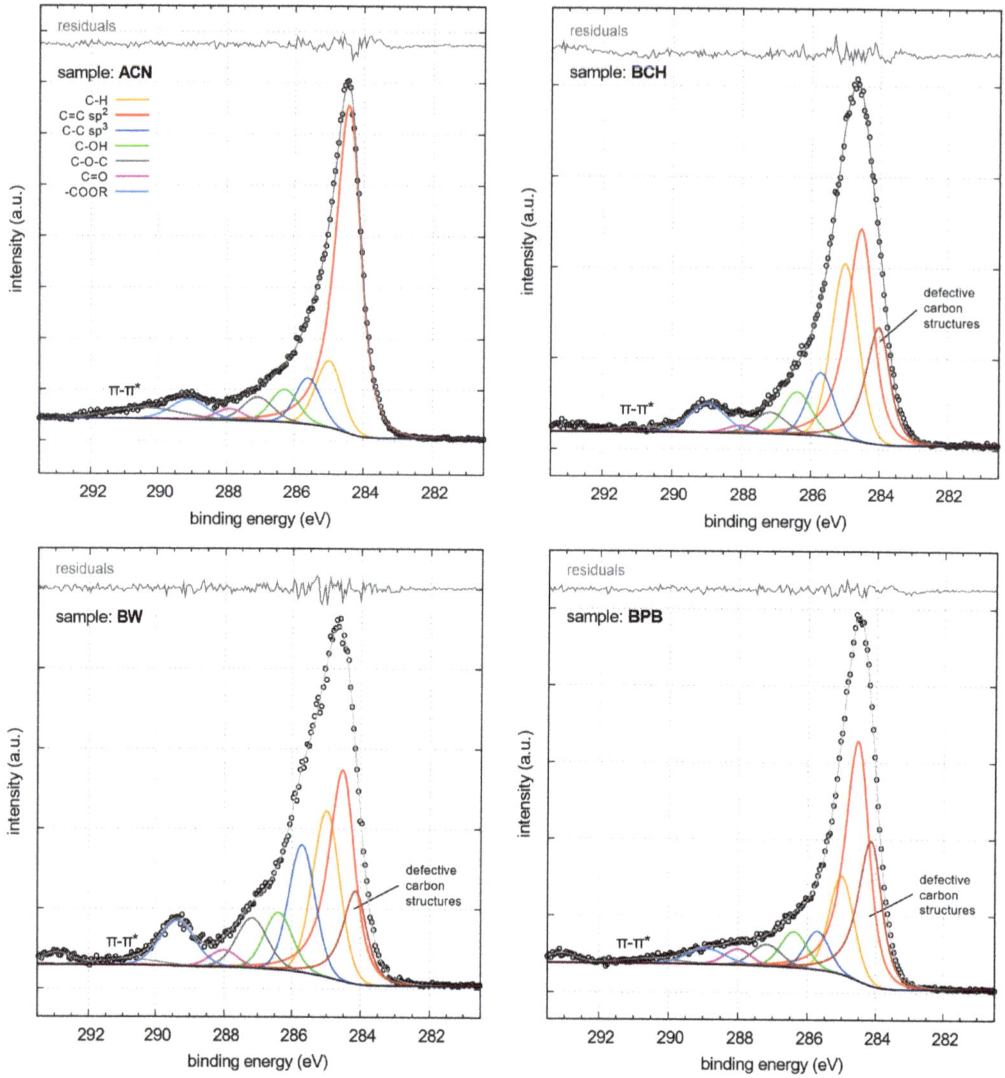

Figure A6. Fitting of XPS C1s spectra. Eight main components were used: C–H, C=C sp^2, C–C sp^3, C–OH, C–O–C, C=O, –COOR, and π-π* shake-up (i.e., bonding to antibonding transition). Details concerning these moieties and their binding energies are presented in table SI-T5 and in publications [72–74]. Additionally, for BCH, BW, and BPB samples, the component corresponding to defective carbon structures was added [75].

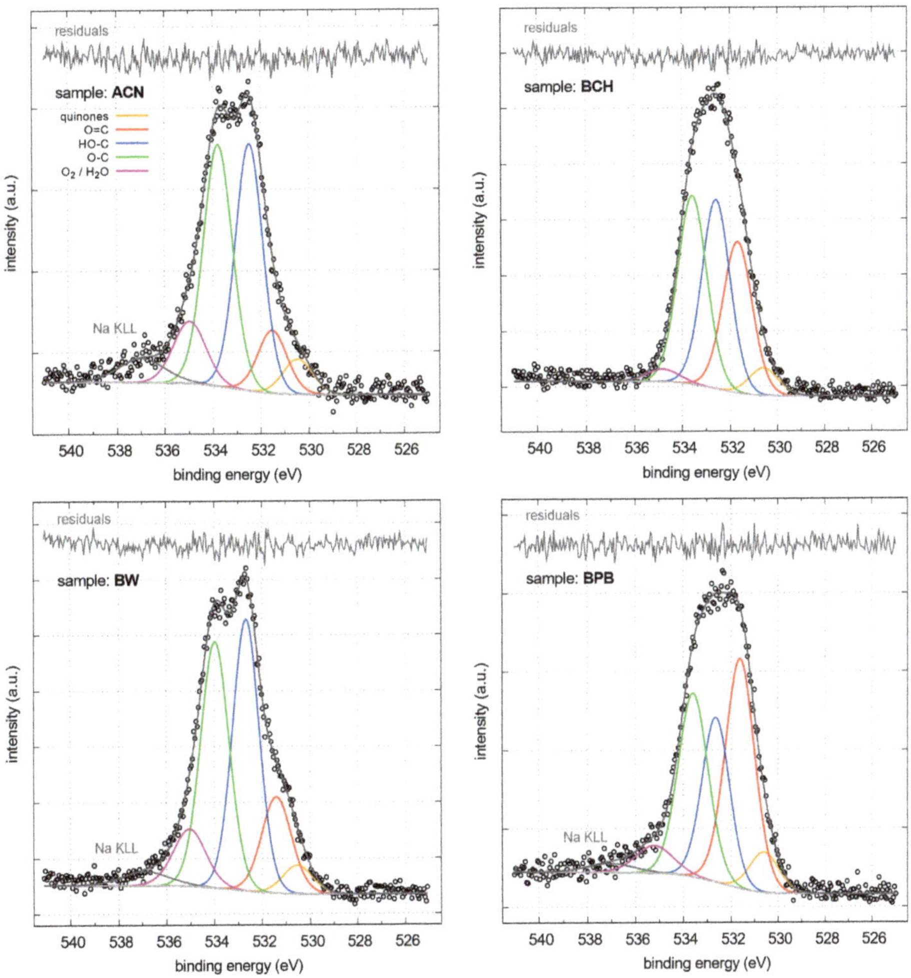

Figure A7. Fitting of XPS O1s spectra. Five peaks for the following components were used: quinones, carbonyl oxygen (O=C), hydroxyl groups (HO–C), oxygen single-bonded to carbon atoms (O–C, other than hydroxyls), and adsorbed molecular oxygen and/or water (O_2/H_2O). Details concerning their binding energies are given in table SI_T6 and publications [75–78].

Table A1. Fitting of XPS C1s spectra: surface atomic concentrations of components, their binding energies, and FWHMs. DCS stands for "defective carbon structures" [75].

Adsorbent	Name	Position	FWHM	%At Conc
BW	C–H	285.00	0.92	20.2
	C=C sp2	284.51	0.83	29.1
	C–C sp3	285.71	0.89	15.1
	C–OH	286.39	0.98	7.4
	C–O–C	287.17	1.08	7.1
	C=O	288.00	0.96	2.1
	–COOR	289.36	1.20	7.2

Table A1. *Cont.*

Adsorbent	Name	Position	FWHM	%At Conc
BPB	C–H	285.00	0.85	14.6
	C=C sp2	284.51	0.76	40.3
	C–C sp3	285.70	0.82	5.5
	C–OH	286.38	0.90	5.6
	C–O–C	287.17	0.99	3.7
	C=O	288.00	0.89	2.5
	–COOR	288.91	1.28	3.5
	DCS	284.13	0.76	24.3
BCH	DCS	284.13	0.76	11.8
	C=C sp2	284.52	0.83	33.6
	C–C sp3	285.71	0.89	8.6
	C–OH	286.39	0.97	6.3
	C–O–C	287.18	1.08	3.5
	C=O	288.01	0.96	1.1
	–COOR	288.97	1.19	5.2
	DCS	284.01	0.77	17.2
ACN	C–H	285.00	0.90	11.4
	C=C sp2	284.42	0.81	63.4
	C–C sp3	285.61	0.87	7.7
	C–OH	286.29	0.95	6.0
	C–O–C	287.07	1.05	4.8
	C=O	287.91	0.94	2.0
	–COOR	289.09	1.23	4.7

Table A2. Fitting of XPS O1s spectra: surface atomic concentrations of components, their binding energies, and FWHMs.

Adsorbent	Name	Position	FWHM	%At Conc
BW	quinones	530.52	1.44	4.2
	O=C	531.39	1.46	14.4
	HO–C	532.67	1.31	36.5
	O–C	533.97	1.41	35.4
	O_2/H_2O	535.04	1.61	9.5
BPB	quinones	530.58	1.46	6.2
	O=C	531.61	1.48	36.0
	HO–C	532.63	1.36	23.8
	O–C	533.58	1.47	28.7
	O_2/H_2O	535.20	1.83	5.3
BCH	quinones	530.59	1.45	4.9
	O=C	531.68	1.45	27.1
	HO–C	532.58	1.37	32.3
	O–C	533.56	1.41	33.1
	O_2/H_2O	534.81	1.59	2.5
ACN	quinones	530.46	1.45	5.5
	O=C	531.51	1.44	9.9
	HO–C	532.47	1.36	36.5
	O–C	533.78	1.43	37.5
	O_2/H_2O	534.97	1.56	10.6

Table A3. Parameters of the intraparticle diffusion model of metal adsorption.

Metal	Phase	Parameter	BW	BPB	BCH	ACN
Cd	1	$k_{i,1}$ (mg/g·min$^{0.5}$)	8×10^{-5}	0.0001	0.0002	4×10^{-5}
		$C_{i,1}$ (mg/g)	0.032	0.0024	**−0.0002**	0.0037
		R^2	0.9578	0.9297	0.8731	0.9270
	2	$k_{i,2}$ (mg/g·min$^{0.5}$)	4×10^{-6}	8×10^{-7}	2×10^{-5}	2×10^{-5}
		$C_{i,2}$ (mg/g)	0.0041	0.0039	0.002	0.0041
		R^2	1.0000	1.0000	1.0000	1.0000
Cu	1	$k_{i,1}$ (mg/g·min$^{0.5}$)	0.0008	0.0023	0.0016	0.003
		$C_{i,1}$ (mg/g)	0.0266	0.0167	**−0.0016**	0.0617
		R^2	0.9556	0.9995	0.9612	0.7872
	2	$k_{i,2}$ (mg/g·min$^{0.5}$)	7×10^{-5}	0.0001	0.0005	7×10^{-5}
		$C_{i,2}$ (mg/g)	0.0353	0.0931	0.0128	0.0924
		R^2	1.0000	1.0000	1.0000	0.6663
Ni	1	$k_{i,1}$ (mg/g·min$^{0.5}$)	0.0006	0.0006	0.0001	0.0003
		$C_{i,1}$ (mg/g)	0.007	0.0033	0.0038	0.0204
		R^2	0.7360	0.9855	1.0000	0.8977
	2	$k_{i,2}$ (mg/g·min$^{0.5}$)	2×10^{-5}	1×10^{-5}	4×10^{-5}	0.0002
		$C_{i,2}$ (mg/g)	0.0126	0.0121	0.0049	0.0213
		R^2	0.5853	1.0000	0.9111	0.9966
Pb	1	$k_{i,1}$ (mg/g·min$^{0.5}$)	0.0002	0.0013	0.0006	0.00007
		$C_{i,1}$ (mg/g)	0.0637	0.0522	0.0557	0.0706
		R^2	0.8727	0.9599	0.9942	0.8881
	2	$k_{i,2}$ (mg/g·min$^{0.5}$)	1×10^{-5}	2×10^{-5}	3×10^{-5}	-6×10^{-6}
		$C_{i,2}$ (mg/g)	0.067	0.0686	0.0629	0.0773
		R^2	1.0000	1.0000	1.0000	1.000
Zn	1	$k_{i,1}$ (mg/g·min$^{0.5}$)	0.0003	0.0068	0.0032	0.0008
		$C_{i,1}$ (mg/g)	0.114	**−0.0132**	0.0083	0.1227
		R^2	0.7261	0.9746	0.9663	0.9193
	2	$k_{i,2}$ (mg/g·min$^{0.5}$)	4×10^{-7}	0.0009	0.0004	0.0001
		$C_{i,2}$ (mg/g)	0.1188	0.0615	0.0485	0.1297
		R^2	1.0000	1.0000	1.0000	0.9999

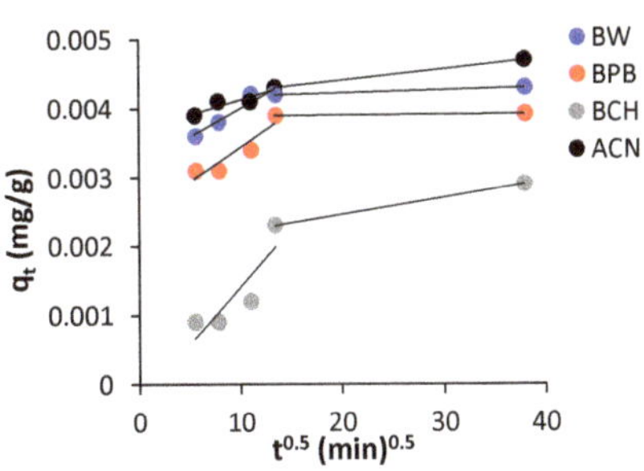

(a)

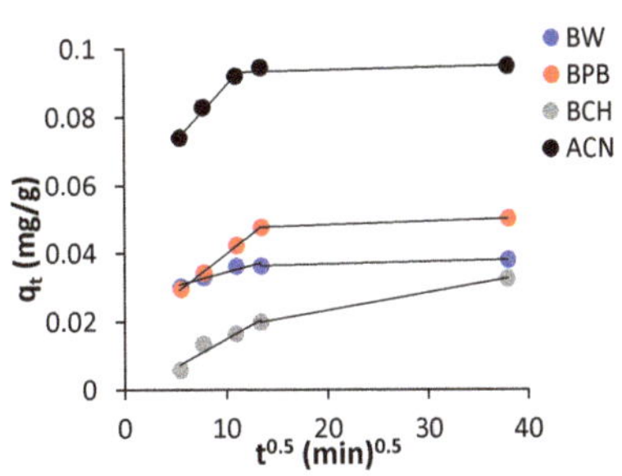

(b)

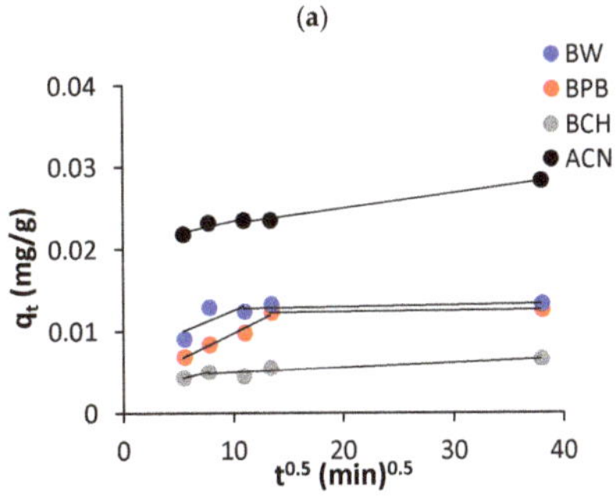

(c)

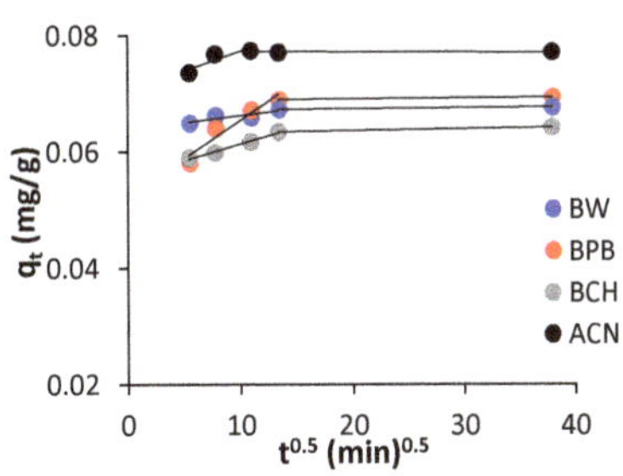

(d)

Figure A8. *Cont.*

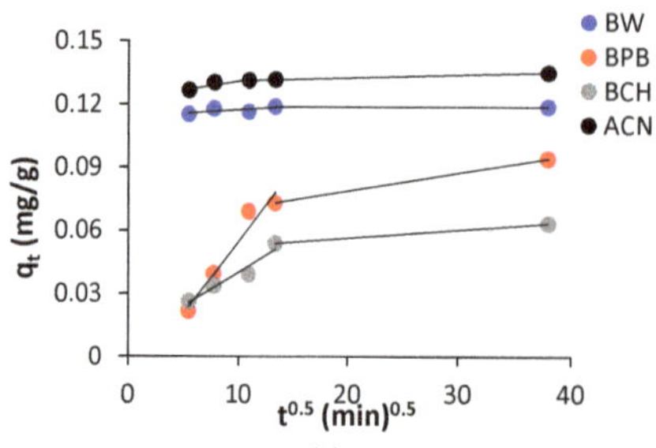

(e)

Figure A8. Kinetics of metal adsorption according to the intraparticle diffusion model: (**a**) Cd, (**b**) Cu, (**c**) Ni, (**d**) Pb, and (**e**) Zn. The first line refers to phase 1, while the second line refers to phase 2.

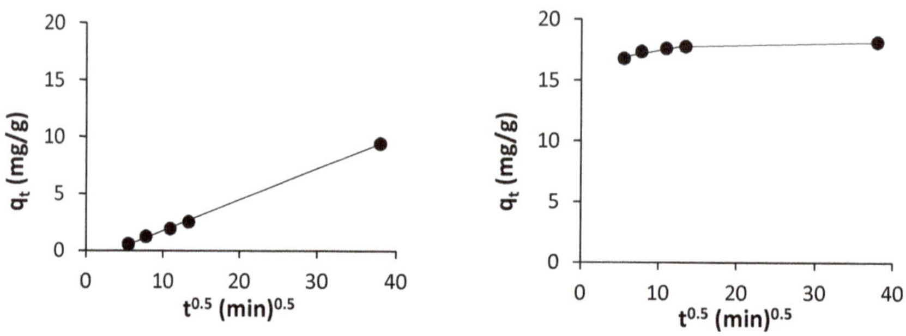

Figure A9. Kinetics of soluble organics (as TOC) adsorption according to the intraparticle diffusion model: (**a**) BW and (**b**) ACN. The first line refers to phase 1, while the second line refers to phase 2.

References

1. Andreozzi, R.; Fabbricino, M.; Ferrraro, A.; Lerza, S.; Marotta, R.; Pirozzi, F.; Race, M. Simultaneous removal of Cr(III) from high contaminated soil and recovery of lactic acid from the spent solution. *J. Environ. Manag.* **2020**, *268*, 110584. [CrossRef] [PubMed]
2. Wang, L.; Rinklebe, J.; Tack, F.M.G.; Hou, D. A review of green remediation strategies for heavy metal contaminated soil. *Soil Use Manag.* **2021**, *37*, 936–963. [CrossRef]
3. Yang, Z.; Wang, D.; Wang, G.; Zhang, S.; Cheng, Z.; Xian, J.; Pu, Y.; Li, T.; Jia, Y.; Li, Y.; et al. Removal of Pb, Zn, Ni and Cr from industrial sludge by biodegradable washing agents: Caboxyethylthiosuccinic acid and itaconic-acrylic acid. *J. Environ. Chem. Eng.* **2021**, *9*, 105846. [CrossRef]
4. Gusiatin, Z.M.; Kumpiene, J.; Janiszewska, S.; Kasiński, S.; Pecio, M.; Piec, R.; Radziemska, M. A mineral by-product from gasification of poultry feathers for removing Cd from highly contaminated synthetic wastewater. *Minerals* **2020**, *10*, 1048. [CrossRef]
5. Klik, B.; Gusiatin, Z.M.; Kulikowska, D. A holistic approach to remediation of soil contaminated with Cu, Pb and Zn with sewage sludge-derived washing agents and synthetic chelator. *J. Clean Prod.* **2021**, *311*, 127664. [CrossRef]
6. Kulikowska, D.; Klik, B.K.; Gusiatin, Z.M.; Hajdukiewicz, K. Characteristics of humic substances from municipal sewage sludge: A case study. *Desalination Water Treat.* **2019**, *144*, 57–64. [CrossRef]
7. Futalan, C.M.; Kim, J.; Yee, J.J. Adsorptive treatment via simultaneous removal of copper, lead and zinc from soil washing wastewater using spent coffee grounds. *Water Sci. Technol.* **2019**, *79*, 1029–1041. [CrossRef]
8. Wang, Q.; Chen, J. Recovery of EDTA from soil-washing wastewater with calcium-hydroxide-enhanced sulfide precipitation. *Chemosphere* **2019**, *237*, 124286. [CrossRef]
9. Ferraro, A.; van Hullebusch, E.D.; Huguenot, D.; Fabbricino, M.; Esposito, G. Application of an electrochemical treatment for EDDS soil washing solution regeneration and reuse in a multi-step soil washing process: Case of a Cu contaminated soil. *J. Environ. Manag.* **2015**, *163*, 62–69. [CrossRef]
10. Futalan, C.M.; Phatai, P.; Kim, J.; Maulana, A.Y.; Yee, J.J. Treatment of soil washing wastewater via adsorption of lead and zinc using graphene oxide. *Environ. Sci. Pollut. Res.* **2019**, *26*, 17292–17304. [CrossRef]
11. de Oliveira Silva, K.N.; Rodrigo, M.A.; dos Santos, E.V. Electrochemical treatment of soil-washing effluent with boron-doped diamond electrodes: A review. *Curr. Opin. Solid State Mater. Sci.* **2021**, *25*, 100962. [CrossRef]

12. Zhan, L.; Chen, M. The improvement effects of different treatment methods of soil wastewater washing on environmental pollution. *Water* **2020**, *12*, 2329. [CrossRef]
13. Reynier, N.; Coudert, L.; Blais, J.F.; Mercier, G.; Besner, S. Treatment of contaminated soil leachate by precipitation, adsorption and Ion exchange. *J. Environ. Chem. Eng.* **2015**, *3*, 977–985. [CrossRef]
14. Fabbricino, M.; Ferraro, A.; Luongo, V.; Pontoni, L.; Race, M. Soil washing optimization, recycling of the solution, and ecotoxicity assessment for the remediation of Pb-contaminated sites using EDDS. *Sustainability* **2018**, *10*, 636. [CrossRef]
15. Makino, T.; Takano, H.; Kamiya, T.; Itou, T.; Sekiya, N.; Inahara, M.; Sakurai, Y. Restoration of cadmium-contaminated paddy soils by washing with ferric chloride: Cd extraction mechanism and bench-scale verification. *Chemosphere* **2008**, *70*, 1035–1043. [CrossRef]
16. Gusiatin, Z.M.; Kulikowska, D.; Klik, B. New-generation washing agents in remediation of metal-polluted soils and methods for washing effluent treatment: A review. *Int. J. Environ. Res. Public Health* **2020**, *17*, 6220. [CrossRef]
17. Park, S.M.; Kim, T.; Kim, Y.H.; Baek, K. Hybrid process of combined soil washing and selective adsorption to treat Cs-contaminated soil. *Chem. Eng. J.* **2021**, *423*, 129921. [CrossRef]
18. Xu, Z.; Guo, H.; Liu, T.; Zhang, W.; Ma, X. Regeneration of washing effluents for remediation of petroleum-hydrocarbons-contaminated soil by corncob-based biomass materials. *ACS Omega* **2019**, *4*, 18711–18717. [CrossRef]
19. Chen, X.; Yin, L.; Zhou, H.; Liu, J.; Ai, X.; Huang, K. Efficient removal of lead from washing effluent of lead-contaminated soil with garlic peel. *Chem. Res. Chin. Univ.* **2018**, *34*, 1020–1027. [CrossRef]
20. Hughes, D.L.; Afsar, A.; Laventine, D.M.; Shaw, E.J.; Harwood, L.M.; Hodson, M.E. Metal removal from soil leachates using DTPA-functionalised maghemite nanoparticles, a potential soil washing technology. *Chemosphere* **2018**, *209*, 480–488. [CrossRef] [PubMed]
21. Li, H.; Dong, X.; da Silva, E.B.; de Oliveira, L.M.; Chen, Y.; Ma, L.Q. Mechanisms of metal sorption by biochars: Biochar characteristics and modifications. *Chemosphere* **2017**, *178*, 466–478. [CrossRef]
22. Jjagwe, J.; Olupot, P.W.; Menya, E.; Kalibbala, H.M. Synthesis and application of granular activated carbon from biomass waste materials for water treatment: A review. *J. Bioresour. Bioprod.* **2021**, *6*, 292–322. [CrossRef]
23. Obey, G.; Adelaide, M.; Ramaraj, R. Biochar derived from non-customized matamba fruit shell as an adsorbent for wastewater treatment. *J. Bioresour. Bioprod.* **2022**, *7*, 109–115. [CrossRef]
24. Xiang, W.; Zhang, X.; Chen, J.; Zou, W.; He, F.; Hu, X.; Tsang, D.C.W.; Ok, Y.S.; Gao, B. Biochar technology in wastewater treatment: A critical review. *Chemosphere* **2020**, *252*, 126539. [CrossRef] [PubMed]
25. Kulikowska, D.; Gusiatin, Z.M.; Bułkowska, K.; Klik, B. Feasibility of using humic substances from compost to remove heavy metals (Cd, Cu, Ni, Pb, Zn) from contaminated soil aged for different periods of time. *J. Hazard. Mater.* **2015**, *300*, 882–891. [CrossRef]
26. EBC. *(2012–2022) European Biochar Certificate—Guidelines for a Sustainable Production of Biochar*; Carbon Standards International (CSI): Frick, Switzerland, 2022; Available online: http://European-Biochar.Org (accessed on 1 April 2023).
27. Pueyo, M.; Mateu, J.; Rigol, A.; Vidal, M.; López-Sánchez, J.F.; Rauret, G. Use of the modified BCR three-step sequential extraction procedure for the study of trace element dynamics in contaminated soils. *Environ. Pollut.* **2008**, *152*, 330–341. [CrossRef]
28. Damian, G.E.; Micle, V.; Sur, I.M. Mobilization of Cu and Pb from multi-metal contaminated soils by dissolved humic substances extracted from leonardite and factors affecting the process. *J. Soils Sediments* **2019**, *19*, 2869–2881. [CrossRef]
29. Piccolo, A.; Spaccini, R.; De Martino, A.; Scognamiglio, F.; di Meo, V. Soil washing with solutions of humic substances from manure compost removes heavy metal contaminants as a function of humic molecular composition. *Chemosphere* **2019**, *225*, 150–156. [CrossRef] [PubMed]
30. Marzeddu, S.; Décima, M.A.; Camilli, L.; Bracciale, M.P.; Genova, V.; Paglia, L.; Marra, F.; Damizia, M.; Stoller, M.; Chiavola, A.; et al. Physical-chemical characterization of different carbon-based sorbents for environmental applications. *Materials* **2022**, *15*, 7162. [CrossRef]
31. Ahmedna, M.; Marshall, W.E.; Husseiny, A.A.; Rao, R.M.; Goktepe, I. The use of nutshell carbons in drinking water filters for removal of trace metals. *Water Res.* **2004**, *38*, 1062–1068. [CrossRef]
32. Hsieh, C.T.; Teng, H. Influence of mesopore volume and adsorbate size on adsorption capacities of activated carbons in aqueous solutions. *Carbon* **2000**, *38*, 863–869. [CrossRef]
33. Lawal, A.A.; Hassan, M.A.; Zakaria, M.R.; Yusoff, M.Z.M.; Norrrahim, M.N.F.; Mokhtar, M.N.; Shirai, Y. Effect of oil palm biomass cellulosic content on nanopore structure and adsorption capacity of biochar. *Bioresour. Technol.* **2021**, *332*, 125070. [CrossRef]
34. Jindo, K.; Mizumoto, H.; Sawada, Y.; Sanchez-Monedero, M.A.; Sonoki, T. Physical and chemical characterization of biochars derived from different agricultural residues. *Biogeosciences* **2014**, *11*, 6613–6621. [CrossRef]
35. Décima, M.A.; Marzeddu, S.; Barchiesi, M.; Di Marcantonio, C.; Chiavola, A.; Boni, M.R. A review on the removal of carbamazepine from aqueous solution by using activated carbon and biochar. *Sustainability* **2021**, *13*, 11760. [CrossRef]
36. Mukome, F.N.; Zhang, X.; Silva, L.C.; Six, J.; Parikh, S.J. Use of chemical and physical characteristics to investigate trends in biochar feedstocks. *J. Agric. Food Chem.* **2013**, *61*, 2196–2222. [CrossRef] [PubMed]
37. Hmid, A.; Mondelli, D.; Fiore, S.; Fanizzi, F.P.; Al Chami, Z.; Dumontet, S. Production and characterization of biochar from three-phase olive mill waste through slow pyrolysis. *Biomass Bioenergy* **2014**, *71*, 330–339. [CrossRef]
38. Askeland, M.; Clarke, B.; Paz-Ferreiro, J. Comparative characterization of biochars produced at three selected pyrolysis temperatures from common woody and herbaceous waste streams. *PeerJ* **2019**, *7*, e6784. [CrossRef]

39. Al-Wabel, M.I.; Al-Omran, A.; El-Naggar, A.H.; Nadeem, M.; Usman, A.R.A. Pyrolysis temperature induced changes in characteristics and chemical composition of biochar produced from conocarpus wastes. *Bioresour. Technol.* **2013**, *131*, 374–379. [CrossRef]
40. Chu, G.; Zhao, J.; Liu, Y.; Lang, D.; Wu, M.; Pan, B.; Steinberg, C.E.W. The relative importance of different carbon structures in biochars to carbamazepine and bisphenol a sorption. *J. Hazard. Mater.* **2019**, *373*, 106–114. [CrossRef]
41. Socrates, G. *Infrared and Raman Characteristic Group Frequencies: Tables and Charts*; John Wiley & Sons: Hoboken, NJ, USA, 2001.
42. Keiluweit, M.; Nico, P.S.; Johnson, M.G.; Kleber, M. Dynamic molecular structure of plant biomass-derived black carbon (biochar). *Environ. Sci. Technol.* **2010**, *44*, 12471253. [CrossRef]
43. Xiaoli, C.; Yongxia, H.; Guixiang, L.; Xin, Z.; Youcai, Z. Spectroscopic studies of the effect of aerobic conditions on the chemical characteristics of humic acid in landfill leachate and its implication for the environment. *Chemosphere* **2013**, *91*, 1058–1063. [CrossRef] [PubMed]
44. Morterra, C.; Low, M.J.D. The nature of the 1600 cm^{-1} band of carbons. *Spectrosc. Lett.* **1982**, *15*, 689–697. [CrossRef]
45. Yang, C.Q.; Simms, J.R. Comparison of photoacoustic, diffuse reflectance and transmission infrared spectroscopy for the study of carbon fibres. *Fuel* **1995**, *74*, 543–548. [CrossRef]
46. Lu, H.; Zhang, W.; Yang, Y.; Huang, X.; Wang, S.; Qiu, R. Relative distribution of Pb2+ sorption mechanisms by sludge-derived biochar. *Water Res.* **2012**, *46*, 854–862. [CrossRef]
47. Qiu, B.; Tao, X.; Wang, H.; Li, W.; Ding, X.; Chu, H. Biochar as a low-cost adsorbent for aqueous heavy metal removal: A review. *J. Anal. Appl. Pyrolysis* **2021**, *155*, 105081. [CrossRef]
48. Shan, R.; Shi, Y.; Gu, J.; Wang, Y.; Yuan, H. Single and competitive adsorption affinity of heavy metals toward peanut shell-derived biochar and its mechanisms in aqueous systems. *Chin. J. Chem. Eng.* **2020**, *28*, 1375–1383. [CrossRef]
49. Chen, X.; Chen, G.; Chen, L.; Chen, Y.; Lehmann, J.; McBride, M.B.; Hay, A.G. Adsorption of copper and zinc by biochars produced from pyrolysis of hardwood and corn straw in aqueous solution. *Bioresour. Technol.* **2011**, *102*, 8877–8884. [CrossRef]
50. Ambaye, T.G.; Vaccari, M.; van Hullebusch, E.D.; Amrane, A.; Rtimi, S. Mechanisms and adsorption capacities of biochar for the removal of organic and inorganic pollutants from industrial wastewater. *Int. J. Environ. Sci. Technol.* **2021**, *18*, 3273–3294. [CrossRef]
51. Zhao, M.; Dai, Y.; Zhang, M.; Feng, C.; Qin, B.; Zhang, W.; Zhao, N.; Li, Y.; Ni, Z.; Xu, Z.; et al. Mechanisms of Pb and/or Zn adsorption by different biochars: Biochar characteristics, stability, and binding energies. *Sci. Total Environ.* **2020**, *717*, 136894. [CrossRef] [PubMed]
52. Fei, Y.H.; Zhang, Z.; Ye, Z.; Wu, Q.; Tang, Y.T.; Xiao, T. Roles of soluble minerals in Cd sorption onto rice straw biochar. *J. Environ. Sci.* **2022**, *113*, 64–71. [CrossRef]
53. Li, J.; Zheng, L.; Wang, S.L.; Wu, Z.; Wu, W.; Niazi, N.K.; Shaheen, S.M.; Rinklebe, J.; Bolan, N.; Ok, Y.S.; et al. Sorption mechanisms of lead on silicon-rich biochar in aqueous solution: Spectroscopic investigation. *Sci. Total Environ.* **2019**, *672*, 572–582. [CrossRef]
54. Zhang, F.; Wang, X.; Yin, D.; Peng, B.; Tan, C.; Liu, Y.; Tan, X.; Wu, S. Efficiency and mechanisms of Cd removal from aqueous solution by biochar derived from water hyacinth (eichornia crassipes). *J. Environ. Manag.* **2015**, *153*, 68–73. [CrossRef] [PubMed]
55. Zhao, J.J.; Shen, X.J.; Domene, X.; Alcañiz, J.M.; Liao, X.; Palet, C. Comparison of biochars derived from different types of feedstock and their potential for heavy metal removal in multiple-metal solutions. *Sci. Rep.* **2019**, *9*, 9869. [CrossRef] [PubMed]
56. Yan, X.; Du, W.; Ma, C.; Cheng, S.; Li, X. Humic acid adsorption behavior and mechanism comparison between biochars and activated carbon. *Desalination Water Treat.* **2020**, *173*, 213–222. [CrossRef]
57. Liu, H.; Ye, M.; Dong, X.; Ren, Z.; Long, S.; Lichtfouse, E. Removal of humic substances by the synergistic effect of biochar adsorption and activation of persulfate. *J. Water Process Eng.* **2021**, *44*, 102428. [CrossRef]
58. Liu, Z.; Zhang, F.S. Removal of lead from water using biochars prepared from hydrothermal liquefaction of biomass. *J. Hazard. Mater.* **2009**, *167*, 933–939. [CrossRef] [PubMed]
59. Mohan, D.; Pittman, C.U.; Bricka, M.; Smith, F.; Yancey, B.; Mohammad, J.; Steele, P.H.; Alexandre-Franco, M.F.; Gómez-Serrano, V.; Gong, H. Sorption of arsenic, cadmium, and lead by chars produced from fast pyrolysis of wood and bark during bio-oil production. *J. Colloid Interface Sci.* **2007**, *310*, 57–73. [CrossRef]
60. Anoop Krishnan, K.; Sreejalekshmi, K.G.; Baiju, R.S. Nickel(II) adsorption onto biomass based activated carbon obtained from sugarcane bagasse pith. *Bioresour. Technol.* **2011**, *102*, 10239–10247. [CrossRef]
61. Manjuladevi, M.; Anitha, R.; Manonmani, S. Kinetic study on adsorption of Cr(VI), Ni(II), Cd(II) and Pb(II) Ions from aqueous solutions using activated carbon prepared from cucumis melo peel. *Appl. Water Sci.* **2018**, *8*, 36. [CrossRef]
62. Zhou, R.; Zhang, M.; Shao, S. Optimization of target biochar for the adsorption of target heavy metal Ion. *Sci. Rep.* **2022**, *12*, 13662. [CrossRef]
63. Liu, J.; Wang, H.; Ma, N.; Zhou, B.; Chen, H.; Yuan, R. Optimization of the raw materials of biochars for the adsorption of heavy metal Ions from aqueous solution. *Water Sci. Technol.* **2022**, *85*, 2869–2881. [CrossRef] [PubMed]
64. Hameed, B.H. Evaluation of papaya seeds as a novel non-conventional low-cost adsorbent for removal of methylene blue. *J. Hazard. Mater.* **2009**, *162*, 939–944. [CrossRef] [PubMed]
65. Fierro, V.; Torné-Fernández, V.; Montané, D.; Celzard, A. Adsorption of phenol onto activated carbons having different textural and surface properties. *Microporous Mesoporous Mater.* **2008**, *111*, 276–284. [CrossRef]

66. Ofomaja, A.E. Intraparticle diffusion process for lead(II) biosorption onto mansonia wood sawdust. *Bioresour. Technol.* **2010**, *101*, 5868–5876. [CrossRef]
67. Chen, J.; Lu, X. Equilibrium and kinetics studies of Cd(II) sorption on zeolite nax synthesized from coal gangue. *J. Water Reuse Desalin.* **2018**, *8*, 94–101. [CrossRef]
68. Zhu, Q.; Moggridge, G.D.; D'Agostino, C. Adsorption of pyridine from aqueous solutions by polymeric adsorbents MN 200 and MN 500. Part 2: Kinetics and diffusion analysis. *Chem. Eng. J.* **2016**, *306*, 1223–1233. [CrossRef]
69. Zhang, J.; Lu, W.; Zhan, S.; Qiu, J.; Wang, X.; Wu, Z.; Li, H.; Qiu, Z.; Peng, H. Adsorption and mechanistic study for humic acid removal by magnetic biochar derived from forestry wastes functionalized with Mg/Al-LDH. *Sep. Purif. Technol.* **2021**, *276*, 119296. [CrossRef]
70. Omri, A.; Benzina, M.; Trabelsi, W.; Ammar, N. Adsorptive removal of humic acid on activated carbon prepared from almond shell: Approach for the treatment of industrial phosphoric acid solution. *Desalination Water Treat.* **2014**, *52*, 2241–2252. [CrossRef]
71. Tan, K.L.; Hameed, B.H. Insight into the adsorption kinetics models for the removal of contaminants from aqueous solutions. *J. Taiwan Inst. Chem. Eng.* **2017**, *74*, 25–48. [CrossRef]
72. Koinuma, M.; Tateishi, H.; Hatakeyama, K.; Miyamoto, S.; Ogata, C.; Funatsu, A.; Taniguchi, T.; Matsumoto, Y. Analysis of reduced graphene oxides by X-ray photoelectron spectroscopy and electrochemical capacitance. *Chem. Lett.* **2013**, *42*, 924–926. [CrossRef]
73. Rabchinskii, M.K.; Ryzhkov, S.A.; Kirilenko, D.A.; Ulin, N.V.; Baidakova, M.V.; Shnitov, V.V.; Pavlov, S.I.; Chumakov, R.G.; Stolyarova, D.Y.; Besedina, N.A.; et al. From graphene oxide towards aminated graphene: Facile synthesis, its structure and electronic properties. *Sci. Rep.* **2020**, *10*, 6902. [CrossRef] [PubMed]
74. Maddi, C.; Bourquard, F.; Barnier, V.; Avila, J.; Asensio, M.C.; Tite, T.; Donnet, C.; Garrelie, F. Nano-architecture of nitrogen-doped graphene films synthesized from a solid CN source. *Sci. Rep.* **2018**, *8*, 3247. [CrossRef] [PubMed]
75. Barinov, A.; Malcioğlu, O.B.; Fabris, S.; Sun, T.; Gregoratti, L.; Dalmiglio, M.; Kiskinova, M. Initial stages of oxidation on graphitic surfaces: Photoemission study and density functional theory calculations. *J. Phys. Chem. C* **2009**, *113*, 9009–9013. [CrossRef]
76. Oh, Y.J.; Yoo, J.J.; Kim, Y., II; Yoon, J.K.; Yoon, H.N.; Kim, J.H.; Park, S. Bin oxygen functional groups and electrochemical capacitive behavior of incompletely reduced graphene oxides as a thin-film electrode of supercapacitor. *Electrochim. Acta* **2014**, *116*, 118–128. [CrossRef]
77. Arrigo, R.; Hävecker, M.; Wrabetz, S.; Blume, R.; Lerch, M.; McGregor, J.; Parrott, E.P.J.; Zeitler, J.A.; Gladden, L.F.; Knop-Gericke, A.; et al. Tuning the acid/base properties of nanocarbons by functionalization via amination. *J. Am. Chem. Soc.* **2010**, *132*, 9616–9630. [CrossRef]
78. Stobinski, L.; Lesiak, B.; Malolepszy, A.; Mazurkiewicz, M.; Mierzwa, B.; Zemek, J.; Jiricek, P.; Bieloshapka, I. Graphene oxide and reduced graphene oxide studied by the XRD, TEM and electron spectroscopy methods. *J. Electron. Spectros. Relat. Phenomena* **2014**, *195*, 145–154. [CrossRef]

energies

Article

Phosphorus Removal Rate and Efficiency in an Electrochemical Sequencing Reactor for the Treatment of Wastewater with Low Organic Carbon Content

Artur Mielcarek [1,*], Kamil Łukasz Bryszewski [2], Joanna Rodziewicz [1], Karolina Kłobukowska [1] and Wojciech Janczukowicz [1]

[1] Faculty of Geoengineering, University of Warmia and Mazury in Olsztyn, Warszawska St. 117a, 10-719 Olsztyn, Poland; joanna.rodziewicz@uwm.edu.pl (J.R.); karolina.klobukowska@uwm.edu.pl (K.K.); jawoj@uwm.edu.pl (W.J.)

[2] Hanna Instruments Sp. z.o.o., Al. J. Piłsudskiego St. 73, 10-449 Olsztyn, Poland; r1@hanna-polska.com

* Correspondence: artur.mielcarek@uwm.edu.pl; Tel.: +48-89-523-38-46

Abstract: Energy is essential for the operation of wastewater treatment systems. Simultaneously, it can be a factor facilitating the electrochemical purification processes. Previous studies have shown that under specific conditions, there is no technological justification for using bio-electrochemical reactors designed for the simultaneous removal of both phosphorus and nitrogen compounds. This is because similar dephosphatation process effects can be achieved in an electrochemical reactor. Additionally, in a bio-electrochemical reactor, a portion of the organic substrate introduced for biological treatment is lost due to the electrocoagulation process. The aim of the research was to determine the influence of low direct current densities (0.4–2.0 A/m^2) on the rate and efficiency of phosphorus and other compound removal in a sequencing electrochemical reactor treating real wastewater from a greenhouse with low organic compound content. In the conducted studies, an increase in electric current density resulted in an increase in the removal rates of phosphorus from 26.45 to 34.79 mg/L·h, nitrogen from 2.07 to 6.58 mg/L·h, and organic compounds from 0.44 to 1.50 mg/L·h. This corresponded to maximum removal efficiencies of 88.6 ± 2.5% for phosphorus, 7.4 ± 2.5% for nitrogen, and 51.1 ± 8.3% for organic compounds. As a result of electrocoagulation, sludge rich in phosphorus was obtained, ranging from 347 ± 38 mg/L (18.1% P) to 665 ± 36 mg/L (11.7% P). The obtained results can be utilized in the future for the development of two-stage systems for wastewater treatment with a low content of organic compounds, aiming at the removal of phosphorus and nitrogen.

Keywords: electrocoagulation; phosphorus recovery; iron electrode; electrocoagulation kinetics; soilless cultivation; quality of electrocoagulation sludge

1. Introduction

Soilless cultivation of tomatoes is often conducted in open systems, where the excess nutrient-rich medium (wastewater) is discharged into the environment in an uncontrolled manner. The concentrations of nutrients in wastewater, such as nitrate reaching up to 614 mg N/L and total phosphorus up to 370 mg P/L, pose a serious threat to both aquatic and groundwater environments [1–3]. On the other hand, this represents a case of wasting biogenic substances that could potentially be reused for fertilizing specific types of crops [4–6].

There are methods that allow for the recirculation of wastewater generated in soilless plant cultivation and their reuse. These methods primarily rely on the analysis of electrical conductivity and disinfection. However, these technologies do not solve the ultimate problem of managing drainage water by discharging it into the environment, as well as the challenge of properly balancing nutrient solutions or the accumulation of chloride and sodium ions [7,8]. Therefore, their practical application is mainly limited to soilless seedling cultivation, where changes in the concentration of individual nutrient components undergo minimal variations due to low transpiration and nutrient uptake by the plants. It should be emphasized that even with the application of recirculation, a portion of the wastewater must still be directed outside the system. Hence, there is a need to explore methods for their proper disposal [9].

The unique composition of wastewater led researchers to hypothesize that effective purification could be achieved by combining physicochemical processes such as electrocoagulation with biological processes. Bio-electrochemical reactors were considered a potential solution [10,11]. A previously tested solution for greenhouse wastewater was the flow-through aerobic rotating electrobiological disc contactor (REBDC), where the flow of electric current was expected to create favorable conditions for the simultaneous removal of phosphorus through electrocoagulation, nitrate through hydrogenotrophic denitrification, and electrochemical reduction [12]. To enhance nitrate removal efficiency, an external source of organic carbon was introduced into the REBDC, assuming that this would increase the participation of heterotrophic denitrification and biomass growth in nitrate removal [12]. The cited studies demonstrated that in the case of the aerobic bio-electrochemical reactor like REBDC, the introduction of an external carbon source created conditions not only for heterotrophic and hydrogenotrophic denitrification but also for nitrogen consumption in biofilm growth. This led to higher nitrogen removal effectiveness in REBDC compared to the rotating electrochemical disk contactor (RECDC). Despite this technological intervention, the nitrogen concentrations in the effluent from REBDC did not meet the required standards for wastewater discharged into the environment. Simultaneously, for phosphorus, its concentration in wastewater treated with electric current densities of 2.50, 5.00, and 10.00 A/m^2 and retention times of 8, 12, and 24 h was below 1.0 mg P/L [12]. It is worth noting that similarly low phosphorus concentrations were observed in the effluent from RECDC for an electric current density of 10.00 A/m^2, where phosphorus removal was solely a result of the electrocoagulation process [12]. These studies revealed that under specific conditions, there is no technological justification for using bio-electrochemical reactors designed for the simultaneous removal of both phosphorus and nitrogen compounds, as similar dephosphatation effects can be achieved in an electrochemical reactor. The external source of organic carbon did not significantly contribute to the efficiency of phosphorus removal [12].

This hypothesis was confirmed by research results using sequential anaerobic reactors: a bio-electrochemical and an electrochemical reactor of the SBBR type (sequencing batch biofilm reactor) [13]. Higher efficiencies of nitrogen and phosphorus removal were achieved in reactors subjected to high current densities (4.4 A/m^2, 8.8 A/m^2, and 13.3 A/m^2) compared to aerobic RECDC and REBDC reactors. At the highest current density, the phosphorus concentration in the effluent from the electrochemical reactor was lower than in the bio-electrochemical reactor. The studies also demonstrated that in the bio-electrochemical reactor, a portion of the introduced organic substrate was not utilized by heterotrophic denitrifying bacteria but was removed through the electrochemical process of electrocoagulation [13,14].

The separation of electrochemical and biological processes and their implementation in sequentially connected SBBR-type reactors (electrochemical and biological) could potentially impact electricity consumption and the amount of organic carbon required to support denitrification. It is also expected to result in lower concentrations of biogenic and organic compounds in treated wastewater. Validating these assumptions requires further research, which should consider the possibility of applying lower current densities than those commonly used [15–17]. This would provide additional support for this technological solution. It would also allow for the powering of such reactors with electrical energy from sources such as photovoltaic cells [18] or microbial fuel cells [19] without the need for converting direct current (DC) to alternating current (AC). This could facilitate the diversification of energy sources and reduce dependence on electricity from the grid.

The aim of this research was to determine the influence of electric current density on the kinetics (1) and efficiency (2) of phosphorus removal through electrochemical processes in a sequential reactor treating real wastewater from soilless tomato cultivation. Additionally, the scope of the study included the quantitative (3) and qualitative (4) analysis of the generated sludge. The impact of the applied electric current density on the nitrogen concentration (5) and organic compounds (6) in treated wastewater was also assessed. The obtained results may be utilized in the future to design two-stage systems for wastewater treatment with low organic compound content, targeting the removal of phosphorus and nitrogen.

2. Materials and Methods

2.1. Experimental Setup and Research Organization

The research was conducted on a sequential electrochemical reactor model. The reactor's construction and the research scheme are presented in Figure 1. To carry out the process, an anode made of iron with an electrochemical surface area of 0.023 m^2 was utilized, while the cathode consisted of discs with a total electrochemical surface area of 0.113 m^2. The difference in the size of electrode surfaces resulted from the technological solution applied in the reactor, which, in addition to electrochemical processes, can be used for biological treatment. In such a configuration, a biofilm can develop on the cathode. The discs were made of stainless steel and mounted on a vertically rotating shaft with a speed of 14 rpm. The electrodes were connected to a DC power supply (Rohde & Schwarz HMP 4040, Munich, Germany). The reactor's effective volume was 2 L. The hydraulic retention time (HRT) of the wastewater was set at 24 h based on previous studies. At the end of each cycle, the entire reactor volume was replaced, introducing a new batch of wastewater (2 L). The kinetics of pollutant removal were determined from samples collected at 0, 3, 6, 12, and 24 h. The following electric current densities (J) were applied: 0.4, 1.2, and 2.0 A/m^2, corresponding to current intensities of 0.045, 0.136, and 0.226 A, respectively. These values were selected based on the literature data and previous studies [12,20–22].

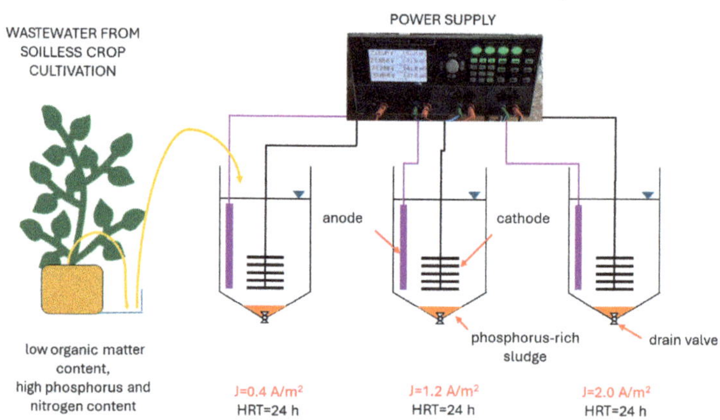

Figure 1. Research scheme.

The obtained results from kinetic studies allowed for the determination of reaction orders, the rate of organic compound utilization, and the removal of nitrogen and phosphorus compounds. Reaction rate constants and statistical analysis of the results were performed using the STATISTICA 13.3 PL computer program. A significance level of $\alpha = 0.05$ was adopted.

2.2. Wastewater

The study utilized wastewater (drainage water) from a greenhouse where tomatoes were cultivated using mineral wool as a substrate. The greenhouse, located in northern Poland, covers an area of approximately 20 hectares. Wastewater for the study was collected in July and August 2023 and exhibited the following physicochemical indicators: pH—6.21; electrolytic conductivity—6.213 $\pm$ 0.202 mS/cm; total organic carbon (TOC)—15.6 $\pm$ 3.1 mg C/L; total nitrogen (TN)—536.9 $\pm$ 10.7 mg N/L; nitrate—535.71 $\pm$ 7.40 mg N/L; nitrite—0.07 $\pm$ 0.05 mg N/L; ammonium nitrogen—0.02 $\pm$ 0.01 mg N/L; and total phosphorus (TP)—103.8 $\pm$ 5.3 mg P/L. The wastewater was stored in a tank with a volume of 1 m^3 made of synthetic material. Based on the measurement of wastewater parameters during retention, it can be concluded that they do not exhibit changes in composition and can be retained in this way for at least 6 months.

2.3. *Physicochemical Analysis of Wastewater*

The following parameters of wastewater were measured: pH value, electrolytic conductivity, and temperature using an HQ4300 multimeter (HACH Company, Loveland, CO, USA); total nitrogen (TN) and total organic carbon (TOC) using a TOC-L CPH/CPN device (Shimadzu Corporation, Kyoto, Japan; oxidative combustion-chemiluminescence method—TN; oxidizing incineration—infrared analysis—TOC); total phosphorus (TP; HACH Lange LCK 348–350 method), ammonia nitrogen ($N\text{-}NH_4$; HACH Lange LCK303-305), nitrite ($N\text{-}NO_2$; HACH Lange LCK341-342), and iron (Fe; HACH Lange LCK 321 and LCK 521) using a DR5000 HACH Lange Spectrophotometer (Malente, Germany).

2.4. *Quantity and Quality of Sludge*

The quantity of generated sludge in the form of total suspended solids (TSS) and volatile suspended solids (VSS) was determined according to APHA 2012 [20]. For qualitative analysis of the sludge, the content of nitrogen (N), phosphorus (P), carbon (C), and iron (Fe) was determined. For this purpose, the sludge was dried at 105 °C. The determination of elements (Fe, P) involved mineralizing 0.2 g of dried sludge in a Teflon container (easy prep) with the addition of 7.5 mL of 69–70% nitric acid and 2.5 mL of 36% hydrochloric acid. Microwave-assisted mineralization was carried out using a MARS CEM 5 device (CEM Corporation, Matthews, NC, USA) at a temperature of 210 °C for 15 min. After the completion of sample dissolution, the samples were quantitatively transferred to a Falcon-type container and diluted to a volume of 20 mL using 2% nitric acid [21,22]. The content of elements (Fe, P) in the prepared sample was determined using inductively coupled plasma optical emission spectrometry (ICP-OES; Avio 220 ICP-OES; Perkin Elmer, Waltham, MA, USA). The amount of N and C in the dried sludge was determined using a Flash 2000 elemental analyzer (Thermo Fisher Scientific, Waltham, MA, USA).

3. Results and Discussion

The conducted research aimed to determine the impact of electrical current density on the kinetics and efficiency of treating wastewater generated in a greenhouse during soilless tomato cultivation. These wastewaters are characterized by low organic compound content, high phosphorus and nitrogen compound content, and electrolytic conductivity. The electrocoagulation (EC) process employed integrates the benefits of coagulation, flotation, and electrochemistry into a single system [17]. Throughout the electrocoagulation process, both reduction and oxidation processes take place at both the anode and cathode, influencing the characteristics of pollutants in the treated wastewater, as well as the interactions among them and with the dissolved metal [23]. Due to the flow of electric current, the anode undergoes dissolution (Equations (1)–(3)), and water electrolysis occurs at the cathode, leading to the release of hydroxyl ions and hydrogen (Equation (4)) [24].

Anode reactions:

$$\mathrm{Fe(s)} \rightarrow \mathrm{Fe^{n+}(aq)} + \mathrm{ne^{-1}} \tag{1}$$

$$\mathrm{4Fe^{2+}(aq)} + \mathrm{10H_2O} + \mathrm{O_2(aq)} \rightarrow \mathrm{4Fe(OH)_3(s)} + \mathrm{8H^+} \tag{2}$$

$$\mathrm{Fe^{2+}(aq)} + \mathrm{2OH^-} \rightarrow \mathrm{Fe(OH)_2(s)} \tag{3}$$

Cathode reaction:

$$\mathrm{2e^{-1}} + \mathrm{2H_2O} \rightarrow \mathrm{H_2} + \mathrm{2OH^-} \tag{4}$$

The research demonstrated an increase in the rate of pollutant removal with the rise in applied electrical current density. Simultaneously, the applied solution ensured high efficiency.

3.1. pH, Electrolytic Conductivity, Iron Content

The pH of treated wastewater significantly impacts both physicochemical and biological processes. Depending on the pH, the electrode dissolves and forms metal ions. The formation of Fe ions is more complex than Al ions when using aluminum anodes. The ferric ion predominates in highly acidic solutions, while the ferrous ion is more common at pH < 6.5. Most studies indicate that Fe^{2+} is initially formed by iron electrolysis and then oxidized to Fe^{3+} by dissolved oxygen under suitable conditions (e.g., pH 5–9). Subsequently, it hydrolyzes to form insoluble $Fe(OH)_3(s)$ and FeOOH(s) [23,25]. Fe^{2+} is highly soluble but a weak coagulant, with a lower pollutant adsorption capacity compared to Fe^{3+}. Simultaneously, in the electrocoagulation reactor, an uneven pH distribution occurs during the flow of electric current. The solution around the anode becomes slightly acidic due to the formation of H^+, while the solution around the cathode becomes slightly alkaline due to the production of OH^- and the release of H_2. This may affect the efficiency of electrocoagulation in removing pollutants [26]. In most studies using electrocoagulation, pH increases due to the production of OH^- ions. However, in the conducted research, no such relationship was observed (Figure 2A). With an initial pH of 6.1, as the electric current intensity increased, the pH decreased. For the highest current density (2.0 A/m^2), the pH of the treated wastewater was 2.8, possibly due to high sulfate concentrations. Greenhouse wastewater contains 58 to 1425 mg SO_4^{2-} [27]. Nariyan et al. [28], treating sulfur-rich mine waters with a concentration of 13,000 mg SO_4^{2-}/L using electrocoagulation at a density of 25 mA/cm^2, achieved removal efficiencies of 83.75% and 84.38% for sulfates in monopolar and bipolar electrode configurations, including aluminum and stainless steel electrodes. In their study, the initial pH was 11.3, and after the purification process, it decreased to pH 6.8 and 6.6 for monopolar and bipolar electrode configurations, respectively. In Lei et al.'s research [29], the pH of municipal wastewater decreased during purification when treating wastewater with a pH below 6. Purifying wastewater with a pH of 5 resulted in a pH drop to 3.2 after 24 h using a platinum anode and titanium cathode. Under these conditions, phosphorus was removed in the form of struvite with an efficiency of 50% at a current density of 8.3 A/m^2. The pH decrease in the cited studies was attributed to the consumption of OH^- ions by Mg^{2+} and NH_4^+, leading to the accumulation of H^+ ions and a subsequent drop in pH.

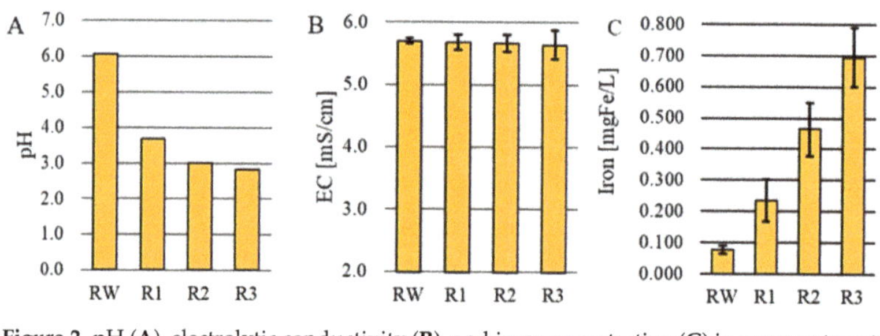

Figure 2. pH (**A**), electrolytic conductivity (**B**), and iron concentration (**C**) in raw wastewater (RW) and treated wastewater (R1–R3).

The electrolytic conductivity of greenhouse wastewater depends on the quality and quantity of ions in the aqueous solution, varying widely based on weather conditions and plant growth phases [3]. Electrolytic conductivity ensures the free flow of electric current in treated wastewater and reduces electricity consumption [30]. The type of cations and anions influences the electrocoagulation kinetics, affecting the ability of coagulants to form flocs. The electrolytic conductivity of the applied wastewater was 5.704 $\pm$ 0.049 mS/cm

(Figure 2B). Different electric current densities did not significantly affect the electrolytic conductivity of treated wastewater despite achieving higher pollutant removal efficiencies, as described in subsequent chapters. This may be because the removed ions did not play a major role in creating electrolytic conductivity, and/or their removal was compensated by an increase in iron ion concentration in the treated wastewater compared to raw wastewater.

Iron is a commonly used electrode material due to its reliability and availability [16]. Iron electrodes are less efficient than aluminum ones but are significantly cheaper, which is crucial for wastewater treatment. Additionally, greenhouse wastewater naturally contains iron ions essential for plant development, and their presence, unlike aluminum ions, does not negatively impact plants [31]. The overall iron concentration in raw wastewater was 0.077 ± 0.012 mg Fe/L.

With an increase in electric current density, there was an increase in iron ion concentration in treated wastewater, reaching 0.695 ± 0.095 mg Fe/L at J = 2.0 A/m^2 (Figure 3). This value is sometimes observed in raw wastewater as well [13]. This is also the permissible value for wastewater discharged into the environment (maximum allowable concentration of 10.0 mgFe/L) [32]. Excessive electric current density can cause secondary pollution of wastewater with ions from the electrode and shorten its lifespan [33]. Behbahani et al. [34], treating synthetic wastewater with an initial concentration of 400 mg P/L and pH 3 using electrocoagulation, removed 84.7% (iron electrode) and 94.15% (aluminum electrode) of orthophosphates. However, there was an increase in iron and aluminum concentrations in the solution to 2 mg Fe/L and 14.5 mg Al/L, respectively. In the cited study, despite a shorter electrocoagulation time (40 min), higher concentrations of iron and aluminum were noted than in the presented research (Figure 3). This is due to the much higher current density used in the cited study, which was 250 A/m^2. It also suggests the use of lower electric current densities when planning the subsequent use of treated wastewater, for example, in plant irrigation.

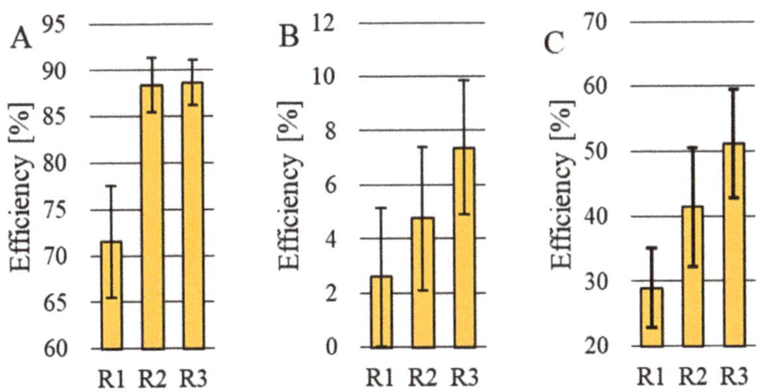

Figure 3. Efficiency of total phosphorus removal (**A**), total nitrogen removal (**B**), and total organic carbon removal (**C**).

3.2. Removal of Phosphorus Compounds, Nitrogen and Organic Compounds

The concentration and efficiency of phosphorus removal are directly associated with the concentration of iron ions available in the solution and the pH of the treated wastewater. The concentration of total phosphorus in the greenhouse wastewater subjected to purification was 88.0 ± 4.9 mg P/L. The average efficiency of total phosphorus removal is presented in Figure 3A.

The efficiency of phosphorus removal in electrochemical reactors through electrocoagulation increased with the applied current density, ranging from 71.5 ± 6.0% in R1 (pH = 3.7) to 88.6 ± 2.5% in R3 (pH = 2.8). The concentration of TP in the treated effluents in R2 and R3 was similar, measuring 10.2 ± 2.6 mg P/L and 10.0 ± 2.2 mg P/L, respectively (no statistically significant differences, $\alpha = 0.05$). In reactor R1, where the lowest current density was applied, the phosphorus concentration was significantly higher, reaching 25.1 ± 2.3 mg P/L. Analyzing the obtained results, the increase in electrical current density from 0.4 A/m^2 to 1.2 A/m^2, along with a simultaneous rise in the concentration of iron ions and a decrease in the pH of the treated wastewater, had a significant impact on the purification efficiency. However, the subsequent increase in electrical current density from 1.2 A/m^2 to 2.0 A/m^2 did not have a substantial effect. Simultaneously, the concentrations of phosphorus in the treated wastewater exceeded the permissible value. For the wastewater to be discharged into the environment, the phosphorus concentration should be below 3.0 mgP/L [32]. Electrocoagulation does not cause microbiological contamination of the wastewater. Therefore, these effluents can be used in re-fertigation or directed to further purification processes.

A fivefold increase in electric current density resulted in only a 32% increase in the dephosphatation rate. This corresponded to a phosphorus removal rate of 26.45 mg P/L·h at J = 0.4 A/m^2 and 34.79 mg P/L·h at an electric current density of J = 2.0 A/m^2 (Table 1). The obtained results allow for a precise estimation of the required time for phosphorus removal from greenhouse effluents, given the knowledge of phosphorus concentration in raw effluents. Meanwhile, the use of significantly higher electric current densities seems unjustified.

Table 1. Kinetic parameters of total phosphorus, total nitrogen, and organic compounds removal (k—rate constant—[1/h], r—removal rate—[mg/L·h], R^2—coefficient of determination).

	Total Phosphorus			Total Nitrogen			Organic Compounds		
	k	r	R^2	k	r	R^2	k	r	R^2
R1	0.421	26.45	0.996	0.137	2.07	0.998	0.130	0.44	0.988
R2	0.396	30.65	0.982	0.109	2.78	0.980	0.125	0.61	0.983
R3	0.448	34.79	0.992	0.190	6.58	0.969	0.262	1.50	0.999

Under the conditions of electric current flow, phosphorus was mainly removed through the electrocoagulation process. The precipitation of phosphorus ions with other ions, including calcium and magnesium ions, was limited due to the low pH of the treated effluents. In the pH range of 2.2–7.2, most orthophosphates exist in the form of an anion $H_2PO_4^-$ [35,36]. These ions, along with calcium and magnesium ions, form water-soluble compounds [4]. Due to the pH, direct precipitation of PO_4^{3-} with Fe^{2+} and Fe^{3+} ions was also limited. As a result of the electric current flow, monomeric and polymeric iron complexes with OH^- ions were most likely formed [37]. Generally, the pH of the point of zero charge (pHpzc) for iron oxides is 5.4–8.6 [38]. Therefore, when the solution pH < pHpzc, the surface of iron oxides occupies a positive charge due to the protonation of surface hydroxyl. Thus, it can adsorb P anions by electrostatic adherence and further immobilize them by ligand exchange (Equation (5)) [39,40]. Iron oxides could also capture P via electrostatic attraction (Equation (6)) [41]. Phosphate can be adsorbed on the surface of iron oxides through surface complexation, including monodentate mononuclear, bidentate mononuclear, and bidentate binuclear [42].

$$FeOH + H_2PO_4^- \rightarrow FeH_2PO_4 + OH^- \quad (5)$$

$$FeOH_2^+ + H_2PO_4^- \rightarrow FeOH_2 \ldots H_2PO_4 \quad (6)$$

The efficiency of removal increased with the current density from 0.4 A/m^2 to 1.2 A/m^2. However, a subsequent increase in current density from 1.2 A/m^2 to 2.0 A/m^2 did not result in a significant improvement in phosphorus removal efficiency. The amount of coagulant increases with the rise in current density, but likely, the quantity of coagulant generated from the increase in current density from 1.2 A/m^2 to 2.0 A/m^2 did not cause a significant enhancement in phosphorus removal efficiency, potentially due to the retention time of the effluent in the reactor. The obtained results confirm that the removal of phosphate ions was not caused by direct precipitation with iron ions, whose quantity increased with the rise in electrical current density, but rather due to various types of interactions. Most likely, the dominant process, influenced by the low pH resulting in the protonation of iron-based materials, was the sorption of phosphorus through electrostatic adsorption [36]. This underscores the need to test the electrocoagulation process using actual wastewater under different conditions to determine its real efficiency and optimize the process. Similar dependencies were reported by Shalaby et al. [43]. The cited authors argue that an increase in density beyond the optimal level does not lead to an increase in the efficiency of pollutant removal since the quantity of generated coagulant is no longer a limiting factor. Kłodowska et al. [20], in experiments with a hydraulic retention time of 24 h in a sequencing batch biofilm reactor with an aluminum electrode, removed between 71.01% and 86.99% of total phosphorus, increasing the density from 0.053 mA/m^2 to 0.210 mA/m^2. In these studies, an increase in current density from 0.105 mA/m^2 to 0.210 mA/m^2 resulted in a slight efficiency increase of only 4.19%. The cited research involved treating municipal wastewater with a low carbon-to-nitrogen ratio and a total phosphorus concentration of 5.16 mg P/L.

The results obtained from the electrical current-based studies suggest that the most favorable current density for efficient phosphorus removal is 1.2 A/m^2. Beyond this value, a significant increase in the efficiency of total phosphorus removal from wastewater was not observed. Simultaneously, the achieved effectiveness using a DC was higher than when using AC. In the research conducted by Bryszewski et al. [13] with a hydraulic retention time of 24 h and a current density of 4.4 A/m^2, but with AC, the phosphorus removal efficiency was 17.2%. Meanwhile, in the study by Rodziewicz et al. [12], the use of a DC in the RECDC and a DC along with an external source of organic carbon in the REBDC for synthetic wastewater with parameters similar to those from greenhouse effluents provided slightly higher efficiency and lower phosphorus concentration in the treated effluents.

Similarly to phosphorus, the conducted studies demonstrated that the rate and efficiency of nitrogen removal increased with the applied electric current density. However, the obtained values for nitrogen were several times lower compared to those achieved for phosphorus. The concentration of TN in the treated effluents was 467.0 $\pm$ 12.3 mg N/L, 456.6 $\pm$ 12.8 mg N/L, and 444.1 $\pm$ 11.9 mg N/L in R1–R3, respectively. This corresponded to nitrogen removal efficiencies ranging from 2.6 $\pm$ 2.6% to 7.4 $\pm$ 2.5%. The nitrogen removal rate in electrochemical reactors increased with the rise in current density, ranging from 2.78 mg N/L·h (R1) to 6.58 mg N/L·h (R3).

During electrocoagulation, electrochemical reduction occurs at the cathode, and electrochemical oxidation takes place at the anode (direct oxidation) or the generation of oxidants such as free radicals, ozone, or chloric acid, which then oxidize compounds present in wastewater (indirect oxidation). In the case of greenhouse wastewater, reduction processes play a significant role. Reducing nitrates to gaseous nitrogen can be challenging because it is one of the possible products (N_2, NO_2^-, NH_3, NH_2OH, N_2O, NO). A crucial indicator that most influences the final product of electrochemical nitrate reduction is the pH. At neutral and alkaline pH, by-products of this process are nitrites, ammonia, hydroxylamine, and hydrazine. In acidic conditions, undesired products of the reduction process are nitrites and ammonia [44]. If chloride ions are present in the treated wastewater, they undergo anodic oxidation, forming hypochlorous acid. Ammonium ions and nitrites react with hypochlorous acid to form nitrates and molecular nitrogen [45]. These reactions affect the overall efficiency of total nitrogen removal.

In the conducted studies, the predominant form of nitrogen was nitrates, both in raw and treated wastewater. After a 24-h retention time, the percentage of nitrates was 99.78, 99.76, and 99.70%, and the percentage of nitrites was 0.10, 0.09, and 0.09%, while the remainder consisted of ammonium and organic nitrogen, respectively, in R1–R3. This corresponded to nitrite concentrations in the treated effluents of 0.48 $\pm$ 0.13, 0.40 $\pm$ 0.14, and 0.40 $\pm$ 0.12 mg N/L, respectively, in R1–R3.

By analyzing the obtained results and literature data, it can be concluded that higher electrical current density leads to a faster reduction of nitrates on the electrode surface and, consequently, higher efficiency. However, the applied low electrical current densities did not result in a significant decrease in the nitrogen concentration in the treated wastewater.

The composition of treated wastewater, including the presence of other pollutants and the process operation, can significantly influence the final efficiency. Benekos et al. [46], in the current density range of 20–40 mA/cm^2, removed 100% of nitrates from real nitrate-contaminated groundwater (initial nitrate concentration of 25 mg N/L). However, with the increase in current density, ammonium nitrogen accumulation occurred, ranging from 4.5 mg N/L to 6.5 mg N/L. Ghanim and Ajjam [47], electrochemically treating synthetic wastewater with a nitrate concentration of 150 mg N/L, achieved efficiency ranging from 40% (2 mA/cm^2) to 88% (10 mA/cm^2). Ghazouani et al. [48], treating synthetic wastewater with an electric current density from 1 mA/cm^2 to 40 mA/cm^2, observed an initial nitrate concentration of 350 mg N/L and a phosphate concentration of 50 mg P/L. The efficiency of nitrogen and phosphorus removal in the investigated range of current density was 37–99% and 51–99%, respectively. Despite the high efficiency of nitrate removal, the main product of the process was ammonium ions, constituting approximately 90% at the highest current density. In the case of greenhouse wastewater treatment, previous studies have shown that the addition of an external organic carbon source improves denitrification efficiency. However, this solution is ineffective due to the loss of organic substrate during electrocoagulation and the formation of sludge containing phosphorus and a high proportion of organic compounds [12,13]. This complicates their storage and subsequent utilization. There is also no justification for nutrient solution recirculation [9].

Wastewater from soilless tomato cultivation is characterized by a low content of organic compounds. This is due to the infiltration of root exudates into the wastewater, as well as the presence of plant fragments [3]. In the case of electrochemical processes applied to such wastewater, the removal of organic compounds is considered an added value and is not the primary objective. The low content of organic compounds also prevents the use of conventional biological methods for removing biogenic compounds based, for example, on activated sludge [49]. When electrical current is applied, the mechanism for removing organic compounds is complex and involves processes such as complexation, charge neutralization or destabilization, coagulation, flocculation, oxidation, entrapment, and adsorption [50]. The initial content of organic compounds in raw wastewater was 11.3 ± 2.1 mg C/L. The efficiency of organic compound removal increased with the electrical current density, ranging from $28.9 \pm 6.1\%$ (R1) to $51.1 \pm 8.3\%$ (R3) (Figure 3C). This corresponded to an organic carbon concentration ranging from 8.0 ± 0.7 mg C/L (R1) to 5.5 ± 0.9 mg C/L (R3). The higher efficiency of pollutant removal also correlated with a higher removal rate observed during the study (Table 1). Statistical analysis indicated a significant impact of electrical current density on the concentration of organic compounds in the treated wastewater. The absence of external organic carbon supplementation prevented secondary pollution of the treated wastewater, a situation that can occur, especially in anaerobic systems characterized by lower efficiency in utilizing organic compounds. This is due to the lack of electron acceptors such as oxygen or oxidized forms of compounds, e.g., nitrates, after complete denitrification [13,51].

3.3. *Quantity and Quality of Sludge*

The application of DC and the dissolution of the anode result in the formation of sludge. This sludge not only contains dissolved metal from the anode but also the pollutants removed from the wastewater.

In the conducted studies, the amount of generated sludge expressed as TSS increased with the applied electric current density, ranging from 347 ± 38 to 665 ± 36 mg/L in R1–R3 (Figure 4A). A fivefold increase in electric current density resulted in approximately a twofold increase in the amount of generated sludge. This results in the formation of sludge ranging from 0.347 kgTSS/m^3 to 0.665 kgTSS/m^3 for 1 m^3 of greenhouse wastewater, depending on the process parameters. Assuming that an average of 45 m^3 to 120 m^3 of wastewater is discharged daily from 1 hectare of soilless tomato cultivation [52], the amount of generated sludge would be between 16–42 kgTSS/d·ha and 30–80 kgTSS/d·ha. The formed sludge was mainly characterized by the content of the mineral fraction, which ranged from 96.7% (R1) to 97.8% (R3; Figure 4B). Simultaneously, the amount of milligrams of TSS per milligram of removed phosphorus ranged from 5.5 to 8.5 (Figure 4C). The decrease in the phosphorus content in the discharged sediment corresponded to an increase in the iron content with the applied current density (Table 2). Additionally, the obtained sludge had a phosphorus content at a similar level to sludge obtained by precipitation with the use of alkalizing agents in the treatment of greenhouse wastewater [4]. Electrocoagulation led to a decrease in pH rather than an increase, which is crucial for the further utilization of treated wastewater. The applied electric current density of 1.2 A/m^2 proved to be the most efficient in terms of dephosphatation, as well as the quality and quantity of obtained sludge. Using both electrocoagulation and biologically enhanced treatment with an external carbon source in REBDC results in a sediment amount that is an order of magnitude higher. In the studies by Rodziewicz et al. [53], the amount of sediment ranged from 3944 mg/L to 4783 mg/L for HRT = 24 h and 4 h, and J = 1.25 A/m^2. Additionally, in the cited studies, for electric current densities ranging from 0.63 A/m^2 to 10 A/m^2, the sediment amount ranged from 3873 mg/L (HRT = 24 h; J = 0.63 A/m^2) to 5215 mg/L (HRT = 4 h; J = 10 A/m^2). This sludge also had a much higher content of organic compounds (77–88%) and a lower phosphorus content (0.68–0.84%) [53]. In the studies by Bryszewski et al. [13], using alternating current and an external organic carbon source, they recorded the amount of generated sludge ranging from 416.0 $\pm$ 267.3 mg/L (HRT = 24 h; J = 4.4 A/m^2) to 1079.4 $\pm$ 251.3 mg/L (HRT = 24 h; J = 13.3 A/m^2). The share of organic compounds was 28.6% and 22.4%, and phosphorus was 1.3–2.1%. The amount and quality of sludge resulting from electrocoagulation can vary widely and depend primarily on the quality of the treated wastewater, the type of electrode used, and the applied electric current density and electrocoagulation time [54,55]. Akyol [56], using electric current density in the range of 5–10 A/m^2, noted the amount of sludge resulting from electrocoagulation in the range of 4570–5330 mg/L. The studies concerned the treatment of wastewater from paint manufacturing, and aluminum electrodes were used. In the studies by Mansoorian et al. [57], where the electric current density ranged from 2 mA/cm^2 to 10 mA/cm^2, the amount of generated sludge ranged from 81 mg/L to 84 mg/L. The research focused on treating wastewater from battery production, and iron electrodes were used.

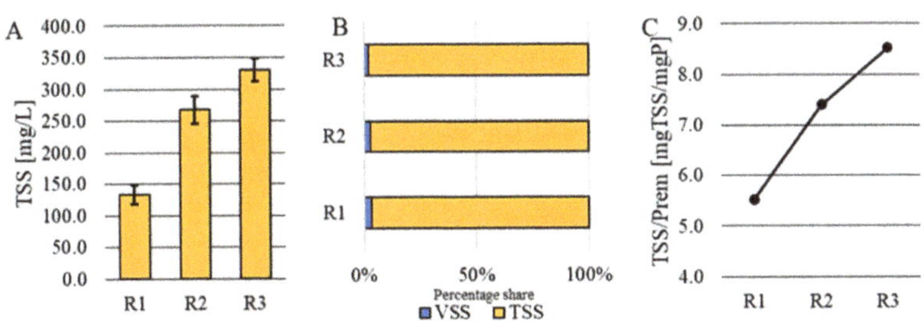

Figure 4. Amount of sediment discharged from the reactor (**A**), the proportion of volatile suspended solids (VSS) in total suspended solids (TSS) (**B**), and the amount of mgTSS per mg of removed phosphorus (**C**).

Table 2. Percentage content of selected elements in the sediment.

W	Average Elemental Content [%]			
	N	P	C	Fe
R1	4.18	18.1	0.65	8.08
R2	2.28	13.5	0.43	8.67
R3	2.68	11.7	0.55	13.1

Sludge containing over 30% organic compounds is not stabilized and should be directly managed or subjected to stabilization processes [58]. In the conducted studies, the sludge had a much lower organic compound content, making it easy to store and utilize in the future. The choice of a method for managing sediments after electrocoagulation depends on the type of pollutants, sediment composition, as well as local regulations and norms. Sediments from EC can be utilized as fertilizer [59,60], pigment [61], construction material (mainly in the form of blocks) [62], adsorbent [60], or catalyst [63]. Analyzing the composition of sediments generated during the electrocoagulation of wastewater from soilless tomato cultivation, it can be inferred that these sediments are rich in phosphorus and iron. Therefore, their management should focus on the reuse of these elements.

The presence of iron and iron oxides may facilitate the reuse of sediments as sorbents for phosphorus ions or as catalysts in reactions such as pyrolysis [63] or the Fenton reaction [64]. This could extend the so-called "product life cycle", but at the same time, it might make phosphorus in the sediments unsuitable for agricultural use. Due to the significant role of phosphorus in human activities, this element should play a decisive role in sediment management, where agricultural utilization currently appears to be the best solution.

Simultaneously, greenhouse wastewater subjected to electrocoagulation has no contact with domestic wastewater or other biological factors, making it safe in terms of hygiene and not posing a threat to cultivation.

4. Conclusions

The research presented in the paper complements existing studies on the impact of various treatment methods on the quality of wastewater from soilless tomato cultivation. The obtained results indicate that electrocoagulation allows for the purification of wastewater, which can subsequently be used to prepare fresh nutrient solutions or subjected to processes such as biological treatment. Electrocoagulation is a method that does not cause microbiological contamination, a crucial aspect of reusing purified wastewater in cultivation. Bacteria and fungi present in biological treatment systems could potentially induce diseases in cultivated plants. Simultaneously, the sediment generated during electrocoagulation is characterized by a low content of organic compounds and a high phosphorus content, making these sediments suitable as a substitute for mineral fertilizers. Greenhouse wastewater, due to its lack of contact with domestic wastewater, is safe in terms of hygiene and can be utilized in various branches of agriculture. The resulting sediments do not require a stabilization process; they only need to be dewatered.

In addition, it was found that:

1. A threefold increase in electrical current density (from 0.4 to 1.2 A/m^2) resulted in a 16% increase in dephosphatation rate, while a fivefold increase in electrical current density (from 0.4 to 2.0 A/m^2) resulted in a 32% increase.
2. The rate of TP removal ranged from 26.45 to 34.79 mg/L·h, TN from 2.07 to 6.58 mg/L·h, and organic compounds from 0.44 to 1.50 mg/L·h, respectively, for electrical current densities of 0.4 and 2.0 A/m^2.
3. Increasing the electrical current density above 1.2 A/m^2 did not significantly increase the efficiency of TP removal. The observed effectiveness was 71.5 ± 6.0%, 88.0 ± 4.9%, and 88.6 ± 2.5%, respectively, for electrical current densities of 0.4, 1.2 and 2.0 A/m^2.
4. Due to the low pH of the treated wastewater, phosphorus was likely removed through electrostatic adsorption on the surface of iron oxides via surface complexation.

5. The use of DC and iron electrodes also allowed for the removal of a maximum of 7.4 ± 2.5% TN and 51.1 ± 8.3% organic compounds for an electrical current density of 2.0 A/m^2.
6. The amount of sludge resulting from electrocoagulation using an iron electrode and HRT = 24 h increased with the increase in electrical current density, ranging from 347 ± 38 mg/L (J = 0.4 A/m^2) to 665 ± 36 mg/L (J = 2.0 A/m^2). This sludge was characterized by a high percentage of the mineral fraction, ranging from 96.7% to 97.8%.
7. With the increase in applied electrical current density and the increase in the amount of generated sludge, the percentage of phosphorus in the sludge decreased, ranging from 18.1% (J = 0.4 A/m^2) to 11.7% (J = 2.0 A/m^2).
8. Electrocoagulation led to a decrease in the pH of the treated wastewater from 6.1 to 2.8 (J = 2.0 A/m^2), while no significant change in electrolytic conductivity was observed.
9. Dissolving the iron electrode led to an increase in the iron concentration in the treated wastewater from 0.077 ± 0.012 mg Fe/L to 0.695 ± 0.095 mg Fe/L (J = 2.0 A/m^2).

Author Contributions: Conceptualization, A.M.; methodology, A.M., K.Ł.B. and J.R.; validation, W.J. and J.R.; formal analysis, A.M. and J.R.; investigation, K.Ł.B. and K.K.; resources, A.M.; data curation, K.Ł.B.; writing—original draft preparation, A.M. and W.J.; writing—review and editing, A.M. and J.R.; visualization, K.Ł.B. and K.K.; supervision, A.M.; project administration, A.M. and J.R.; funding acquisition, A.M. All authors have read and agreed to the published version of the manuscript.

Funding: The study was part of the project "Development a precise treatment of wastewater from soilless tomato cultivation technology using electro biological hybrid reactor" as part of the LIDER X program, financed by The National Centre for Research and Development No. LIDER/4/0019/L-10/18/NCBR/2019. Co-financing amount: PLN 1,492,500.00.

Data Availability Statement: The data presented in this study are available in the article.

Conflicts of Interest: Author Bryszewski K.Ł. was employed by the company Hanna Instruments Sp. z.o.o. The remaining authors declare that the research was conducted in the absence of any commercial or financial relationships that could be construed as a potential conflict of interest.

References

1. Vermeulen, T.; van Os, E.A.; van der Linden, A.M.A.; Wipfler, E.L. Need for Clean Water and Recirculation to Reduce Emissions of Plant Protection Products from Soilless Cultivation. *Acta Hortic.* **2017**, *1176*, 87–94. [CrossRef]
2. Elvanidi, A.; Benitez Reascos, C.; Gourzoulidou, E.; Kunze, A.; Max, J.; Katsoulas, N. Implementation of the Circular Economy Concept in Greenhouse Hydroponics for Ultimate Use of Water and Nutrients. *Horticulturae* **2020**, *6*, 83. [CrossRef]
3. Mielcarek, A.; Rodziewicz, J.; Janczukowicz, W.; Dobrowolski, A. Analysis of Wastewater Generated in Greenhouse Soilless Tomato Cultivation in Central Europe. *Water* **2019**, *11*, 2538. [CrossRef]
4. Mielcarek, A.; Jóźwiak, T.; Rodziewicz, J.; Bryszewski, K.; Janczukowicz, W.; Kalisz, B.; Tavares, J.M.R. Recovery of Phosphorus and Other Minerals from Greenhouse Wastewater Generated during Soilless Tomato Cultivation by Means of Alkalizing Agents. *Sci. Total Environ.* **2023**, *892*, 164757. [CrossRef]
5. García-Caparrós, P.; Llanderal, A.; Maksimovic, I.; Lao, M. Cascade Cropping System with Horticultural and Ornamental Plants under Greenhouse Conditions. *Water* **2018**, *10*, 125. [CrossRef]
6. Karatsivou, E.; Elvanidi, A.; Faliagka, S.; Naounoulis, I.; Katsoulas, N. Performance Evaluation of a Cascade Cropping System. *Horticulturae* **2023**, *9*, 802. [CrossRef]
7. Richa, A.; Fizir, M.; Touil, S. Advanced Monitoring of Hydroponic Solutions Using Ion-Selective Electrodes and the Internet of Things: A Review. *Environ. Chem. Lett.* **2021**, *19*, 3445–3463. [CrossRef]
8. Thompson, R.B.; Incrocci, L.; van Ruijven, J.; Massa, D. Reducing Contamination of Water Bodies from European Vegetable Production Systems. *Agric. Water Manag.* **2020**, *240*, 106258. [CrossRef]
9. Mielcarek, A.; Kłobukowska, K.; Rodziewicz, J.; Janczukowicz, W.; Bryszewski, K.Ł. Water Nutrient Management in Soilless Plant Cultivation versus Sustainability. *Sustainability* **2023**, *16*, 152. [CrossRef]
10. Tong, S.; Zhang, B.; Feng, C.; Zhao, Y.; Chen, N.; Hao, C.; Pu, J.; Zhao, L. Characteristics of Heterotrophic/Biofilm-Electrode Autotrophic Denitrification for Nitrate Removal from Groundwater. *Bioresour. Technol.* **2013**, *148*, 121–127. [CrossRef]

11. Mousset, E.; Trellu, C.; Olvera-Vargas, H.; Pechaud, Y.; Fourcade, F.; Oturan, M.A. Electrochemical Technologies Coupled with Biological Treatments. *Curr. Opin. Electrochem.* **2021**, *26*, 100668. [CrossRef]
12. Rodziewicz, J.; Mielcarek, A.; Janczukowicz, W.; Jóźwiak, T.; Struk–Sokołowska, J.; Bryszewski, K. The Share of Electrochemical Reduction, Hydrogenotrophic and Heterotrophic Denitrification in Nitrogen Removal in Rotating Electrobiological Contactor (REBC) Treating Wastewater from Soilless Cultivation Systems. *Sci. Total Environ.* **2019**, *683*, 21–28. [CrossRef]
13. Bryszewski, K.Ł.; Rodziewicz, J.; Mielcarek, A.; Janczukowicz, W.; Jóźwiakowski, K. Investigation on the Improved Electrochemical and Bio-Electrochemical Treatment Processes of Soilless Cultivation Drainage (SCD). *Sci. Total Environ.* **2021**, *783*, 146846. [CrossRef] [PubMed]
14. Koyuncu, S.; Arıman, S. Domestic Wastewater Treatment by Real-Scale Electrocoagulation Process. *Water Sci. Technol.* **2020**, *81*, 656–667. [CrossRef] [PubMed]
15. Zhang, Z.; Han, Y.; Xu, C.; Han, H.; Zhong, D.; Zheng, M.; Ma, W. Effect of Low-Intensity Direct Current Electric Field on Microbial Nitrate Removal in Coal Pyrolysis Wastewater with Low COD to Nitrogen Ratio. *Bioresour. Technol.* **2019**, *287*, 121465. [CrossRef] [PubMed]
16. Mousazadeh, M.; Niaragh, E.K.; Usman, M.; Khan, S.U.; Sandoval, M.A.; Al-Qodah, Z.; Khalid, Z.B.; Gilhotra, V.; Emamjomeh, M.M. A Critical Review of State-of-the-Art Electrocoagulation Technique Applied to COD-Rich Industrial Wastewaters. *Environ. Sci. Pollut. Res.* **2021**, *28*, 43143–43172. [CrossRef] [PubMed]
17. Mao, Y.; Zhao, Y.; Cotterill, S. Examining Current and Future Applications of Electrocoagulation in Wastewater Treatment. *Water* **2023**, *15*, 1455. [CrossRef]
18. Chae, K.-J.; Kang, J. Estimating the Energy Independence of a Municipal Wastewater Treatment Plant Incorporating Green Energy Resources. *Energy Convers. Manag.* **2013**, *75*, 664–672. [CrossRef]
19. Mohyudin, S.; Farooq, R.; Jubeen, F.; Rasheed, T.; Fatima, M.; Sher, F. Microbial Fuel Cells a State-of-the-Art Technology for Wastewater Treatment and Bioelectricity Generation. *Environ. Res.* **2022**, *204*, 112387. [CrossRef]
20. Kłodowska, I.; Rodziewicz, J.; Janczukowicz, W. The Influence of Electrical Current Density and Type of the External Source of Carbon on Nitrogen and Phosphorus Efficiency Removal in the Sequencing Batch Biofilm Reactor. *J. Ecol. Eng.* **2018**, *19*, 172–179. [CrossRef]
21. Tong, S.; Liu, H.; Feng, C.; Chen, N.; Zhao, Y.; Xu, B.; Zhao, J.; Zhu, M. Stimulation Impact of Electric Currents on Heterotrophic Denitrifying Microbial Viability and Denitrification Performance in High Concentration Nitrate-Contaminated Wastewater. *J. Environ. Sci.* **2019**, *77*, 363–371. [CrossRef]
22. Tong, S.; Chen, N.; Wang, H.; Liu, H.; Tao, C.; Feng, C.; Zhang, B.; Hao, C.; Pu, J.; Zhao, J. Optimization of C/N and Current Density in a Heterotrophic/Biofilm-Electrode Autotrophic Denitrification Reactor (HAD-BER). *Bioresour. Technol.* **2014**, *171*, 389–395. [CrossRef]
23. Moussa, D.T.; El-Naas, M.H.; Nasser, M.; Al-Marri, M.J. A Comprehensive Review of Electrocoagulation for Water Treatment: Potentials and Challenges. *J. Environ. Manag.* **2017**, *186*, 24–41. [CrossRef]
24. Kabdaşlı, I.; Arslan-Alaton, I.; Ölmez-Hancı, T.; Tünay, O. Electrocoagulation Applications for Industrial Wastewaters: A Critical Review. *Environ. Technol. Rev.* **2012**, *1*, 2–45. [CrossRef]
25. Almukdad, A.; Hafiz, M.; Yasir, A.T.; Alfahel, R.; Hawari, A.H. Unlocking the Application Potential of Electrocoagulation Process through Hybrid Processes. *J. Water Process Eng.* **2021**, *40*, 101956. [CrossRef]
26. Ingelsson, M.; Yasri, N.; Roberts, E.P.L. Electrode Passivation, Faradaic Efficiency, and Performance Enhancement Strategies in Electrocoagulation—A Review. *Water Res.* **2020**, *187*, 116433. [CrossRef]
27. Kwon, M.J.; Hwang, Y.; Lee, J.; Ham, B.; Rahman, A.; Azam, H.; Yang, J.-S. Waste Nutrient Solutions from Full-Scale Open Hydroponic Cultivation: Dynamics of Effluent Quality and Removal of Nitrogen and Phosphorus Using a Pilot-Scale Sequencing Batch Reactor. *J. Environ. Manag.* **2021**, *281*, 111893. [CrossRef]
28. Nariyan, E.; Wolkersdorfer, C.; Sillanpää, M. Sulfate Removal from Acid Mine Water from the Deepest Active European Mine by Precipitation and Various Electrocoagulation Configurations. *J. Environ. Manag.* **2018**, *227*, 162–171. [CrossRef] [PubMed]
29. Lei, Y.; Hidayat, I.; Saakes, M.; van der Weijden, R.; Buisman, C.J.N. Fate of Calcium, Magnesium and Inorganic Carbon in Electrochemical Phosphorus Recovery from Domestic Wastewater. *Chem. Eng. J.* **2019**, *362*, 453–459. [CrossRef]
30. Sahu, O.; Mazumdar, B.; Chaudhari, P.K. Treatment of Wastewater by Electrocoagulation: A Review. *Environ. Sci. Pollut. Res.* **2014**, *21*, 2397–2413. [CrossRef] [PubMed]
31. Connolly, E.L.; Guerinot, M. Lou Minireview Iron Stress in Plants. *Plant J.* **2002**, *3*, reviews1024.1.
32. *Decree of the Minister of Maritime Affairs and Inland Navigation of 12 July 2019 on Substances Particularly Hazardous to Water and Conditions to Be Complied with When Discharging Waste Water into Water Bodies or into the Ground and When Discharging Rainwater or Snowmelt into Water Bodies or into Water Installations*; Ministry of Maritime Affairs and Inland Navigation: Warsaw, Poland, 2019.
33. Tahreen, A.; Jami, M.S.; Ali, F. Role of Electrocoagulation in Wastewater Treatment: A Developmental Review. *J. Water Process Eng.* **2020**, *37*, 101440. [CrossRef]
34. Behbahani, M.; Alavi Moghaddam, M.R.; Arami, M. A Comparison between Aluminum and Iron Electrodes on Removal of Phosphate from Aqueous Solutions by Electrocoagulation Process. *Int. J. Environ. Res.* **2011**, *5*, 403–412.
35. Wang, Y.; Kuntke, P.; Saakes, M.; van der Weijden, R.D.; Buisman, C.J.N.; Lei, Y. Electrochemically Mediated Precipitation of Phosphate Minerals for Phosphorus Removal and Recovery: Progress and Perspective. *Water Res.* **2022**, *209*, 117891. [CrossRef]

36. Wang, Q.; Liao, Z.; Yao, D.; Yang, Z.; Wu, Y.; Tang, C. Phosphorus Immobilization in Water and Sediment Using Iron-Based Materials: A Review. *Sci. Total Environ.* **2021**, *767*, 144246. [CrossRef]
37. Mao, Y.; Ninh Pham, A.; Xin, Y.; David Waite, T. Effects of PH, Floc Age and Organic Compounds on the Removal of Phosphate by Pre-Polymerized Hydrous Ferric Oxides. *Sep. Purif. Technol.* **2012**, *91*, 38–45. [CrossRef]
38. Jang, M.; Min, S.-H.; Kim, T.-H.; Park, J.K. Removal of Arsenite and Arsenate Using Hydrous Ferric Oxide Incorporated into Naturally Occurring Porous Diatomite. *Environ. Sci. Technol.* **2006**, *40*, 1636–1643. [CrossRef] [PubMed]
39. Omwene, P.I.; Kobya, M.; Can, O.T. Phosphorus Removal from Domestic Wastewater in Electrocoagulation Reactor Using Aluminium and Iron Plate Hybrid Anodes. *Ecol. Eng.* **2018**, *123*, 65–73. [CrossRef]
40. Reza, A.; Chen, L. Electrochemical Treatment of Livestock Waste Streams. A Review. *Environ. Chem. Lett.* **2022**, *20*, 1863–1895. [CrossRef]
41. Gamshadzehi, E.; Nassiri, M.; Ershadifar, H. One-Pot Synthesis of Microporous $Fe_2O_3/g\text{-}C_3N_4$ and Its Application for Efficient Removal of Phosphate from Sewage and Polluted Seawater. *Colloids Surf. A Physicochem. Eng. Asp.* **2019**, *567*, 7–15. [CrossRef]
42. Weng, Y.; Vekeman, J.; Zhang, H.; Chou, L.; Elskens, M.; Tielens, F. Unravelling Phosphate Adsorption on Hydrous Ferric Oxide Surfaces at the Molecular Level. *Chemosphere* **2020**, *261*, 127776. [CrossRef] [PubMed]
43. Shalaby, A.; Nassef, E.; Mubark, A.; Hussein, M. Phosphate Removal from Wastewater by Electrocoagulation Using Aluminium Electrodes. *Am. J. Environ. Eng. Sci.* **2014**, *1*, 90–98.
44. Majlesi, M.; Mohseny, S.M.; Sardar, M.; Golmohammadi, S.; Sheikhmohammadi, A. Improvement of Aqueous Nitrate Removal by Using Continuous Electrocoagulation/Electroflotation Unit with Vertical Monopolar Electrodes. *Sustain. Environ. Res.* **2016**, *26*, 287–290. [CrossRef]
45. Mook, W.T.; Chakrabarti, M.H.; Aroua, M.K.; Khan, G.M.A.; Ali, B.S.; Islam, M.S.; Abu Hassan, M.A. Removal of Total Ammonia Nitrogen (TAN), Nitrate and Total Organic Carbon (TOC) from Aquaculture Wastewater Using Electrochemical Technology: A Review. *Desalination* **2012**, *285*, 1–13. [CrossRef]
46. Benekos, A.K.; Tziora, F.E.; Tekerlekopoulou, A.G.; Pavlou, S.; Qun, Y.; Katsaounis, A.; Vayenas, D.V. Nitrate Removal from Groundwater Using a Batch and Continuous Flow Hybrid Fe-Electrocoagulation and Electrooxidation System. *J. Environ. Manag.* **2021**, *297*, 113387. [CrossRef] [PubMed]
47. Ghanim, A.N.; Ajjam, S.K. Kinetic Modelling of Nitrate Removal from Aqueous Solution during Electrocoagulation. *Civ. Environ. Res.* **2013**, *3*, 64–73.
48. Ghazouani, M.; Bousselmi, L.; Akrout, H. Combined Electrocoagulation and Electrochemical Treatment on BDD Electrodes for Simultaneous Removal of Nitrates and Phosphates. *J. Environ. Chem. Eng.* **2020**, *8*, 104509. [CrossRef]
49. Puig, S.; Corominas, L.; Balaguer, M.D.; Colprim, J. Biological Nutrient Removal by Applying SBR Technology in Small Wastewater Treatment Plants: Carbon Source and C/N/P Ratio Effects. *Water Sci. Technol.* **2007**, *55*, 135–141. [CrossRef]
50. Ryan, D.R.; Maher, E.K.; Heffron, J.; Mayer, B.K.; McNamara, P.J. Electrocoagulation-Electrooxidation for Mitigating Trace Organic Compounds in Model Drinking Water Sources. *Chemosphere* **2021**, *273*, 129377. [CrossRef]
51. Erses, A.S.; Onay, T.T.; Yenigun, O. Comparison of Aerobic and Anaerobic Degradation of Municipal Solid Waste in Bioreactor Landfills. *Bioresour. Technol.* **2008**, *99*, 5418–5426. [CrossRef]
52. Prystay, W.; Lo, K.V. Treatment of Greenhouse Wastewater Using Constructed Wetlands. *J. Environ. Sci. Health Part B* **2001**, *36*, 341–353. [CrossRef]
53. Rodziewicz, J.; Mielcarek, A.; Janczukowicz, W.; Tavares, J.M.R.; Jóźwiakowski, K. Characteristics of Sludge from the Treatment of Soilless Plant Cultivation Wastewater in a Rotating Electrobiological Disc Contactor (REBDC). *Energies* **2023**, *16*, 1022. [CrossRef]
54. Rajaniemi, K.; Tuomikoski, S.; Lassi, U. Electrocoagulation Sludge Valorization—A Review. *Resources* **2021**, *10*, 127. [CrossRef]
55. Boinpally, S.; Kolla, A.; Kainthola, J.; Kodali, R.; Vemuri, J. A State-of-the-Art Review of the Electrocoagulation Technology for Wastewater Treatment. *Water Cycle* **2023**, *4*, 26–36. [CrossRef]
56. Akyol, A. Treatment of Paint Manufacturing Wastewater by Electrocoagulation. *Desalination* **2012**, *285*, 91–99. [CrossRef]
57. Mansoorian, H.J.; Mahvi, A.H.; Jafari, A.J. Removal of Lead and Zinc from Battery Industry Wastewater Using Electrocoagulation Process: Influence of Direct and Alternating Current by Using Iron and Stainless Steel Rod Electrodes. *Sep. Purif. Technol.* **2014**, *135*, 165–175. [CrossRef]
58. Tchobanoglous, G.; Burton, F.L.; Stensel, D.H. *Metcalf &Eddy: Wastewater Engineering: Treatment and Reuse*; McGraw Hill Companies, Inc.: New York, NY, USA, 2014.
59. Lin, X.; Han, Z.; Yu, H.; Ye, Z.; Zhu, S.; Zhu, J. Struvite Precipitation from Biogas Digestion Slurry Using a Two-Chamber Electrolysis Cell with a Magnesium Anode. *J. Clean. Prod.* **2018**, *174*, 1598–1607. [CrossRef]
60. Golder, A.K.; Samanta, A.N.; Ray, S. Removal of Phosphate from Aqueous Solutions Using Calcined Metal Hydroxides Sludge Waste Generated from Electrocoagulation. *Sep. Purif. Technol.* **2006**, *52*, 102–109. [CrossRef]
61. Tezcan Un, U.; Onpeker, S.E.; Ozel, E. The Treatment of Chromium Containing Wastewater Using Electrocoagulation and the Production of Ceramic Pigments from the Resulting Sludge. *J. Environ. Manag.* **2017**, *200*, 196–203. [CrossRef] [PubMed]
62. Sharma, P.; Joshi, H. Utilization of Electrocoagulation-Treated Spent Wash Sludge in Making Building Blocks. *Int. J. Environ. Sci. Technol.* **2016**, *13*, 349–358. [CrossRef]

63. Tezcan Un, U.; Topal, S.; Ates, F. Electrocoagulation of Tissue Paper Wastewater and an Evaluation of Sludge for Pyrolysis. *Desalination Water Treat.* **2016**, *57*, 28724–28733. [CrossRef]
64. Samy, M.; Alalm, M.G.; Mossad, M. Utilization of Iron Sludge Resulted from Electro-Coagulation in Heterogeneous Photo-Fenton Process. *Water Pr. Technol.* **2020**, *15*, 1228–1237. [CrossRef]

energies MDPI

Article

Performance Assessment of Venturi-Assisted Confined Tube Aerators with Varying Diameter

Roohany Mahmud [1], Joseph Carpenter [2] and David W. MacPhee [3,*]

1 Department of Civil and Environmental Engineering, Michigan State University, East Lansing, MI 48824, USA; mahmudr2@msu.edu
2 Faculty of Mechanical Engineering, The University of Alabama, Tuscaloosa, AL 35487, USA; jcarpenter@eng.ua.edu
3 Department of Mechanical and Aerospace Engineering, Carleton University, Ottawa, ON K1S 5B6, Canada
* Correspondence: davidwmacphee@cunet.carleton.ca

Abstract: A significant amount of energy in wastewater treatment plants is spent on aeration to treat the organic matter with microorganisms in an oxygen-enriched environment. In this study, a novel and simplistic aeration concept known as Confined Tube Aeration (CTA) is proposed, in which the main elements are a Venturi injector and a coiled tube at its outlet. Two Venturi injector diameters were chosen for evaluation in this study, measuring 1 inch (25.4 mm) and 4 inch (101.6 mm). In this study, a relationship was developed between air suction rate and pressure differential across the injector. Then, a numerical model was developed to analyze hydrodynamic conditions and evaluate system performance. The main findings are that the larger diameter aerator performs 20% better than the smaller injector in terms of standard aeration efficiency (SAE), with a maximal value of 0.74 kgO_2/kWh found for the larger diameter system. These results suggest that future SAE improvements may be made for larger diameter systems in full-scale wastewater treatment applications with suitably designed injectors.

Keywords: Venturi; injector; efficiency; wastewater

1. Introduction

Aeration is the process of bringing air and water into close contact to promote mass transfer of oxygen. Aeration is required for the proper growth of microorganisms in a wastewater treatment aeration basin for timely decomposition of organic materials. Aeration accounts for 30% to 75% of total energy consumption of wastewater treatment plants [1,2]. Oxygenation of water is accomplished using devices called aerators, whose primary purpose is to enhance mass transfer down a concentration gradient by increasing gas–liquid surface area. These aerators, which come in a variety of designs, are typically compared using two indicators: standard oxygen transfer rate (SOTR, in kgO_2/h) and standard aeration efficiency (SAE, in kgO_2/kWh).

There are generally two classes of aerators: surface aerators, which discharge water into the air, and diffused aerators, which release gaseous oxygen into the water column. The latter is by far the most common in industry due to higher SAE [3]. Here, blowers or compressors move ambient air through a network of pipes, terminating in porous pipes or disks arranged at the bottom of the basin [4]. Diffusers with fine pores are capable of producing microbubbles in the range of 10 μm to 60 μm [3] and are very efficient in diffusing gas to the liquid because of the large gas–liquid interfacial area, longer residence time, and hence higher mass transfer rates [5]. Such diffusers are generally categorized into fine and coarse bubble diffusers based on the bubble sizes produced [6]; fine-bubble diffusers produce smaller bubbles (2–5 mm in diameter) than coarse-bubble diffusers (6–10 mm) [7]. Since bubbles require residence time to transfer oxygen effectively, aeration tanks must be sufficiently (up to 10 m) deep to optimize oxygen transfer. Furthermore, a deep aeration tank provides higher hydrostatic pressure, which increases the local saturation of dissolved oxygen concentration, which, in turn, increases the oxygen deficit to promote increased mass transfer. However, diffused aeration has its own drawbacks. For starters, generating microbubbles requires a higher compression cost due to the significant pressure drop through the small pores. Diffuser systems can also easily be clogged and require higher upfront costs due to the complexity of installations [8]. However, their high mass transfer efficiency makes them nonetheless quite energy efficient and can achieve SAE values up to 4.56 kgO_2/kWh [9] for fine pore diffusers, whereas coarse bubble diffusers enjoy SAE values typically in the 1.22–2.13 kgO_2/kWh range.

Another type of diffused aerator is the Venturi aerator. Here, a submersible pump is located near the base of the aeration basin with a piping system where a Venturi device is attached and through which liquid is circulated. Venturi aerators can draw air from the surface naturally in a throat based on Bernoulli's principle; therefore, no compressor is required. Venturi aerators generally consume less energy than conventional airlift reactors, bubble columns, and stirred tanks [3,10]. A Venturi bubble generator consists of a nozzle section, a suction chamber, and a divergent section [11]. Venturi aerators are capable of producing bubbles with a mean diameter below 100 μm, thus providing high interfacial areas for mass transfer [12,13]. Additionally, the higher kinetic energy of liquid and gas flow creates intense turbulence and promotes mixing. Venturi aerators require higher liquid flow at the entrance of the injector to produce enough negative static pressure at the throat to entrain air and, therefore, can suffer from low efficiency. However, Venturi aeration systems are less energy intensive and less expensive to install and maintain as they have no moving parts that may break or fail. Moreover, they are less prone to clogging compared to coarse and fine bubble diffusers [14]. Venturi aerators generally have SAE values between 0.5 and 2.3 kgO_2/kWh [15].

Several recent studies in the literature have been conducted to evaluate the performance of Venturi aerators for aquaculture and wastewater purposes. Yadav et al. [16] used a dimensionless technique to optimize performance of such an aeration system on the basis of Venturi geometric properties. Here, the maximal SAE was found to be 0.611 kgO_2/kWh, and the nondimensional SAE was found to rely only on the (water) Reynolds number and Froude number. Dong et al. [17] analyzed commercially available injectors in series and parallel, with differing injection depths. It was found here that better aeration efficiency could be realized by connecting aerators in parallel, with a maximal SAE reported to be 0.306 kgO_2/kWh. Zhu et al. [18] conducted a similar study and found that parallelizing injectors resulted in SAE values that doubled the series configuration (0.14 kgO_2/kWh versus 0.07 kgO_2/kWh). Dange and Warkhedkar [19] assessed the effectiveness of a Venturi aeration for the purposes of pond/river oxygenation on the basis of injector depth. It was found here that, within the injector depths investigated, deeper injection resulted in increased air flow rates in the system. The maximal reported SAE value was 0.2936 kgO_2/kWh.

In this study, a novel Venturi bubble aeration system termed the Confined Tube Aerator (CTA) is proposed, previously introduced by the authors [20,21]. Figure 1 shows a schematic of the proposed system. The basic components of this aeration system are a pump, a Venturi injector, and a coiled tube. The pump draws water from the aeration basin and circulates it through the Venturi injector and coiled tube.

With a horizontally oriented helical axis, the CTA may reduce bubble coalescence widely observed in horizontal pipes [22], maintaining a higher interfacial area necessary for increased oxygen transfer. This device has the potential to revolutionize wastewater treatment for several reasons:

- Pressure requirements are decreased as there is no need to discharge at the bottom of basins. CTA pumps may draw from the bottom of basins and discharge to the surface, enhancing mixing.
- Thermodynamic work is decreased since pumps are used in lieu of compressors or blowers. Membrane pressure losses experienced in bubble diffusers are also avoided.
- Maintenance costs, which can be substantial for bubble diffusers as they are prone to fouling from debris and dissolved surfactants [9], are decreased as VAs may be cleaned/replaced

easily. Furthermore, the CTA system allows for placement of Venturi injectors on land, making replacement easier than if placed at the bottom of aeration basins.

- CTA systems do not require large, deep basins and may drastically reduce construction costs. They may also be particularly well suited for water treatment in remote settings and for oceanic vessels, where traditional aeration techniques are not feasible.

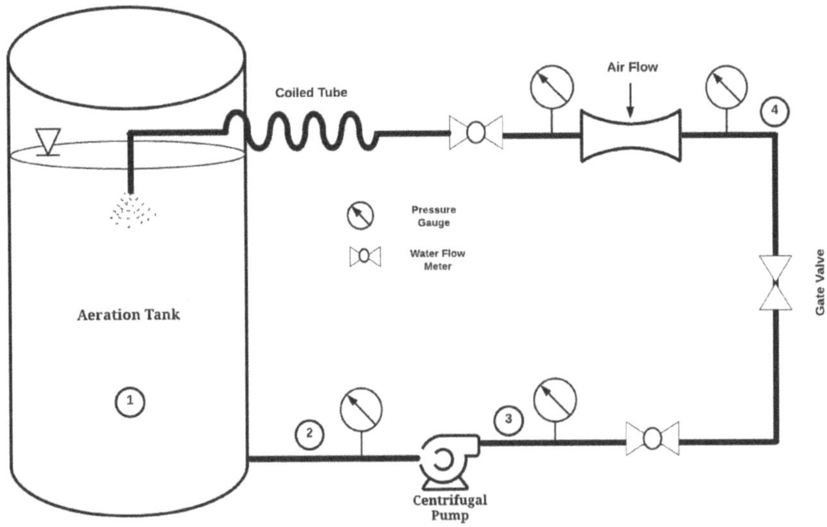

Figure 1. Confined Tube Aerator (CTA) System Schematic. 1: Aeration tank. 2: Pump suction. 3: Pump discharge. 4: Venturi inlet. 5: Venturi discharge (CTA inlet).

Furthermore, they are a potentially disruptive technology in the aquaculture aeration industry, as the current industry standard paddle wheel is of comparably low efficiency. The frequent use of dredge nets for fish harvesting make the use of bubble diffusers impractical.

In the current study, the performance of the proposed system is analyzed with respect to differing aerator sizes at varying hydrodynamic conditions. The discrete bubble model concept is used to model the mass transfer from bubbles to water in the tube, with the goal of improving aeration efficiency through parametric analysis of the results. This is the first study to estimate performance of CTA systems using commercially available Venturi injectors larger than 2.54 cm in diameter.

2. Methodology

The following is divided into five sections. The first discusses calculation of two-phase pressure drop in the coiled tube, the second explains aeration theory and performance parameters, the third outlines the theory underlying aeration prediction, the fourth discusses model validation with experiments, and the fifth gives an overview of the simulation procedure.

2.1. Two-Phase Pressure Drop

Flow inside the CTA tube is two-phase by nature. Accurate prediction of the two-phase flow pressure drop is important in designing and optimizing the CTA system's performance. The total pressure loss gradient for a steady two-phase flow consists of three components—frictional pressure drop (Δp_f), acceleration (momentum) pressure drop (Δp_a), and gravitational pressure drop (Δp_{grav}) [23]. Therefore, the total pressure drop is derived from the following equation:

$$\Delta P = \Delta p_f + \Delta p_a + \Delta p_{grav}$$
$$\left(\frac{dP}{dx}\right) = \left(\frac{dp}{dx}\right)_f + \left(\frac{dp}{dx}\right)_a + \left(\frac{dp}{dx}\right)_g \tag{1}$$

In the above, Δp_a and Δp_{grav} are ignored since the flow velocity of the two phases (air bubbles and water) is relatively constant along the length of the pipe and there are no changes in gravitational potential energy between the inlet and outlet. For simplicity, it is common to treat the two-phase mixture as a single phase with average properties of the two phases considering the homogeneously mixed flow assumption. However, a more accurate prediction can be made using a friction multiplier method [24]. Here, the frictional pressure drop ΔP_L in a pipe with two-phase flow can be described by the Darcy–Weisbach equation, as shown below:

$$\left(\frac{dp}{dx}\right)_f = \Delta P_L \cdot \phi_{fr}^2 \tag{2}$$

where ϕ_{fr}^2 is a two-phase flow multiplier that accounts for the additional pressure losses due to the presence of both liquid and gas phases. The liquid-only frictional pressure drop, ΔP_L, is calculated using the following relation:

$$\Delta P_L = f_l \frac{L}{D_{pipe}} \frac{\dot{m}_{total}^2}{2 \cdot \rho_l \cdot A_{pipe}^2} \tag{3}$$

The two-phase flow multiplier, ϕ_{fr}^2, is further defined as

$$\phi_{fr}^2 = E_{fr} + \frac{3.24 \cdot F_{fr} \cdot H_{fr}}{Fr_m^{0.045} \cdot We_l^{0.035}} \tag{4}$$

with E_{fr}, F_{fr} and H_{fr} given by

$$E_{fr} = (1-\chi)^2 + \chi^2 \cdot \frac{\rho_l \cdot f_g}{\rho_g \cdot f_l} \tag{5}$$
$$F_{fr} = \chi^{0.78} \cdot (1-\chi)^{0.224} \tag{6}$$
$$H_{fr} = \left(\frac{\rho_l}{\rho_g}\right)^{0.91} \cdot \left(\frac{\mu_g}{\mu_l}\right)^{0.19} \cdot \left(1 - \frac{\rho_l}{\rho_g}\right)^{0.7} \tag{7}$$

here, ρ, μ, and σ are the density, viscosity, and surface tension of the liquid. The subscripts l, g, and m represent the liquid phase, gas phase, and mixture, respectively. f_l and f_g are the Darcy–Weisbach friction factors for the liquid and gas phases, respectively, and χ is the gas quality:

$$\chi = \frac{\dot{m}_g}{\dot{m}_l + \dot{m}_g} \tag{8}$$

The mixture's Froude number (Fr_m) and Weber number (We_m) for the combined flow are calculated as follows:

$$Fr_m = \frac{\dot{m}_{total}^2}{g \cdot D_{pipe} \cdot \rho_m^2 \cdot A_{pipe}^2} \tag{9}$$

$$We_m = \frac{\dot{m}_{total}^2 \cdot D_{pipe}}{\rho_m^2 \cdot \sigma \cdot A_{pipe}^2} \tag{10}$$

In the above relationship, L, D_{pipe}, and A_{pipe} represent the length, diameter, and cross-sectional area of the pipe, respectively. The mixture density (ρ_m) is calculated as volume-weighted average of the gas (ρ_g) and liquid (ρ_l) densities as follows:

$$\rho_m = \epsilon_g \cdot \rho_g + (1 - \epsilon_g)\rho_l \tag{11}$$

here, ϵ_g is the volume fraction of gas and is calculated as the ratio of the volumetric flow rate of the gas phase to the total volume flow rate of liquid (Q_l) and gas phase (Q_g).

$$\varepsilon_g = \frac{Q_g}{Q_l + Q_g} \tag{12}$$

The Colebrook–White equation is used for friction factor calculation in turbulent flow [25], as there are no instances of laminar flow in this study:

$$\frac{1}{\sqrt{f}} = -2log_{10}\left[\frac{e}{3.7D_{pipe}} + \frac{2.51}{Re\sqrt{f}}\right] \tag{13}$$

here, e is the relative roughness, defined as the pipe diameter divided by tubing roughness (0.0015 mm for standard PVC piping utilized in this study). Additionally, since the CTA assembly is designed in this study to discharge at the surface of the water, the (gauge) pressure of the Venturi discharge in Figure 1 becomes the total pressure differential (ΔP) in Equation (1).

2.2. Aeration Theory

The following equation describes the rate of oxygen transfer in an aeration system [26].

$$\frac{dC_L}{dt} = K_L a(C_s - C_L) \tag{14}$$

The left-hand side of the above represents the rate of change in the dissolved oxygen concentration in the liquid phase over time. Here, C_S refers to the saturation concentration of the gas (O_2 or N_2) in the liquid, and C_L represents the actual concentration at any given moment. $K_L a$ is the overall mass transfer coefficient, which is a product of the liquid-phase mass transfer coefficient (K_L) and the specific interfacial area ($a = A/V$). In practice, it is impractical to measure the interfacial area; as a result, the value of $K_L a$ is measured.

Upon integration of Equation (14), considering initial concentration of $C = C_0$ at time $t = 0$ and $C = C_L$ at any subsequent time $t = t$, the following integrated form is derived:

$$\ln\left(\frac{C_S - C_0}{C_S - C_L}\right) = K_L a \cdot t \tag{15}$$

This integral form allows experimental determination of $K_L a$ by measuring C_L over time. As C_L approaches C_S, the transfer rate diminishes, representing a near equilibrium state. Plotting the left-hand side of Equation (15) with respect to time helps determine $K_L a$ from the slope.

$K_L a$ is often normalized to a standard temperature of 20 °C, $(K_L a)_{20}$, to ensure consistency and comparability between different aeration system performances. The following standardized equation is used, where θ is the temperature correction factor (1.024 for pure water), and T is the water temperature during the experiment (°C) [27]:

$$(K_L a)_{20} = \frac{K_L a}{\theta^{T-20}} \tag{16}$$

The Standard Oxygen Transfer Rate (SOTR) is defined as the mass of oxygen transferred per unit of time into a fixed volume of water at standard conditions of 20 °C and 1 atm [28].

$$SOTR = V(\frac{dC_L}{dt})_{std} = (K_L a)_{20} C_{s,20} V \tag{17}$$

where $C_{s,20}$ denotes the saturation concentration of dissolved oxygen at 20 °C, V is the volume of water aerated, and the dimension of SOTR is kgO_2/h. To quantify the efficiency of an aeration system, the Standard Aeration Efficiency (SAE) is often used, which has dimensions kgO_2/kWh:

$$SAE = \frac{SOTR}{DP}$$
$$SAE = \frac{SOTR}{WP} \tag{18}$$

The power input can be either delivered power (DP) or wired power (WP) in the above equation.

2.3. Discrete Bubble Model (DBM)

In this study, a discrete bubble model (DBM) method was adopted to predict mass transfer inside the confined tube. This model is based on the one-dimensional time-dependent motion of spherical gas bubbles in the coiled pipe system. In this analysis, the effects of turbulence and agglomeration/breakup of bubbles are neglected, and the overall methodology is based on that found in [29]. The mass transfer flux (J_i) of a gas species across the surface of a single bubble is then written as follows:

$$J_i = K_{L,i}(C_{s,i} - C_{b,i}) \tag{19}$$

where $K_{L,i}$ is the mass transfer coefficient, $C_{s,i}$ is the saturation concentration of the gas i in the liquid at the gas–liquid interface, and $C_{b,i}$ is the concentration of the gas i at the bulk liquid. Applying Henry's law at the gas–liquid interface, the equilibrium concentration becomes

$$C_{s,i} = H_i p_i \tag{20}$$

where H_i is Henry's constant, and p_i is the partial pressure of gas i. The bulk aqueous-phase concentration of the gas i at liquid can be derived from the ideal gas law:

$$C_{b,i} = \frac{n_i M_{w,i}}{V} \tag{21}$$

here, n_i is the molar amount of gas in species i, $M_{w,i}$ is the molecular weight of gas i, and V is the volume of the liquid. If the bubble diameter (d_b) is known, the mass transfer rate for the gas species i from the bubble–liquid surface is expressed as

$$\frac{dm_i}{dt} = J_i A(t) = K_{L,i}(C_{s,i} - C_{b,i}) \times \pi d_b^2 \tag{22}$$

$A(t)$ is the time-dependent interfacial area. Assuming no slip between gas bubbles and liquid, an average velocity v_m of the bubbles and liquid can be considered uniform across the length of the pipe:

$$v_m = \frac{dx}{dt} = \frac{\dot{Q}_g + \dot{Q}_l}{A_{pipe}} \tag{23}$$

here, $\dot{Q}_g$ and $\dot{Q}_l$ are the volumetric flow rates of the gas and liquid, and A_{pipe} is the cross-sectional area of the pipe. Given that mass transfer is a relatively slow process, a pseudo-steady state condition for a small elemental distance traveled by a bubble with velocity v_m in time dt is assumed. This leads to the following:

$$\frac{dm_i}{dx} = K_{L,i}(C_{s,i} - C_{b,i})\frac{\pi d_b^2}{v_m} \tag{24}$$

In order to determine total mass transfer per unit length of pipe, Equation (24) is converted from a single bubble to a total number of bubbles ($N_{bub/L}$) in the elemental length of Δx:

$$\frac{dM_i}{dx} = K_{L,i}(C_{s,i} - C_{b,i})\frac{\pi d_b^2 \times N_{bub/L}}{v_m} \quad (25)$$

Equation (25) is an ordinary differential equation and can be integrated numerically using a first-order explicit scheme. Here, $\frac{dM_i}{dx}$ is the total mass transfer per unit length of the pipe for gas species i. $N_{bub/L}$ can be determined by knowing the initial bubble volume, the volume fraction of bubbles, elemental pipe length Δx, and the pipe area:

$$N_{bub/L} = \varepsilon_g \cdot \frac{A_{pipe}}{V_{b,0}} \cdot \Delta x \quad (26)$$

Furthermore, bubbles are assumed to be spherical in shape, and the initial bubble volume ($V_{b,0}$) of a single bubble can be calculated knowing the initial bubble diameter ($d_{b,0}$) at the outlet of the Venturi injector:

$$V_{b,0} = \frac{\pi}{6} d_{b,0}^3 \quad (27)$$

The bubbles move with average velocity of v_m; therefore, their residence time (t_{res}) is calculated for the pipe length L as

$$t_{res} = \frac{L}{v_m} \quad (28)$$

The mass transfer coefficient, K_L is determined through the Sherwood number (Sh) [30], noting that bubble diameter and diffusion coefficient (D_c) are known:

$$\text{Sh} = \frac{K_L d_b}{D_c} \quad (29)$$

Sh is estimated based on an assumption of an *immobile* surface, valid for bubbles in the range investigated in this study [31]

$$\text{Sh} = c\text{Re}_\text{b}^{1/2}\text{Sc}^{1/3} \quad (30)$$

The constant c is approximated as 0.6 as experimental values are found to be varied between 0.42 and 0.95 [32]. The remaining parameters necessary for calculation of K_L are the Reynolds and Schmidt numbers:

$$\text{Re}_\text{b} = \frac{\rho_l v_b d_b}{\mu_l} \qquad \text{Sc} = \frac{\mu_l}{D_c \rho_l} \quad (31)$$

where $v_b = v_m$ is the bubble velocity and μ_l is the dynamic viscosity of the liquid, respectively.

The internal pressure of a spherical bubble within the pipe depends on the atmospheric P_{atm}, static P_L, and Laplace $4\sigma/d_b$ pressure as follows [33]:

$$P = P_{atm} + P_L + \frac{4\sigma}{d_b} \quad (32)$$

Bubbles with a diameter greater than 0.1 mm usually render the Laplace pressure as negligible. The partial pressure of the gas i within the bubble can then be calculated from the total pressure P by considering the mole fraction (y_i) of the gas component:

$$p_i = y_i \cdot P \quad (33)$$

where y_i is the mole fraction of the gas component i. The total number of moles in a single gas bubble can be described by using the ideal gas law:

$$N = \frac{PV_b}{RT} \tag{34}$$

where V_b is the volume of a single bubble, R is the universal gas constant, and T is the temperature of the gas inside the bubble, assumed to be the same as the liquid temperature.

The bubble size distribution in the Venturi injector is not uniform. As a result, it is common to use a characteristic mean diameter (Sauter mean diameter—denoted SMD or d_{32}) for the analysis of wastewater treatment and aquaculture bubble-based aeration systems [34]. In essence, the SMD represents the diameter of a sphere that has the same volume-to-surface area ratio as a statistically averaged collection of bubbles or particles:

$$d_{32} = \frac{\sum_{i=1}^{m} m_i d_i^3}{\sum_{i=1}^{m} m_i d_i^2} \tag{35}$$

here, m is the number of bubbles observed. In this study, the Sauter mean diameter (d_{32}) is determined from a recent study [35] and depends on water and air flow rates, as well as air inlet diameter for Venturi injectors:

$$\frac{d_{32}}{D_{suc}} = 1215.9 \mathrm{Re_w}^{-1.4767} \mathrm{Re_{air}}^{0.7566} \alpha^{-0.5110} \tag{36}$$

here, D_{suc} is the suction port diameter, α is the air to water (volumetric) ratio, and $\mathrm{Re_w}$ and $\mathrm{Re_{air}}$ are water and air Reynolds numbers (the latter based on the suction diameter). The SMD is then used as the bubble diameter (d_b) in mass transfer calculations.

2.4. Simulation Procedure

The liquid in this study was water, and the gas was assumed to be air consisting of 21% O_2 and 79% N_2. The diffusion coefficient (D_c) of oxygen and nitrogen in water at 25 °C is 2.3×10^{-9} m^2/s and 2.0×10^{-9} m^2/s.

The bubble size varies due to two phenomena: mass transfer of species across the bubble surface and pressure changes due to viscous losses. The initial dissolved oxygen concentration was set to 0 g/m^3, and the dissolved nitrogen concentration in the tank water was assumed to be in equilibrium with the atmosphere. Mass transfer from the the top surface of the water tank was neglected as this is sealed in experiments. The system was assumed to be isothermal at 25 °C due to the high thermal mass of the system. Air injection information was taken for commercially available Venturi injectors from Mazzei® (Mazzei Injector Corp, Bakersfield, CA, USA). The injector's specifications were chosen based on the desired flow rate and pressure conditions as per manufacturer's guidelines.

The computational domain (CTA pipe) was split into elemental lengths Δx (m), and each cell moved with an average mixture velocity of v_m. Once the water returned to the aeration tank, the water was assumed to be homogeneously mixed. If the initial position of the control volume is x_i (m), the final position is x_f (m), and the time to travel one elemental length is Δt (s), then the following relationships hold:

$$\begin{aligned} x_f &= x_i + \Delta x \\ \Delta t &= \frac{\Delta x}{v_m} \end{aligned} \tag{37}$$

Equation (25) was used to calculate the mass transfer of oxygen and nitrogen gas from air bubbles to water. However, mass transfer is usually expressed in molar form; therefore, the saturation concentration and instantaneous aqueous concentration of gas in the simulation procedure were adjusted accordingly. Equation (25) can then be transformed into an explicit integration formula for calculating the molar species transfer per time step:

$$\Delta n_i(\Delta x) = \left[K_{L,i}(x)(C_{s,i}(x) - C_{b,i}(x)) \frac{\pi d_b(x)^2 \times N_{bub/L}}{v_m} \cdot \Delta x \right] \bigg/ M_{w,i} \tag{38}$$

here, $M_{w,i}$ is the molecular weight of gas species i. The saturation concentration for O_2 and N_2 are calculated using Henry's constant (H), as follows [36,37]. Note that these constants are used to calculate saturation concentrations after a brief discussion of pressure calculations (see Equation (45)).

$$\begin{aligned} H_O &= 68.0 - 160.672 \times 10^{-2} T + 184.64 \times 10^{-4} T^2 \\ H_N &= 29.197 - 68.649 \times 10^{-2} T + 88.851 \times 10^{-4} T^2 \end{aligned} \tag{39}$$

Equation (38) represents the number of moles of O_2 and N_2 that leave the gaseous phase during the time step, entering into the liquid phase. At $(x + \Delta x)$, the molar amount of dissolved gas i remaining in the gaseous phase and the total number of moles in the gaseous phase are calculated as follows:

$$\begin{aligned} n_i(x + \Delta x) &= n_i(x) - \Delta n_i(\Delta x) \\ N(x + \Delta x) &= N(x) - \sum_{i=1}^{n} \Delta n_i(\Delta x) \end{aligned} \tag{40}$$

Since all bubble sizes investigated herein are larger than 0.1 mm in diameter, the Laplacian term in Equation (32) is ignored. The total pressure in the bubble also varies along the pipe length due to the two-phase frictional pressure drop, calculated using Equations (2)–(7):

$$P(x + \Delta x) = P_{atm} + P_L(x) - \left(\frac{dP}{dx}\right)\Delta x \tag{41}$$

The change in the bubble size, to be updated in the next time step (control volume), is then calculated by modifying Equation (34) in the following form:

$$d_b(x + \Delta x) = \left(\frac{6}{\pi} \cdot \frac{N(x + \Delta x) \cdot R \cdot (T_a + 273.15)}{P(x + \Delta x) \cdot 10^5}\right)^{\frac{1}{3}} (\text{m}) \tag{42}$$

Due to concentration changes as a result of mass transfer and pressure, the mole fractions (y_i) in the gaseous phase and the saturation concentration ($C_{s,i}$) change as follows:

$$y_i(x + \Delta x) = \frac{n_i(x + \Delta x)}{N(x + \Delta x)} \tag{43}$$

$$p_i(x + \Delta x) = y_i(x + \Delta x) P(x + \Delta x) \tag{44}$$

$$C_{s,i}(x + \Delta x) = H_i p_i(x + \Delta x) \tag{45}$$

To generate initial conditions for the problem, two injectors were considered, Model 1078 (2.54 cm diameter) and Model 4091 (10.16 cm diameter), which have sufficient performance data to calculate inlet bubble sizes from Equation (36) given pump suction and discharge pressures as well as water flow rates. The air suction rate was converted from standard to simulation conditions using the following relationship:

$$Q_a = Q_{std} \frac{P_{std} T_a}{P_a T_{std}} \tag{46}$$

where Q_{std} is the air suction rate at standard temperature (T_{std} =20 °C) and standard pressure (P_{std} = 1.013 bar) from the manufacturer. The actual temperature is the air–water mixture temperature (T_a =25 °C), and the actual pressure (P_a) is the total pressure inside the CTA assembly at the Venturi outlet.

The outlet pressure of the Venturi injector is influenced by factors such as the pipe length and the properties of the two-phase mixture flowing through it. To accurately characterize this relationship, a suction flow rate versus pressure differential equation was developed from manufacturer data. To establish this relationship, manufacturer data were fitted with a 3rd-order polynomial. Figure 2 shows this relationship for an inlet pressure of 1.72 bar and two water flow rates (2.66 m^3/h and 67.23 m^3/h for the two injectors) to demonstrate how air flow rates are calculated.

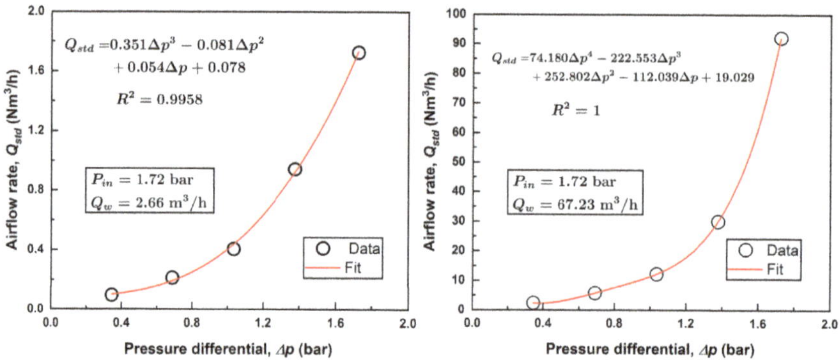

(**a**) (**b**)

Figure 2. Example polynomial fitting of air flow rate for (**a**) Model 1078; (**b**) Model 4091.

As previously mentioned, once a computational volume finishes its journey through the pipe, it is assumed to be homogeneously mixed with the tank, and the dissolved oxygen concentration is updated. This process continues until a steady-state concentration is reached. Figure 3 shows all major steps involved in the simulation.

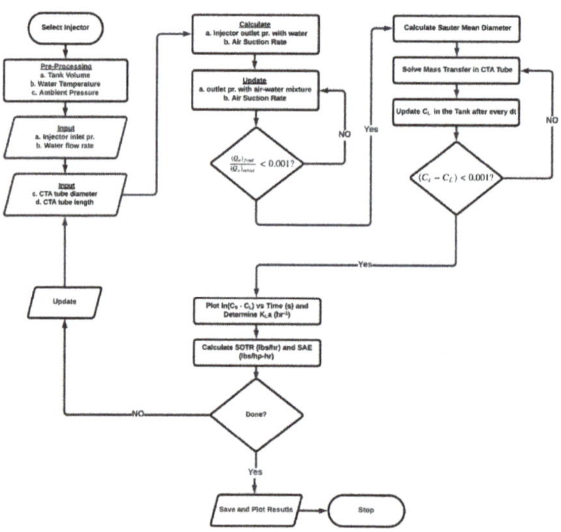

Figure 3. Flowchart of the simulation procedure.

In the current work, only the hydraulic power developed by the pump is considered for the performance analysis of the system. The fluidic power delivered (*DP*) by the pump, in kW, is estimated by the following equation, knowing the differential pressure and flow rate through the pump and noting that the pump pressure is expressed in bar:

$$DP = 100\Delta P_{pump} Q_w \tag{47}$$

In this analysis, water flow rate and pressure at the inlet of the injector at Point 4 in Figure 1 are utilized directly from manufacturer's data. Since the pipe length and diameter are known, the energy equation between Points 3 and 4 in Figure 1 can be solved as in Equation (48):

$$p_3 + \frac{\rho v_3^2}{2} + \rho g h_3 = p_4 + \frac{\rho v_4^2}{2} + \rho g h_4 + \rho g h_{loss(3-4)} \tag{48}$$

Similarly, the pump's inlet pressure can be calculated between Points 1 and 2 in Figure 1 as follows:

$$p_1 + \frac{\rho v_1^2}{2} + \rho g h_1 = p_2 + \frac{\rho v_2^2}{2} + \rho g h_2 + \rho g h_{loss(1-2)} \tag{49}$$

In these equations, elevation is measured from the bottom of the basin or tank. Since the tank is not pressurized and water in the tank is static, p_1 becomes 0 bar and v_1 is 0 m/s. The velocity from Points 3 to 4 is equivalent, and p_4 is the pressure at the injector inlet. h_{loss} includes frictional loss and minor losses from pipe fittings; its general expression is:

$$\Delta h_{loss} = \left(f_l \frac{L}{D_{pipe}} + \sum K \right) \frac{v_m^2}{2g} \tag{50}$$

Thus, the pressure differential across the pump becomes:

$$\Delta P_{pump} = p_3 - p_2 \tag{51}$$

2.5. Model Validation

The simulation procedure outlined in this study was validated with three experiments. Here, a large container was filled with 946 liters of tap water at 25 °C and deoxygenated using anhydrous sodium sulfite (NA_2SO_3). Cobalt chloride ($CoCl_2$) was used as a catalyst to speed up the reactions, and a starting DO content of 0 mg/L was achieved. Next, aeration was initiated using the Model 1078 (2.54 cm) injector, with motive flow provided by a 0.5 hp centrifugal pump. Pressure gauges and flow meters were employed to determine (water and air) flow rates and pressures necessary to conduct DBM simulations; see Figure 1. A 2.54 cm diameter coiled tube with a length of 6.1 m was employed as the confined tube aerator. Two Vernier DO probes (ODO-BTA) measured dissolved oxygen content as water was re-oxygenated. These values were averaged and recorded at one-minute intervals and compared to simulation results using the methodology presented herein, as seen in Figure 4.

Three experimental tests were conducted using valve control to achieve different operational conditions, listed in Table 1.

The results indicate that the methodology utilized in this study is adequate in predicting dissolved oxygen content and in most cases predicts DO values with less than 15% discrepancy between experiments and simulations. More details on the experimental procedure outlined herein may be found in [20].

Table 1. Experimental conditions for validation of simulation methodology.

	Test 1	Test 2	Test 3
Inlet CTA Pressure (gauge, kPa)	37.2	77.2	28.3
Air Flow Rate (mL/s)	142.5	67.0	103.0
Water Flow Rate (mL/s)	594.7	594.7	533.3

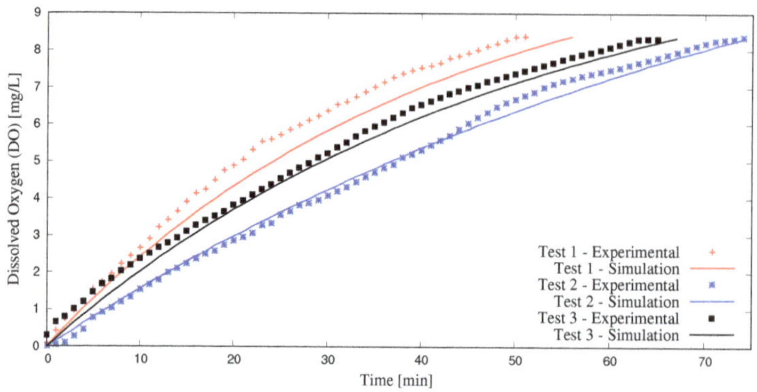

Figure 4. Experimental validation of simulation methodology.

3. Results and Discussion

Simulations were conducted using the aforementioned procedures for various water flow rates, inlet injector pressures, and CTA pipe lengths for both Model 1078 (2.54 cm) and Model 4091 (10.16 cm) injectors. As mentioned previously, air flow rates depend completely on water flow rates and pressure differential across the injector and are available from manufacturer's data; an example of which is shown in Figure 2.

The pipe length, through the associated pressure drop due to viscous effects, affects the outlet pressure of the injector and thereby the pressure differential. Given a Venturi inlet pressure, a longer pipe forces a smaller pressure differential through the injector, which, in turn, decreases the air suction rate, which further affects the two-phase pressure drop through the pipe. Each simulation required an iterative procedure (seen in Figure 3) to converge to an appropriate pressure differential and air flow rate, given an inlet pressure and pipe length. As a result, for every simulation conducted, an increase in pipe length decreased the pressure differential across the injector. This, in turn, decreases the suction air flow rate, as seen in Figure 5.

Furthermore, apparent in Figure 5 is the fact that as inlet injector pressure increases, suction flow rates increase. This is due to the fact that a greater amount of pressure differential becomes available for the injector. Air flow rates increase with increased water flow rates as well due to the increased fluidic energy available to pull air into the injector. The larger (Model 4091) injector draws much more air than the smaller injector mainly for this reason.

Since O_2 transfer is directly proportional to interfacial surface area, which is related to air and water flow rates within the pipe, the two injectors may be further compared according to the volume fraction of air within the pipe, which is calculated using Equation (12). All air volume fractions for simulations in this study are seen in Figure 6.

In Figure 6, similar trends are seen as with the previous discussion of air flow rates. Decreasing air volume fractions can be correlated with increased pipe lengths due to lower air flow. Interestingly, no discernible trend relating the air volume fraction to inlet pressure and flow rate can be discerned. This is perhaps due to the dimensionless nature of the volume fraction—a suitably nondimensionalized set of variables may prove these curves to collapse onto a single relationship. Equally interesting is the comparison between the two injector sizes. The larger model achieves higher air volume fractions than the smaller injector; a maximum of around 60% for Model 4091 versus 41% for Model 1078. This is likely due to the increase in available hydraulic power (larger flow rates are possible) combined with reduced frictional losses (as these scale with the square of velocity, which, for a given flow rate, decreases in an inverse-squared relationship with increased radius).

Of equal importance to the volumetric air flow rates and volume fractions is the bubble sizes; smaller bubble enable a larger surface area per unit air flow rate for mass transfer to occur. Recall that bubble sizes are calculated based on established nondimensional relationships [35] (see Equation (36)) and depend on water and air flow Reynolds numbers as well as the air-to-water ratio, which depends on the volumetric fraction as follows: $\varepsilon_g = \alpha\varepsilon_l$.

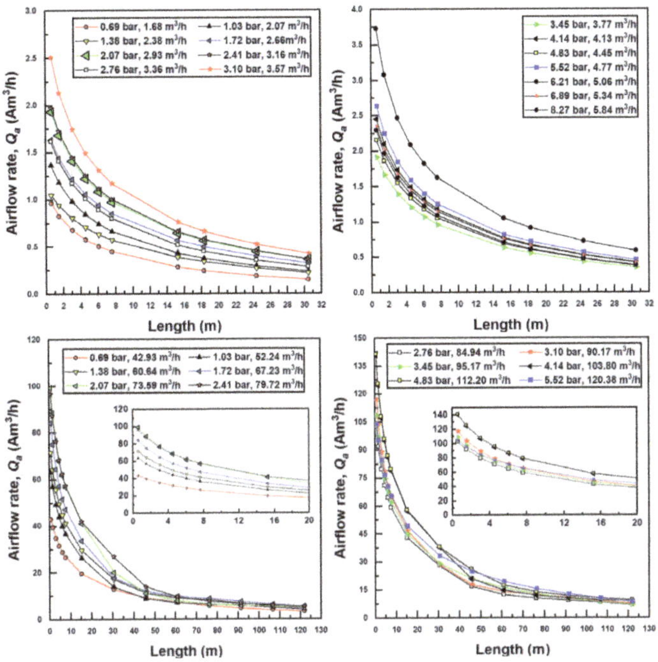

Figure 5. Suction air flow (actual) rate dependence on confined tube length for various inlet pressures and water flow rates; Models 1078 (**top** two figures) and 4091 (**bottom** two figures).

Figure 7 shows the variation in Sauter mean diameter (SMD) for various confined tube length and for differing inlet pressures and flow rates for the two injectors investigated herein. It is observed here that an increase in liquid flow rate generally produces smaller bubbles. This is due to a more substantial pressure recovery (according to Bernoulli's theorem) occurring in the diverging section of the injector, which is known to disassociate bubbles in Venturi injectors. With larger pressure recovery, bubbles generally shrink, and with less severe pressure recovery, bubbles generally grow in size. For this reason, the larger injector generally produces smaller bubbles.

Another interesting aspect of Figure 7 is that bubble sizes decrease with the increase in confined tube length. If inlet pressure and flow rate are held constant, it has already been established that an increase in CTA pipe length decreases air suction rates and gas volume fraction. This, in turn, decreases the air Reynolds number (Re_{air}) as well as the gas fraction (α). Drawing on Equation (36), these changes have the effect of decreasing the SMD (from a reduced Re_{air}) while simultaneously increasing the SMD (from a reduced α). Ultimately, the dependence of SMD on the air Reynolds number is stronger, as noted by the greater magnitude of the exponent in Equation (36), resulting in an overall decreased SMD from the increased CTA pipe length.

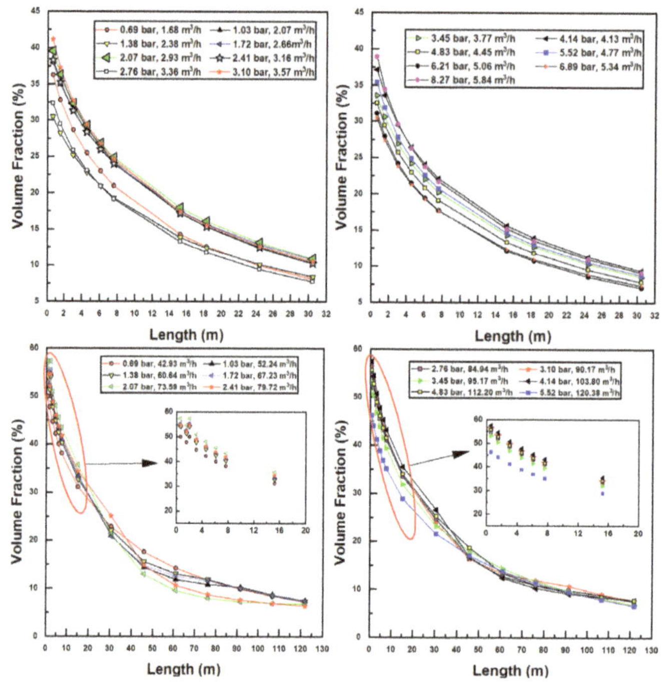

Figure 6. Air volume fraction dependence on confined tube length for various inlet pressures and water flow rates; Models 1078 (**top** two figures) and 4091 (**bottom** two figures).

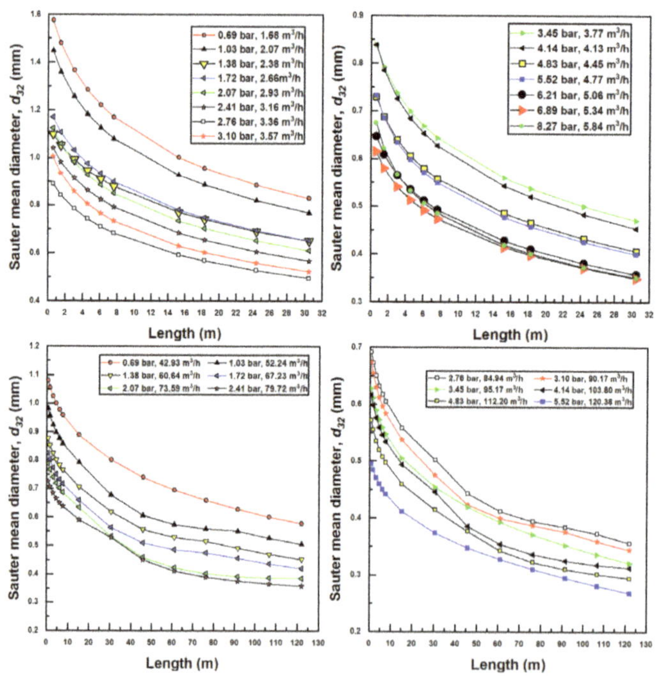

Figure 7. Sauter mean diameter bubble production for various confined tube lengths and water inlet pressure and flow rates; Models 1078 (**top** two figures) and 4091 (**bottom** two figures).

Minimum SMD, experienced for each set of parameters at the longest CTA pipe lengths and at greatest flow rate and inlet pressure, were calculated to be around 0.35 mm for the smaller injector (Model 1078) and around 0.30 mm for the larger injector (Model 4091). Maximum SMD bubbles were experienced at the shortest CTA pipe lengths, and lowest inlet pressures and flow rates were calculated to be around 1.6 mm for the small injector and 1.1 mm for the large injector.

The bubble size, suction air fraction, injector inlet pressure, and flow rate, as well as CTA pipe length, all factor into the performance parameters of the aeration system, namely, the standard oxygen transfer rate (SOTR) and standard oxygen transfer efficiency (SAE). These parameters were calculated using Equations (17) and (18), the latter using developed power in the denominator calculated as the product of flow rate and pressure differential of water across the injector. Figure 8 displays all simulated SOTR values in this investigation.

Some interesting observations can be made from the results in Figure 8. First and foremost, increasing water flow rates and inlet pressures increases SOTR. This is unsurprising, since increasing these two parameters increases the suction air flow rates (Figure 5) and decreases the bubble size (Figure 7), both trends acting to increase interfacial surface area and therefore mass transfer.

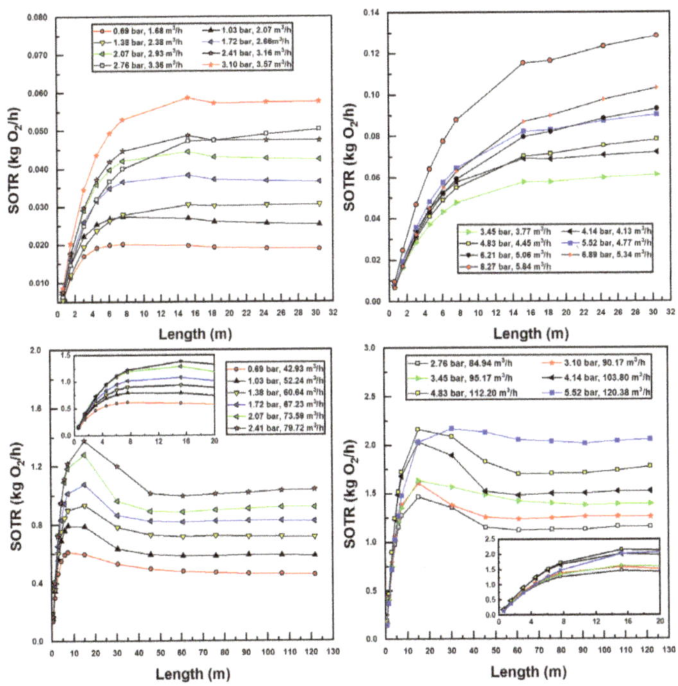

Figure 8. Simulated standard oxygen transfer rate (SOTR) values for various confined tube lengths and water inlet pressure and flow rates; Models 1078 (**top** two figures) and 4091 (**bottom** two figures).

Additionally, it is evident in Figure 8 that, for small pipe lengths, increasing the pipe length increases SOTR. This is due to the fact that, at these small pipe lengths, there is insufficient residence time for entrained bubbles to transfer mass to the water; all plots achieve a SOTR of zero with zero pipe length since mass transfer in the tank is ignored. The rise in SOTR is quite evident at these smaller pipe lengths, due to the high concentration difference between air and water, which increases mass transfer rates. As pipe length increases, concentration differences decrease, while suction air flow rates and air fractions decrease simultaneously, until a maximum SOTR is encountered (in some instances), after which SOTR tends toward a relatively stable value. This trend can be expected, since sufficient increases in pipe length—while increasing bubble residence time—decrease the air suction fraction and therefore the SOTR. Interestingly, for the larger (Model 4091) injector, a clear maximum SOTR is reached for each pipe length with a given injector inlet pressure and flow rate—between 5 m and 30 m length in call cases. Bubble sizes are significantly lower—and air fractions higher—for this injector, as compared to the smaller (Model 1078) injector, making mass transfer considerably greater. Additionally, as the larger pipe size incurs lower frictional pressure drops, larger pipe lengths may be utilized within the operational envelope of these injectors, further increasing SOTR for the larger model. Lastly, the gradual flattening of SOTR data with the increased pipe length in Figure 8 is a result of approaching saturation concentration. In many instances—and especially for the larger injector model—bubble residence times are sufficient so that water becomes saturated with oxygen, preventing any further increases in SOTR.

Standard Aeration Efficiency (SAE) trends are seen in Figure 9. SAE trends generally follow those of SOTR (Figure 8); however, those inlet injectors and pressures with the highest SOTR values achieve the lowest SAE values. This is due to the fact that SAE values are calculated by dividing SOTR by developed (fluidic) power, which is highest for high flow rates and pump differentials. As a result, from an energy efficiency perspective, lower flow rates and injector input pressures are more desirable. Maximum SAE values reported were 0.74 kgO_2/kWh for Model 4091 and 0.62 kgO_2/kWh for Model 1078. These values are generally in the range of those found in recent studies [16–19] with maximal SAE reported herein being slightly higher due to the use of the novel CTA assembly to assist aeration.

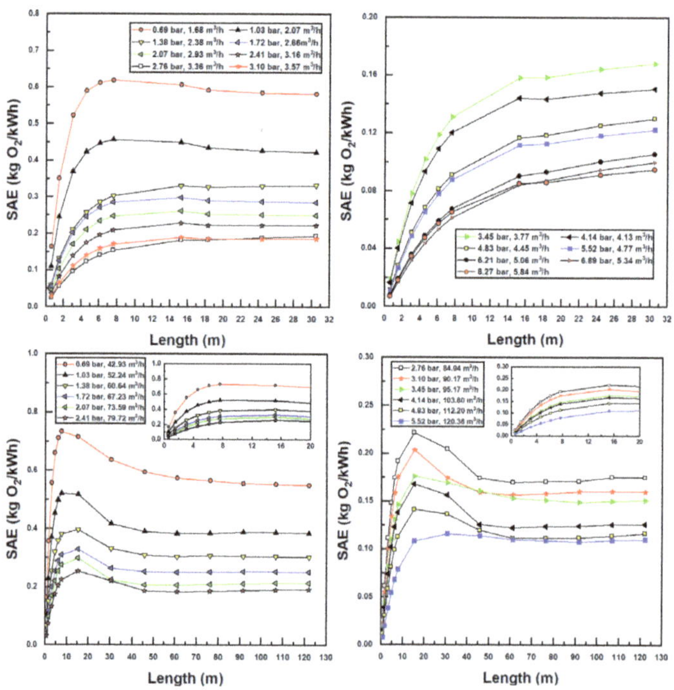

Figure 9. Simulated standard aeration efficiency (SAE) values for various confined tube lengths and water inlet pressure and flow rates; Models 1078 (**top** two figures) and 4091 (**bottom** two figures).

These results indicate that the CTA system scales well with increased diameter and power requirements in terms of system efficiency. Additionally, having knowledge of injector performance metrics allows this methodology to estimate the maximum aeration efficiency of the system with respect to confined tube pipe length, inlet injector pressure, and motive (water) flow rates, without the need for time-consuming and costly aeration experiments.

4. Conclusions

A novel microbubble aeration system is proposed herein, incorporating a Venturi injector and a coiled tube intended to facilitate bubble residence time without bubble coalescence. The proposed aeration system has the benefit of producing the interfacial area required for effective mass transfer without the need for compressors or blowers and deep aeration basins. Using manufacturer's data for two injectors (2.54 cm and 10.16 cm diameter), along with nondimensional correlations available from recent studies to predict bubble sizes, air suction rates, air volumetric fractions, standard oxygen transfer rate (SOTR), and standard aeration efficiency (SAE) calculations were performed based on simulations using the discrete bubble model concept. The general results of this study are as follows:

1. The proposed aeration system is capable of producing bubbles in the range of 0.15–1.58 mm for Model 1078 and 0.14–1.1 mm for Model 4091.
2. Increases in water flow rates effectively increase the volume fraction of injected air. Maximum volume fraction varied between 40 and 60% for the injectors investigated, with the larger-diameter model enjoying larger maximal air volume fractions.
3. Increasing the liquid flow rate improves the oxygen transfer rate through the system, due to decreases in bubble sizes and increases in injected air volume fraction.
4. Increasing the liquid flow rate increases the power requirements. As a result, the system operates with lower energy consumption and higher standard aeration efficiency at lower water and pressure at the injector inlet.
5. The confined tube length has a significant impact on the oxygen transfer rate—longer pipe lengths can effectively improve bubble residence times, increasing mass transfer. However, too long of a pipe decreases the suction air flow rates and increases viscous losses significantly, affecting standard aeration efficiency negatively. For each injector investigated, a maximal standard aeration efficiency was found with respect to pipe length. This was found to be 0.74 kgO_2/kWh for the larger (10.16 cm) injector and CTA assembly, compared to 0.62 kgO_2/kWh. Both of these values were higher than those found in the recent literature concerning traditional Venturi aeration.

Further studies are underway to investigate the relative cost of bubble production in Venturi injectors on the basis of energy (pressure) losses for CTA applications, as they relate to the competing effects of viscous losses in the coiled pipes themselves, at larger diameters in varying configurations. Other future works in this area, some of which are intended to address limitations of this study, are as follows:

- Addressing bubble agglomeration/breakup, which may become especially important as aeration tubes become larger and potentially slower-moving.
- Analysis of effects of radius of curvature, which may affect frictional pressure drop at low radii.
- Parallelization of Venturi injectors and aeration tubes, which may prove beneficial at larger scales necessary for municipal wastewater treatment.
- Analysis of system performance when clean water is replaced with a mixed-liquor fluid typically seen in wastewater aeration basins.

The results of this study are encouraging, especially given that increases in aeration efficiency were found for an increased injector diameter and, hence, the system capacity for delivering oxygen to wastewater. Since traditional Venturi aeration in wastewater applications is currently focused on ejecting the air/water mixture in the bottoms of large basins, relying on smaller bubbles' increased residence time to increase oxygen transfer, this study represents a fundamental shift in the application of Venturi injectors for wastewater treatment applications.

Author Contributions: Methodology and writing, R.M.; review and editing, J.C.; supervision and editing, D.W.M. All authors have read and agreed to the published version of the manuscript.

Funding: This research was partially funded by the United States Department of Energy's Industrial Assessment Centers, grant number DE-FOA-0002452.

Data Availability Statement: Data are contained within the article.

Conflicts of Interest: The authors declare no conflict of interest.

References

1. Henderson, M. Energy Reduction Methods in the Aeration Process at Perth Waste Water Treatment Plant. Renewable Energy Systems and the Environment. Master's Thesis, University of Strathclyde, Glasgow, UK, 2002.
2. Rosso, D.; Stenstrom, M.K. Alpha factors in full-scale wastewater aeration systems. *Proc. Water Environ. Fed.* **2006**, *7*, 4853–4863. [CrossRef]
3. Terasaka, K.; Hirabayashi, A.; Nishino, T.; Fujioka, S.; Kobayashi, D. Development of microbubble aerator for waste water treatment using aerobic activated sludge. *Chem. Eng. Sci.* **2011**, *66*, 3172–3179. [CrossRef]
4. Nadayil, J.; Mohan, D.; Dileep, K.; Rose, M.; Parambi, R.R.P. A study on effect of aeration on domestic wastewater. *Int. J. Interdiscip. Res. Innov.* **2015**, *3*, 10–15.
5. Ahmad, N.; Javed, F.; Awan, J.A.; Ali, S.; Fazal, T.; Hafeez, A.; Aslam, R.; Rashid, N.; Rehman, M.S.U.; Zimmerman, W.B.; et al. Biodiesel production intensification through microbubble mediated esterification. *Fuel* **2019**, *253*, 25–31. [CrossRef]
6. Jenkins, T.E. *Aeration Control System Design: A practical Guide to Energy and Process Optimization*; John Wiley & Sons: Hoboken, NJ, USA, 2013.
7. Shammas, N.K. Fine pore aeration of water and wastewater. In *Advanced Physicochemical Treatment Technologies*; Springer: Berlin/Heidelberg, Germany, 2007; pp. 391–448.
8. Hafeez, A.; Shamair, Z.; Shezad, N.; Javed, F.; Fazal, T.; Ur Rehman, S.; Bazmi, A.A.; Rehman, F. Solar powered decentralized water systems: A cleaner solution of the industrial wastewater treatment and clean drinking water supply challenges. *J. Clean. Prod.* **2021**, *289*, 125717. [CrossRef]
9. Rosso, D.; Larson, L.E.; Stenstrom, M.K. Aeration of large-scale municipal wastewater treatment plants: State of the art. *Water Sci. Technol.* **2008**, *57*, 973–978. [CrossRef]
10. Basso, A.; Hamad, F.; Ganesan, P. Effects of the geometrical configuration of air–water mixer on the size and distribution of microbubbles in aeration systems. *Asia-Pac. J. Chem. Eng.* **2018**, *13*, e2259. [CrossRef]
11. Gourich, B.; Azher, N.E.; Vial, C.; Soulami, M.B.; Ziyad, M.; Zoulalian, A. Influence of operating conditions and design parameters on hydrodynamics and mass transfer in an emulsion loop-venturi reactor. *Chem. Eng. Process. Process Intensif.* **2007**, *46*, 139–149. [CrossRef]
12. Yoshida, A.; Takahashi, O.; Ishii, Y.; Sekimoto, Y.; Kurata, Y. Water purification using the adsorption characteristics of microbubbles. *Jpn. J. Appl. Phys.* **2008**, *47*, 6574. [CrossRef]
13. Zhao, L.; Mo, Z.; Sun, L.; Xie, G.; Liu, H.; Du, M.; Tang, J. A visualized study of the motion of individual bubbles in a venturi-type bubble generator. *Prog. Nucl. Energy* **2017**, *97*, 74–89. [CrossRef]
14. Kadzinga, F. Venturi Aeration of Bioreactors. Master's Thesis, University of Cape Town, Cape Town, South Africa, 2015.
15. West, H.S. Apparatus and Method for Introducing a Gas into a Liquid. U.S. Patent 13/379,441, 28 June 2012.
16. Yadav, A.; Kumar, A.; Sarkar, S. Performance evaluation of venturi aeration system. *Aquac. Eng.* **2021**, *93*, 102156. [CrossRef]
17. Dong, C.; Zhu, J.; Wu, X.; Miller, C.F. Aeration efficiency influenced by venturi aerator arrangement, liquid flow rate and depth of diffusing pipes. *Environ. Technol.* **2012**, *33*, 1289–1298. [CrossRef]
18. Zhu, J.; Miller, C.F.; Dong, C.; Wu, X.; Wang, L.; Mukhtar, S. Aerator module development using venturi air injectors to improve aeration efficiency. *Appl. Eng. Agric.* **2007**, *23*, 661–667. [CrossRef]
19. Dange, A.; Warkhedkar, R. An experimental study of venturi aeration system. *Mater. Today: Proc.* **2023**, *72*, 615–621. [CrossRef]
20. Mahmud, R.; Erguvan, M.; MacPhee, D.W. Performance of closed loop venturi aspirated aeration system: Experimental study and numerical analysis with discrete bubble model. *Water* **2020**, *12*, 1637. [CrossRef]
21. MacPhee, D.W. Confined Tube Aspiration Aeration Devices and Systems. U.S. Patent 11,318,432, 3 May 2022.
22. Baker, O. Design of pipelines for the simultaneous flow of oil and gas. In Proceedings of the SPE Annual Technical Conference and Exhibition, SPE, Dallas, TX, USA, 18–21 October 1953; p. SPE-323.
23. Hetsroni, G. *Handbook of Multiphase Systems*; McGraw Hill: New York, NY, USA, 1982.
24. Friedel, L. Improved friction pressure drop correlations for horizontal and vertical two-phase pipe flow. In Proceedings of the European Two-Phase Flow Group Meeting, Ispra, Italy, 5–8 June 1979.
25. Colebrook, C.F. Turbulent flow in pipes, with particular reference to the transition region between the smooth and rough pipe laws. *J. Inst. Civ. Eng.* **1939**, *11*, 133–156. [CrossRef]
26. Lewis, W.K.; Whitman, W.G. Principles of Gas Absorption. *Ind. Eng. Chem.* **2002**, *16*, 1215–1220. [CrossRef]
27. ASCE. *Standard Guidelines for In-Process Oxygen Transfer Testing*; Technical Report; ASCE: Reston, VA, USA, 1997.
28. ASCE. *Measurement of Oxygen Transfer in Clean Water*; American Society of Civil Engineers: Reston, VA, USA, 1993.

29. McGinnis, D.F.; Little, J.C. Predicting diffused-bubble oxygen transfer rate using the discrete-bubble model. *Water Res.* **2002**, *36*, 4627–4635. [CrossRef]
30. Cengel, Y.; Heat, T.M. *Heat and Mass Transfer: A Practical Approach*, 2nd ed.; McGraw Hill: New York, NY, USA, 2003.
31. Griffith, R. Mass transfer from drops and bubbles. *Chem. Eng. Sci.* **1960**, *12*, 198–213. [CrossRef]
32. Lochiel, A.; Calderbank, P. Mass transfer in the continuous phase around axisymmetric bodies of revolution. *Chem. Eng. Sci.* **1964**, *19*, 471–484. [CrossRef]
33. Guggenheim, E.A. *Thermodynamics: An Advanced Treatise for Chemists and Physicists;* North-Holland Publishing Company: Amsterdam, The Netherlands, 1967.
34. Sauter, J. *Determining Size of Drops in Fuel Mixture of Internal Combustion Engines;* No. NACA-TM-390; NACA: Washington, DC, USA, 1926.
35. Gordiychuk, A.; Svanera, M.; Benini, S.; Poesio, P. Size distribution and Sauter mean diameter of micro bubbles for a Venturi type bubble generator. *Exp. Therm. Fluid Sci.* **2016**, *70*, 51–60. [CrossRef]
36. Weiss, R. The solubility of nitrogen, oxygen and argon in water and seawater. In *Deep Sea Research and Oceanographic Abstracts;* Elsevier: Amsterdam, The Netherlands, 1970; Volume 17, pp. 721–735.
37. Wüest, A.; Brooks, N.H.; Imboden, D.M. Bubble plume modeling for lake restoration. *Water Resour. Res.* **1992**, *28*, 3235–3250. [CrossRef]

energies

MDPI

Article

Uncalcined Zn/Al Carbonate LDH and Its Calcined Counterpart for Treating the Wastewater Containing Anionic Congo Red Dye

Kuppusamy Manjula Rani [1,*], Pachagoundanpalayam Nachimuthugounder Palanisamy [1], Vennila Nagamuthu Kowshalya [1], Ayyasamy Tamilvanan [2], Rajendran Prabakaran [3,*] and Sung Chul Kim [3,*]

[1] Department of Chemistry, Kongu Engineering College, Perundurai, Erode 638060, Tamil Nadu, India; asppavithran@gmail.com (P.N.P.); kowshinila@gmail.com (V.N.K.)

[2] Department of Mechanical Engineering, Kongu Engineering College, Perundurai, Erode 638060, Tamil Nadu, India; tamilvanan.aj@gmail.com

[3] School of Mechanical Engineering, Yeungnam University, 280 Daehak-Ro, Gyeongsan 712-749, Gyeongbuk, Republic of Korea

* Correspondence: chemistrykmr@gmail.com (K.M.R.); praba.auto@gmail.com (R.P.); sungkim@ynu.ac.kr (S.C.K.)

Abstract: In this investigation, Zn/Al carbonate layered double hydroxide (ZAC-LDH) and its derived material on calcination were synthesized for removing the anionic azo dye Congo red (CR) from wastewater. Numerous factors were methodically investigated, including temperature, adsorbent dosage, pH, starting Dye Concentration (DC), and contact time. The CR elimination percentage dropped as the initial DC increased from 25 mg/L to 100 mg/L at 30 °C for uncalcined LDH, and from 97.96% to 89.25% for calcined LDH. The pH analysis indicates that the highest level of dye removal was recorded within the acidic pH range through the electrostatic attraction mechanism. The sorption kinetics analysis results demonstrated that the pseudo-second-order kinetic model exhibited a stronger fit to both uncalcined LDH and CZA-LDH, with the maximum correlation coefficient value. The Van't Hoff plots indicate the spontaneous nature of the physisorption process with a negative ΔG° (<-20 kJ/mol), while the endothermic adsorption process exhibited a positive ΔH°. The X-ray diffraction of calcined LDH reveals a significant intercalation of CR dye molecules, both prior to and following adsorption, showcasing a distinctive memory effect. The Brunauer–Emmett–Teller (BET) gas sorption measurements were performed to support the mesoporous nature of ZAC-LDH and CZA-LDH. The FTIR spectrum confirms the interaction of dye molecules on the surface of uncalcined and calcined LDH. These findings emphasize the efficacy of both the synthesized LDHs in removing CR dye, with CZA-LDH demonstrating superior efficiency compared to uncalcined LDH in the context of CR removal from wastewater.

Keywords: Congo red dye; Langmuir separation factor; layered double hydroxide; kinetics; memory effect; thermodynamics

1. Introduction

Water pollution is a serious threat that the ecosystem is facing nowadays. It is inevitable for hazardous compounds in various forms to contaminate water bodies like lakes, rivers, seas, and groundwater. This contamination can occur through various sources by both natural and human-induced activities, including industrial discharge, agricultural and urban runoff, wastewater disposal, oil spills, plastic pollution, atmospheric deposition, mining activities, etc. [1,2]. A wide array of industries, including but not limited to the textile, paper, culinary, rubber, plastic, and cosmetics sectors, utilize pigments and dyes to impart color to their products [3]. The industries of textile and dyeing utilize a wide range of dye types, including direct, reactive, disperse, acid, basic, vat dye, sulfur, reactive disperse, and azoic dyes, among others [4]. Each of these dyes possesses distinct properties and finds specific applications [5]. Azo dyes, constituting a predominant category among diverse dye types, are widely manufactured and applied to fibers like cotton, rayon, cellulose acetate, and polyester. They encompass over 50% of the entire spectrum of commercial dyes. As a result, a considerable number of industrial effluents contain these azo dyes, which give rise to aesthetic concerns and, more significantly, potential health hazards for aquatic organisms and humans, including reproductive disorders, neurological disorders, and malignancies [6]. The selection of dye is contingent upon various factors, including the intended color, the fiber variety to be dyed, and the method of application. In the textile industry, environmental factors are progressively exerting a greater impact on the dye selection process, resulting in the emergence of dyeing methods that are sustainable or environmentally benign [7]. Dye pollutants released from such industries are the main source of environmental pollution [8]. The release of even a minute amount of dye into an aquatic ecosystem poses a risk to aquatic lifespan [9,10]. This is primarily due to its adverse impact on the ecosystem, as it diminishes sunlight transmission and depletes the crucial dissolved oxygen levels essential for sustaining aquatic life [11].

As an illustration, the widely used diazo dye Congo red (CR) dye breaks down into benzidine, a substance that has been shown to cause cancer in living things [12]. To remove Congo red dye (CRD) from effluents before they are released into the environment, environmental remediation products must be created [13]. This area of study could prove to be an engaging subject, addressing concerns related to public health and environmental protection regulations. Since CRD is known to be resistant to biodegradation, it can persevere in the environment for an extended time, leading to concerns about bioaccumulation in aquatic ecosystems [14]. And it may cause skin and eye irritation upon direct contact. Ingesting or inhaling large amounts of the dye could potentially lead to severe health effects, such as cancer [15]. Dye containing-wastewater removal poses a significant environmental problem, mainly because conventional techniques of treating such effluents are inherently challenging [16]. Established approaches, including biological methods, coagulation, electrochemical techniques, advanced oxidation processes, and membrane processes, often encounter inefficiencies in achieving complete color removal, coupled with elevated capital and operating costs [17–20]. As a result, the development of affordable techniques and materials for effectively eliminating pollutants from the environment is crucial [21].

Adsorption technology is a prevalent innovative method currently being employed to eliminate colorants from aqueous phase due to its user-friendly procedure, efficacy in removal, cost-effectiveness, and absence of secondary pollution [22]. Preceding studies have reported on a number of adsorbents, including carbon [23], natural materials [24], synthetic resins [25,26], banana peel [27,28], lemon peel [29], raw barley straw [30], heavy metal waste, decorated egg shell [31], sawdust [32], rice husk [33], fly ash [34], coir pith carbon [35], and tea detritus [36,37], utilized for the color removal from aqueous phase. Certain adsorbents are prohibitively expensive, challenging to dispose of, or excessively re-generable to be utilized on a large scale. In this context, a category of anionic clays referred to as layered double hydroxides has demonstrated efficacy as adsorbents for the elimination of numerous anionic impurities [38,39].

The general formula of LDHs is $M^{2+}{}_{1-x}M^{3+}{}_{x}(OH)_2]^{x+}(A^{n-})_{x/n}\cdot mH_2O$, where M^{2+} is divalent metal (Zn^{2+}, Mg^{2+}, Fe^{2+}, etc.), M^{3+} is trivalent metal (Al^{3+}, Fe^{3+}, etc.), and A^{n-} is interlayer anions ($CO_3{}^{2-}$, Cl^-, $NO_3{}^-$, $SO_4{}^{2-}$, etc.) [40,41]. Carbonates serve as the interlayer anions within obviously occurring mineral hydroxides, a category to which LDH materials belong. The compositional flexibility of LDH is evident in its ability to incorporate various M^{2+} and M^{3+} ions in the layers, diverse interlayer anions (A^{n-}), and stoichiometric coefficients (x) [42]. This adaptability allows LDH materials to find extensive applications across multiple domains, including catalysis, adsorption, ion exchange, pharmaceuticals, and purification. The exchangeability of interlayer anions and the flexibility in composition make LDH materials versatile in meeting diverse functional requirements [43–45]. When exposed to water or an aqueous solution containing diverse anions, calcined layered double hydroxide (CLDH) displays an impressive capacity to restore the layered structure characteristic of LDH. Importantly, these anions need not be the same as those in the original LDH, and the process demonstrates what is known as the 'structural memory effect' [46,47]. Studies have shown that both LDH and its calcined counterpart serve as effective ion exchangers and adsorbents for the eradication of toxic pollutants water [48,49]. The prime objective of this investigation is to produce calcined LDH (CZA-LDH) and uncalcined LDH, followed by the elucidation of these compounds efficiency as adsorbents enabling the removal of CR in solutions of water.

2. Materials and Methods

2.1. Synthesis of ZAC-LDH and Its Calcined Counterpart CZA-LDH

All the materials utilized in this investigation are of analytical reagent (AR) quality and were employed without undergoing any purification process. The ZAC-LDH was produced using an aqueous solution of $ZnSO_4 \cdot 7H_2O$ and $Al_2(SO_4)_3 \cdot 16H_2O$. The molar ratio of $ZnSO_4$ to $Al_2(SO_4)_3$ in the aqueous solution was 3:1. The solution was then blended using a magnetic stirrer. A precisely measured amount of 0.3 g of CTAB was introduced into the solution and thoroughly mixed using a magnetic stirrer until it formed a uniform mixture. The solution containing sodium hydroxide (1 M) as the precipitating agent and sodium carbonate (0.5 M) as the source of carbonate anions is slowly added until the pH reaches 10. The co-precipitate was transferred into a Teflon walled stainless-steel autoclave and subjected to the hydrothermal method shown in Figure 1. During this procedure, the mixture was subjected to a temperature of around 120 °C and held at that temperature for a duration of approximately 8 h. The collected material is subjected to filtration, followed by multiple washes using double-distilled water until the pH reaches a neutral level. Subsequently, the material is dried at a temperature of 80 °C using a hot-air oven. The product that has been acquired has been recognized as uncalcined LDH. In order to obtain calcined LDH, the uncalcined LDH was subjected to a temperature of approximately 450 °C in a muffle furnace for a duration of two hours. In order to obtain the calcined LDH sample, the rates of heating and cooling were both set to 10 °C per minute. In order to conduct additional analyses and adsorption studies, both the uncalcined and calcined LDH samples were finely ground into a powder.

Figure 1. Synthesis of uncalcined LDH and calcined LDH.

2.2. Characterization

Utilizing the Shimadzu XRD-6000 diffractometer, Shimaddzu analytical (India) Pvt. Ltd, Mumbai, the sample's diffraction of X-rays (XRD) pattern was examined. Filtered copper K-alpha radiation (wavelength, λ = 1.54 angstroms) was used at an X-ray tube voltage of 40 kilovolts and a current of 200 mill amperes. The specimens were then attached to alumina specimen holders, and the spacing between the plans of crystals (d-spacing) was measured using the powder technique. The samples were scanned in the series of 5–80 °C at a scanning speed of 1° per minute. The utilization of scanning electron microscopy with field emission (FESEM) was employed to conduct a morphological analysis. With a secondary electron detector, the researchers created FESEM images of the samples after depositing a layer of gold or palladium.

2.3. Method of Adsorbate Solution Preparation

The dye of interest in this adsorption study was CR. The color index, molecular formula, and molecular weight of the CRD is 22,120, $C_{32}H_{22}N_6Na_2O_6S_2$, and 696.68, respectively. The anionic property of this benzidine-based dye and the molecular structure is shown in Figure 2. A typical solution of CRD was produced by liquefying 500 milligrams of the dye in 1000 milliliters of distilled water, resulting in a concentration of 500 milligrams per liter (mg/L). Subsequently, further dilution was made from stock dye solution for various concentrations of CR that were derived from the stock solution for adsorption studies. Every chemical used in this investigation was of analytical purity. In order to prepare solutions and reagents, double-distilled water was utilized. The preliminary pH was adjusted as needed using 0.1 M sodium hydroxide (NaOH) or 0.1 M hydrogen chloride (HCl).

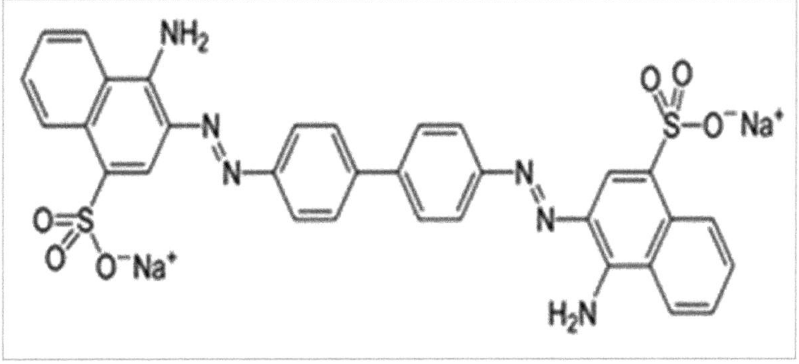

Figure 2. Molecular structure of Congo red.

2.4. Batch Adsorption Experiments

The batch-mode adsorption study was conducted using two synthesized materials, namely uncalcined LDH and calcined LDH, to remove CR. The fixed dose (excluding the dose variation study) and volume for various initial concentrations of CR were subjected to agitation at 170 rpm with the help of REMI make Orbital shaker. The liquid was withdrawn at predetermined time intervals and subsequently subjected to electrical centrifugation at 5000 revolutions per minute for a duration of 20 min. The remaining DC (Dye Concentration) was measured using Elico BL198 Bio-Spectrophotometer at λ_{max} 497 nm with the unadsorbed liquid. To assess the impact of pH for the effective color removal, dilute hydrochloric acid and sodium hydroxide solutions were employed. Furthermore, an investigation was conducted to assess the impact of temperature on the removal of dye at three dissimilar temperatures: 30 °C, 40 °C, and 50 °C. For accuracy, each experiment was replicated. The equation was used to compute the quantities of dye that were adsorbed at equilibrium (q_e) and at a certain time (q_t) on the two adsorbent materials.

$$q_{e,t} = (C_0 - C_e)\, X \frac{V}{m} X\ M \quad (1)$$

where $q_{e,t}$ (mg/g) denotes adsorbed amount at equilibrium (q_e) or at any time (q_t); C_0 (mg/L) indicated Initial Dye Concentration (IDC); $C_{e,t}$ (mg/L) refers to DC at equilibrium

(C_e) or at any time (C_t); V (L) denotes volume of the solution; and m (g) indicates mass of the adsorbent uncalcined LDH/calcined LDH sample.

Both uncalcined and calcined LDH, which were adopted as adsorbent materials in the elimination of CR, underwent regeneration under identical conditions. The regeneration process involved heating them in a muffle furnace at the temperature required for calcination for approximately 2 h, within an atmosphere of air. The capacity for reuse was tested using a thermally regenerated material, including both calcined substances, to eliminate CRD from a liquid solution. The reusability investigation kept the dosage of adsorbent unchanged, probably 0.5 g/L for uncalcined LDH and 0.250 g/L for calcined LDH, while starting with a concentration of dye about 50 mg/L.

3. Result and Discussion

3.1. Characterization of Materials

The characterization of the prepared uncalcined LDH by X-ray diffractogram revealed a phase consistent with pure and original LDH materials, as depicted in Figure 3a. The value of 2θ, which represents the peak's maximal intensity and distinctiveness, indicates that the single phase is well-crystallized and contains significant constituent crystallites [49,50]. Additional components, such as ZnO and $Al(OH)_3$, were found at larger angles (2θ values between 35° and 60°) in the analysis. The crystalline structure of ZnO usually resembled a brukite-type layer [49]. When uncalcined LDH underwent calcination at 450 °C, the initial layered structure disintegrated, causing the removal of the interlayer carbonate anions, OH^-, and H_2O molecules. This transformation led to the creation of metal oxides mixtures. The XRD pattern illustrated in Figure 3b does not contain basal reflections from planes (003) and (006). As stated in Figure 3b, the occurrence of mixed metal oxide formation is indicative of the elimination of the initial layered structure.

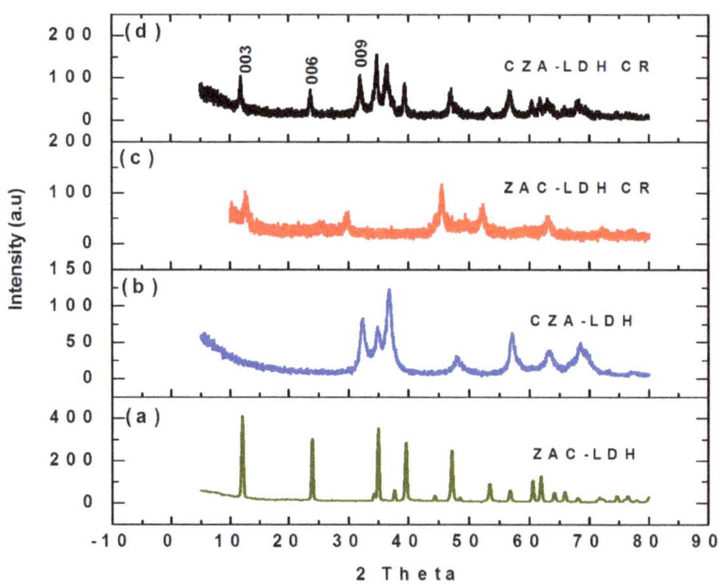

Figure 3. Uncalcined and calcined LDH XRD patterns (**a**,**b**) prior to CR adsorption and (**c**,**d**) subsequent to the process.

The PXRD diffraction patterns depicted in Figure 3a–d provide evidence of the effective synthesis of Zn/Al-LDH, which exhibits a notable level of crystallinity and a pure hydrotalcite phase. Strong peaks at 2θ values of 11.24, 22.68, 34.34, 38.19, 45.17, 60.30, and 61.61, which correspond to the reflection planes (hkl) of (003), (006), (009), (012), (015), (110), and (113), respectively, provide clear evidence of this. The presence of carbonate anions intercalated within the ZA-LDH structure is denoted by these peaks. The XRD patterns of uncalcined LDH and calcined LDH prior to and subsequent to dye adsorption are illustrated in Figure 3a–d. The principal mechanisms responsible for the adsorption of the dye were surface adsorption and intercalation of the dye onto uncalcined and calcined LDH, respectively. The aforementioned conclusion is substantiated by the XRD patterns of the adsorbents prior to and subsequent to dye adsorption (Figure 3a–d). The dye adsorption by calcined LDH is notably associated with the restoration of the original hydrotalcite (HT) assembly through the intercalation of CR, facilitated by a special phenomenon known as the 'memory effect'. This process is additionally validated by the resurgence of basal reflections of planes (003), (006), and (009), as well as the presence of several additional faint peaks as observed in the XRD pattern (Figure 3d). This provides evidence for the uptake of anionic dye CR onto both adsorbents, uncalcined and calcined LDH, through surface adsorption and anion-exchange mechanisms.

Figure 4a,b depict the FESEM images of the adsorbent materials both before and after CRD adsorption. Before adsorption, the rough surface of both the adsorbents has different pores with different sizes. The dye molecule covering the surface of the material was evidenced to have smooth coverage on surface in Figure 4b and flowers, like the structure causing heterogeneity occurrence in Figure 5a,b. This difference in surface further proved the adsorption of CR by both adsorbents.

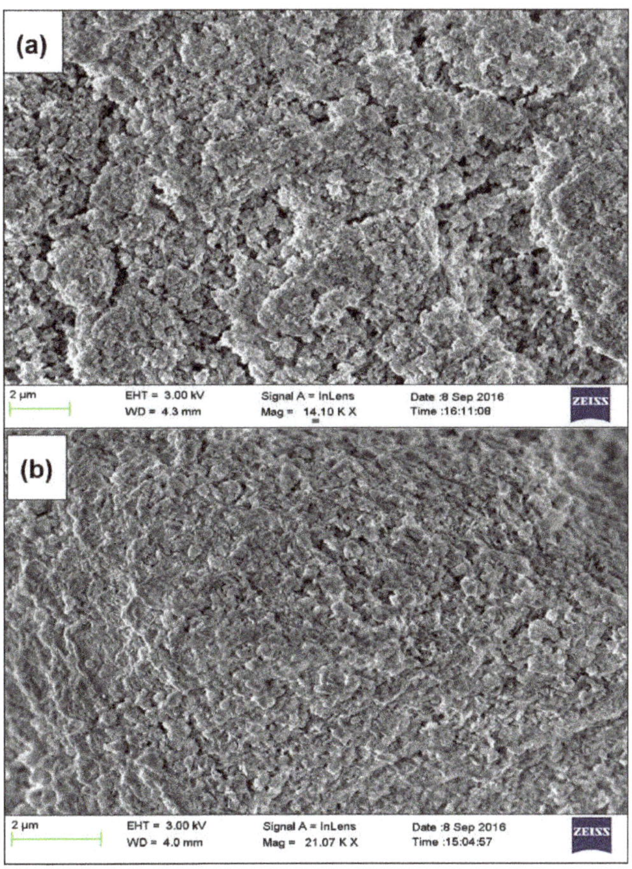

Figure 4. FESEM image of uncalcined LDH (**a**) before and (**b**) after adsorption of CR.

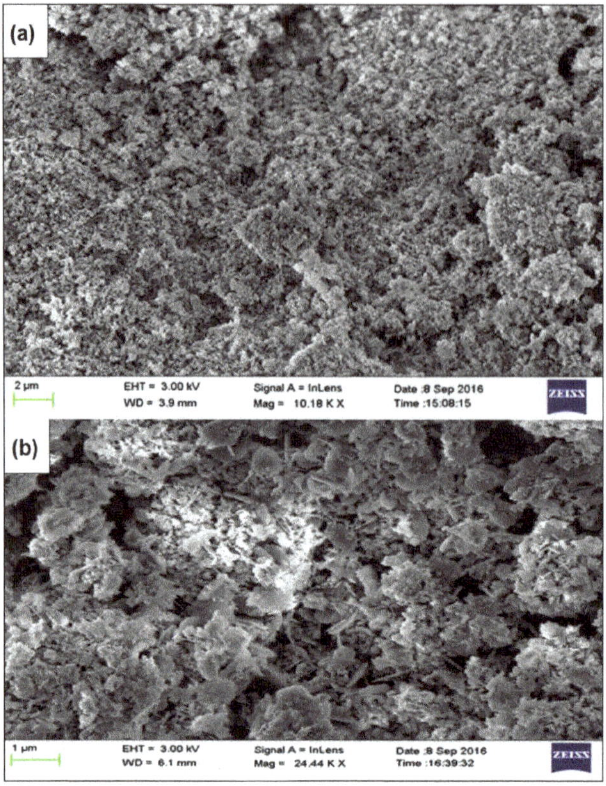

Figure 5. FESEM image of calcined LDH (**a**) before and (**b**) after adsorption of CR.

The particular surface areas and porous characteristics of ZAC-LDH and CZA-LDH were investigated using the Brunauer–Emmett–Teller (BET) gas sorption technique. The samples' BET surface area, pore volume, and average pore width are presented in Table 1. The nitrogen adsorption/desorption isotherms of the ZAC-LDH and CZA-LDH samples, as depicted in Figure 6, exhibited a type IV isotherm with an H3 hysteresis loop. This hysteresis loop is indicative of the creation of plate-like particle aggregates, resulting in the production of slit-shaped pores. The mesoporous nature of the samples was further validated by analyzing the relative pressure graph, which showed an increase in the adsorption and desorption curves between 0.6 and 1.0. The calcined LDH exhibited a significantly greater surface area, total pore volume, and average pore width compared to ZAC-LDH. Furthermore, it was observed that the pore volume and average pore size of CZA-LDH increased due to the removal of interlayer CO_3^{2-} ions during calcination. Furthermore, the relatively large pore size and mesoporosity of these materials make them highly effective as adsorbents for wastewater treatment. This is due to their ability to facilitate the simple diffusion of dye molecules [48].

Table 1. Surface area, pore volume, and pore size of ZAC-LDH and CZA-LDH.

Parameter	ZAC-LDH	CZA-LDH
Single point surface area (P/Po = 0.21) in (m^2/g)	22.0257	22.0469
BET surface area (m^2/g)	22.5610	22.5963
Dubinin–Astakhov—micropore surface area (m^2/g)	23.2209	23.2866
Total pore volume of (P/Po = 0.99) (cm^3/g)	0.132013	0.145968
t-Plot micropore volume (cm^3/g)	0.001020	0.002219
t-Plot mesopore volume (cm^3/g)	0.130993	0.143749
Average pore width (Å)	233.7454	258.7366

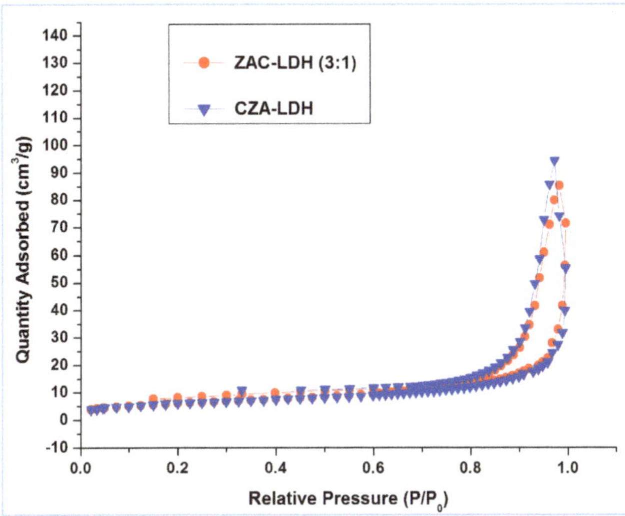

Figure 6. Nitrogen adsorption–desorption isotherm of ZAC-LDH and CZA-LDH.

The FTIR spectra of ZAC LDH and CZA LDH before and after adsorption of CR dye are shown in Figure 7a,b. The band at 3460 cm^{-1} in the FTIR spectra corresponding to ZAC LDH was due to the stretching vibrations of H-bonding of M-OH group in the brucite-like layer structure and interlayer water molecule. The intercalated anion carbonate in ZAC LDH was supported by the appearance of band at 1380 cm^{-1}, 770 cm^{-1}. But in CZA LDH, the disappearance of the peaks at 3460 cm^{-1}, 1380 cm^{-1}, and 770 cm^{-1} confirms the formation of calcined LDH with the loss of hydroxyl group and water molecules. After the adsorption of CR, the dye was evidenced by the decrease in the intensity of peak at 1380 cm^{-1} in ZAC LDH-CR compared with ZAC-LDH (Figure 7a). The reconstruction of the LDH structure by intercalation of anionic dye CR in CZA-LDH was evidenced through the appearance of peak at 1380 cm^{-1} (Figure 7b). Thus, the FTIR spectrum confirms the interaction of dye molecule on the surface of uncalcined and calcined LDH [51].

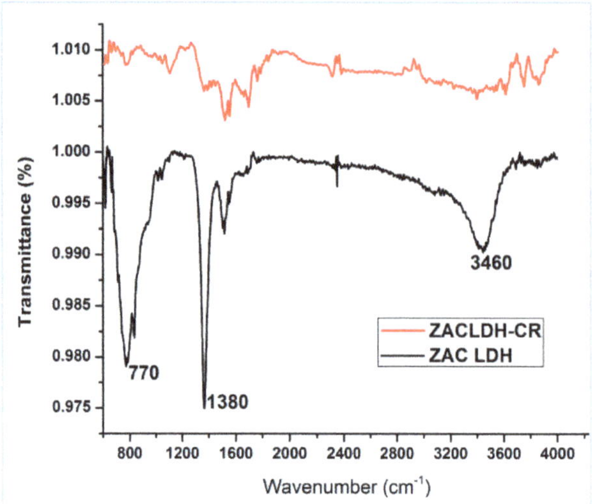

(a)

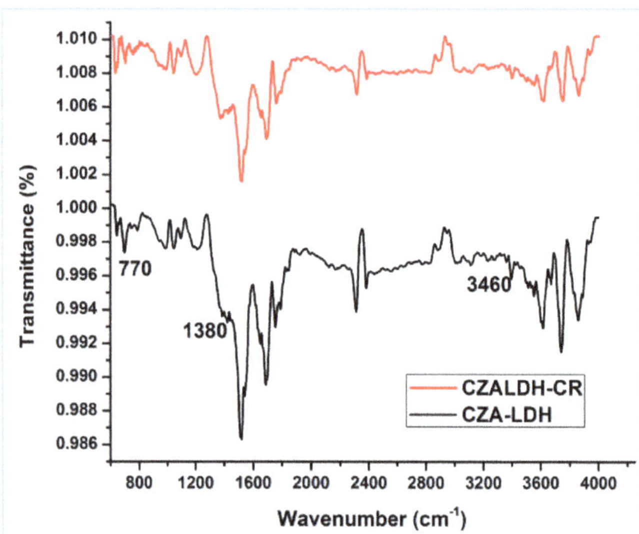

(b)

Figure 7. (**a**) FTIR spectrum of ZAC LDH before and after adsorption of CR. (**b**) FTIR spectrum of CZA LDH before and after adsorption of CR.

3.2. The Influence of IDC and Contact Time

As illustrated in Figure 8, the impact of IDC on CR adsorption by two adsorbents was investigated. With an increase in the IDC from 25 mg/L to 100 mg/L at a temperature of 30 °C, the dye removal efficiency decreased from 95.65% to 89.62% for uncalcined LDH and from 97.96% to 89.25% for calcined LDH. The availability of active spots on the adsorbent is greater for lower concentrations of adsorbate, leading to more adsorption than at the higher concentration. Concurrently, the maximum achievable adsorption rate of CRD per unit mass for both adsorbents rose, corresponding to the initial concentration increase from 25 mg/L to 100 mg/L at 30 °C, from 47.83 mg/g to 179.23 mg/g for uncalcined LDH, and from 97.96 mg/g to 356.99 mg/g for calcined LDH. Both instances demonstrated an initial surge in the contact duration of the dye adsorption during the initial phases. This was attributed to the presence of more active spots on the adsorbent surface, which promoted efficient adsorption. Subsequently, the rate of increase slowed down, indicating a reduction in available sites [52]. After it reached equilibrium nearing at 60 min, there was no significant change in the extent of quantity adsorbed. This was explained by the IDC acting as a driving force between the solid and aqueous phases to surmount the total mass transfer resistance of the CRD. As a result, the adsorption mechanism was enhanced at higher concentrations.

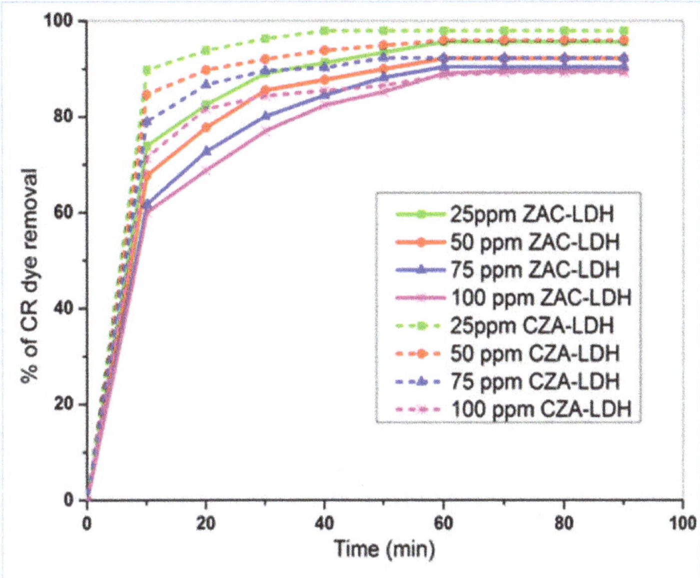

Figure 8. IDC as a factor in CRD elimination by uncalcined and calcined LDH.

3.3. The Influence of Temperature

Experiments were conducted at three distinct temperatures—30 °C, 40 °C, and 50 °C—in order to determine whether temperature affected the removal of CR by two adsorbent materials that were synthesized. Figure 9 describes the temperature dependency of CR onto both uncalcined LDH and calcined LDH. The efficiency of adsorption of CR is significantly influenced by the solution temperature, with a more favorable outcome observed at higher temperatures. Surface diffusion facilitates the adsorption of pigment onto the surface, which is enhanced by a rise in temperature. Consequently, as the temperature rose, the percentage removal of anionic dye CR from uncalcined LDH increased from 89.62% to 92.35%, and from 89.25% to 93.55% for CZA-LDH. The IDC of the dye utilized in these observations was 100 mg/L. The endothermic characteristics of the adsorption process were demonstrated through the proportional increase in CR adsorption and temperature [53].

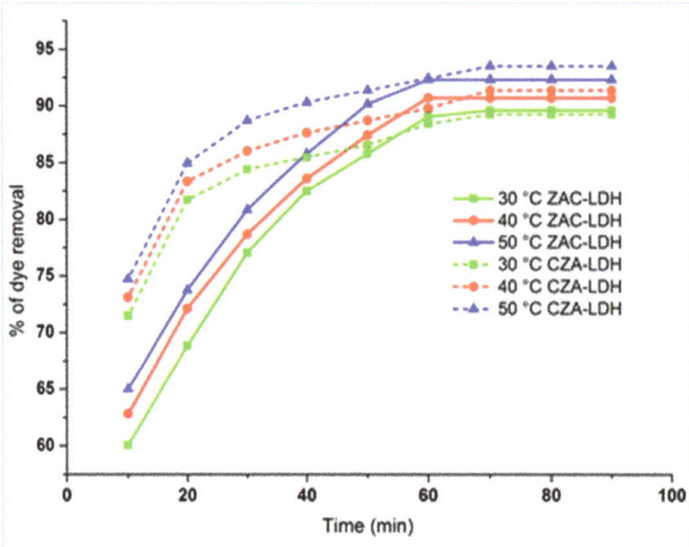

Figure 9. Temperature's effect on the eradication of CRD by uncalcined and calcined LDH.

3.4. The Implication of pH

The solution pH serves as a crucial parameter influencing the adsorption of adsorbate molecules. As illustrated in Figure 10, the consequence of the pH on the uptake of CR by two adsorbents, calcined and uncalcined LDH, was inspected by changing the initial pH of the dye from 2 to 12 for 100 mg/L IDC. The point of zero charge (PZC) of the adsorbent played a substantial role in elucidating the pH effect on dye removal, with the PZC values measured for uncalcined LDH (8.07) and calcined LDH (8.6). Higher dye removal percentages occurred at pH < PZC, and lower percentages were observed at pH > PZC for both uncalcined LDH (PZC = 8.07) and calcined LDH (PZC = 8.6). In contrast to pH > PZC, where the negatively charged anionic dye was repelled by the positively charged LDH surface, electrostatic attraction facilitated dye removal at pH < PZC [54]. The maximum dye removal of 78.38% was observed for uncalcined LDH at pH 6 and further increase in pH decreases the percentage removal due to breaking of LDH structure. Simultaneously, calcined LDH demonstrated complete dye removal, reaching 100%, within a broad pH range spanning from 3 to 7 [55,56]. The increase in the percentage of adsorption can be attributed to the positive charge that develops on the adsorbent surface at an acidic pH. This positive charge enhances the electrostatic force of attraction between the adsorbent surface and the negatively charged dye molecules [57,58].

3.5. The Implications of Adsorbent Dose

The effect of adsorbent incremental dose on CR uptake by CZA-LDH (100 mg/L DC) and uncalcined (50 mg/L DC) was examined by varying the adsorbent dose from 0.20 to 0.50 g/L. The values, which are depicted in Figure 11, indicate that, as the adsorbent dose was upsurge, the amount of dye removed increased. The observed results can be elucidated by the enlarged number of energetic spots and larger surface area of adsorbent, which promote enhanced adsorption until a state of equilibrium is achieved with the adsorbate molecules [59]. Additionally, the adsorption experiments were conducted using 0.50 g/L as the optimal dosage. With this dose, uncalcined LDH achieved a maximum removal of 92.31%, while CZA-LDH achieved 89.25% removal, and these values were fixed for further study [60].

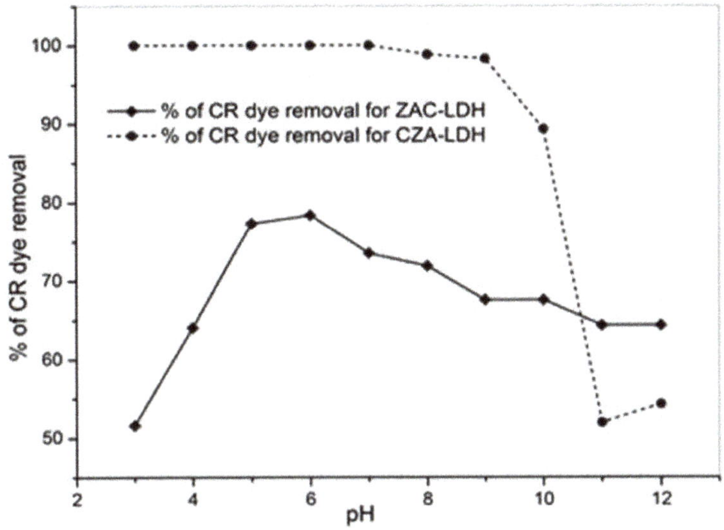

Figure 10. pH effect on the CRD elimination by ZAC-LDH and CZA-LDH (Ci=100 mg/L; T=30 °C).

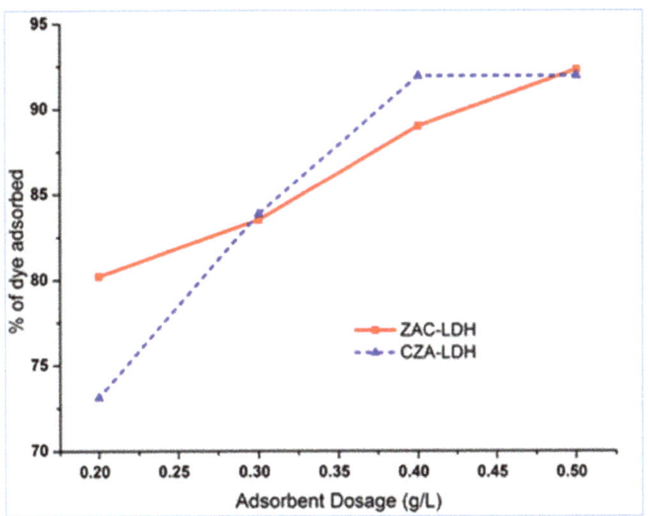

Figure 11. Adsorbent dosage effect on the removal of CRD by calcined and uncalcined LDH.

3.6. Kinetic Studies

3.6.1. Pseudo First Order (PFO) Model

To examine and understand the adsorption processes, including chemical reactions and mass transfer, three kinetic models—namely, PFO, PSO (pseudo-second order) [61], and intra-particle diffusion models—were employed for studying the adsorption of CRD on both uncalcined and calcined materials. Lagergren [62] proposed the pseudo-first-order rate equation, which is denoted as follows:

$$\log(q_e - q_t) = \log q_e - \left(\frac{k_1}{2.303}\right)t \quad (2)$$

where q_e and q_t represent the amounts of dye adsorbed at equilibrium and time, t (min) respectively; and k_1 symbolizes PFO rate constant (min^{-1}).

The values of q_e and k_1 were calculated from intercept and slope of the plot log (q_e-q_t) vs 't' for different 'C_i' and 'temperatures' for both adsorbents. Tables 2 and 3 summarize the calculated kinetic parameters (KPs). It is apparent from the R^2 value that the adsorption characteristics of both adsorbents for the removal of CRD were better described by a pseudo-second-order equation as opposed to a PFO equation.

Table 2. KP for the adsorption of CRD by calcined and uncalcined LDH at a temperature of 30 °C for various IDCs.

Adsorbents	**Uncalcined LDH**				**Calcined LDH**			
	Initial Concentration (mg/L)							
Parameter	25	50	75	100	25	50	75	100
q_e exp (mg/g)	47.83	92.22	135.66	179.23	97.96	191.84	276.92	356.99
	PFO kinetic model							
k_1 (min^{-1})	0.062	0.070	0.068	0.072	0.114	0.081	0.093	0.071
q_e cal (mg/g)	31.21	65.43	111.51	178.11	59.84	92.49	156.24	203.56
R^2	0.9552	0.9694	0.9795	0.9205	0.9114	0.9097	0.9035	0.919
	PSO kinetic model							
k_2 (g/mg min^{-1})	0.005	0.002	0.001	0.001	0.013	0.003	0.002	0.001
h	13.441	22.371	24.390	26.810	123.457	121.951	181.818	135.135
q_e cal (mg/g)	50.25	97.09	147.06	196.08	99.01	196.08	285.71	370.37
R^2	0.9997	0.9996	0.9992	0.9989	0.9999	0.9999	0.9999	0.9999
	IPDM							
k_{id} (mg/g/$min^{½}$)	0.604	0.272	0.149	0.104	0.991	0.331	0.206	0.113
R^2	0.8647	0.8491	0.8784	0.9173	0.6897	0.8282	0.7221	0.7989

Table 3. KP governing the adsorption of CRD by calcined and uncalcined LDH at varying temperatures (100 mg/L IDC).

Adsorbents	**Uncalcined LDH**			**Calcined LDH**		
	Temperature					
Parameter	30 °C	40 °C	50 °C	30 °C	40 °C	50 °C
q_e exp (mg/g)	179.23	181.42	184.70	356.99	365.59	374.19
	PFO kinetic model					
k_1 (min^{-1})	0.073	0.060	0.067	0.066	0.055	0.067
q_e cal (mg/g)	179.80	139.06	150.42	189.37	173.66	196.20
R^2	0.9307	0.9693	0.9665	0.9199	0.8814	0.9367
	PSO kinetic model					
k_2 (g/mg min^{-1})	0.00070	0.00079	0.00083	0.00097	0.00096	0.00097
h	26.954	30.395	33.003	133.333	131.579	142.857
q_e cal (mg/g)	196.08	196.08	200.00	370.37	370.37	384.62
R^2	0.9989	0.999	0.999	0.9999	0.9999	0.9999
	IPDM					
k_{id} (mg/g/$min^{½}$)	0.104	0.111	0.113	0.113	0.109	0.106
R^2	0.9143	0.9123	0.9003	0.8009	0.8158	0.7917

3.6.2. PSO Model

The following is the representation of the PSO kinetic equation,

$$\frac{t}{q_t} = \frac{1}{k_2 {q_e}^2} + \frac{1}{q_e} t \tag{3}$$

The consensus adsorption capacity, q_e, and second order rate constant, k_2, were derived from the relationship between the values of t/q_t and time for various concentrations of Ci (75 mg/L, 50 mg/L, 25 mg/L, and 100 mg/L) and temperatures (30 °C, 40 °C, and 50 °C). To determine the values of q_e and k_2, the slope and intercept of the plots in Figure 12a,b and Figure 13a,b were utilized. The second-order calculated KPs are displayed in Tables 1 and 2. It was observed that the q_e values calculated for both adsorbents were in close proximity to the experimental q_e values. The adsorption behavior of the dye by both adsorbents was highly consistent. For all concentrations, the PSO kinetics model provided a more accurate fit than the PFO kinetics model, as evidenced by the high R^2 values with low standard deviation values (approximately in the range from 0.082 to 0.30) for the adsorptive elimination of CR by both adsorbents [62].

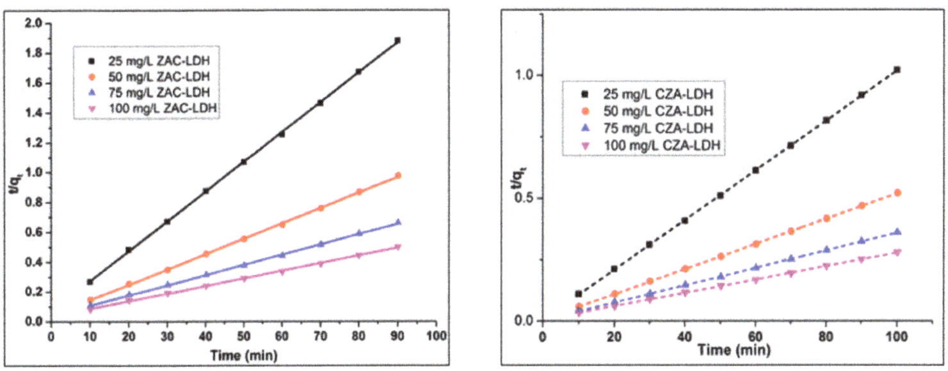

Figure 12. PSO kinetic model for ZAC-LDH and CZA- LDH uptake of CRD.

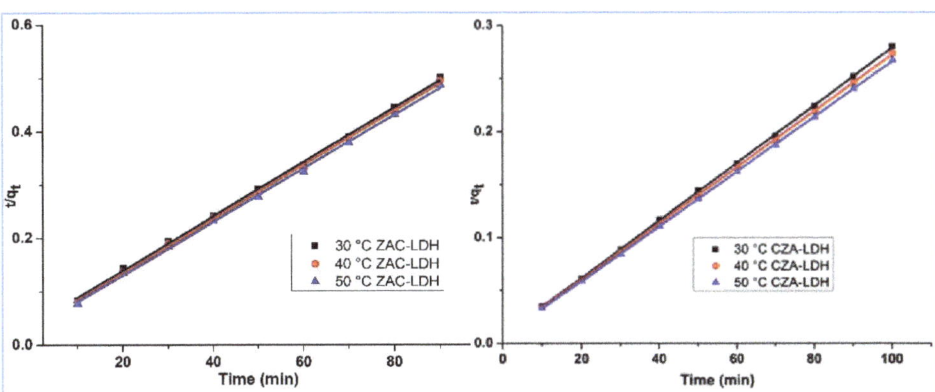

Figure 13. PSO kinetic model for the uptake of CRD at different temperature by ZAC-LDH and CZA- LDH.

3.6.3. IPDM (Intra-Particle Diffusion Model)

The kinetic results were subjected to the intra-particle diffusion model so as to elucidate the mechanism of diffusion. Intense agitation of the solution mixture containing adsorbent and adsorbate during the adsorbent process may result in the adsorbate migration into the pores available on the adsorbent, potentially serving as a tool in the rate-determination step. The intra-particle diffusion model's determination of the relationship among the quantity of dye adsorbed (q_t) and $t^{1/2}$ (k_{id} + C) at various time intermissions for various concentrations is shown in the graph diagram in Figure 14a,b. The values of k_{id} were computed from the gradients for all concentrations at 30 °C and for various temperatures, fixing the IDC as 100 mg/L. Since the intra-particle diffusion plot did not intersect the origin, the adsorption process by both calcined and uncalcined LDH appears to have substantially followed boundary layer diffusion [63].

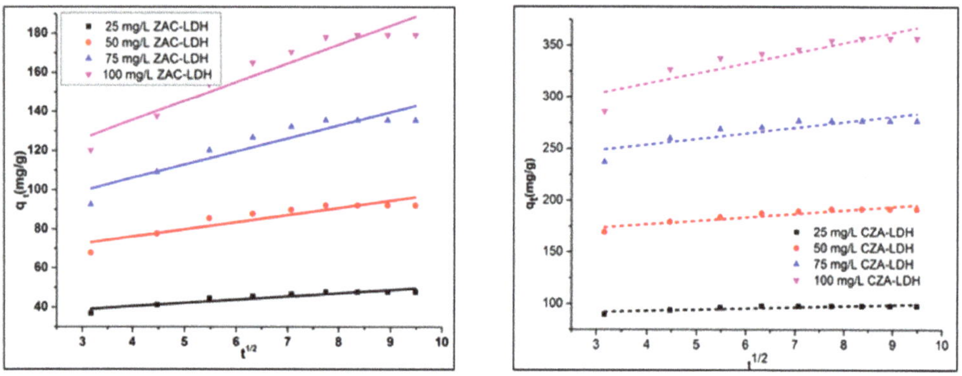

Figure 14. PSO kinetic model for the uptake of CRD at different temperature by ZAC- LDH and CZA-LDH.

3.7. *Adsorption Isotherm*

At constant temperatures, the equilibrium relationship between the adsorbate and the adsorbate adsorbed onto the surface of the adsorbent can be analyzed using the adsorption isotherm. In this investigation, the adsorption equilibrium for CR uptake by the two materials was studied through Langmuir [64] and Freundlich [65] isotherm models.

3.7.1. Langmuir Isotherm

The Langmuir adsorption isotherm, a widely recognized linear method for determining the adsorption of a monolayer onto a homogeneous surface, is represented by the subsequent equation:

$$\frac{C_e}{q_e} = \frac{1}{Q_0 . b_L} + \frac{C_e}{Q_0} \quad (4)$$

where Q_o and b_L are constants representing the energy of adsorption (L/mg) and monolayer adsorption capacity, respectively. The equilibrium concentration of the adsorbate is denoted as C_e in mg/L, while q_e represents the quantity of adsorbate adsorbed/unit mass of adsorbent in g/g. Table 4 provides a summary of the calculated values. A value between 0 and 1, known as the Langmuir dimensionless constant (R_L), denotes the favorable nature of adsorption. The maximal mono-layer converges capacity (Q_o) for uncalcined LDH and calcined LDH at 30 °C was determined to be 769.23 mg/g and 526.32 mg/g, respectively, based on the plot of C_e/q_e against C_e. The low correlation coefficient (R^2) of the Langmuir isotherm model makes it less suitable to explain the nature of CR adsorption.

Table 4. KP governing the adsorption of CRD by calcined and uncalcined LDH at varying temperatures (100 mg/L ICD).

Adsorbent	Temperature (°C)	Langmuir Isotherm				Freundlich Isotherm		
		Q_0	b	R_L	R^2	n	k_f	R^2
Uncalcined LDH	30	769.23	0.0024	0.8914	0.9208	1.133	29.580	0.999
	40	714.29	0.0026	0.8831	0.68	1.122	33.159	0.9915
	50	454.55	0.0042	0.8250	0.8712	1.412	53.407	0.9867
Calcined LDH	30	526.32	0.0073	0.7335	0.9531	1.676	86.298	0.993
	40	588.24	0.0065	0.7543	0.9637	1.745	97.521	0.9983
	50	625.00	0.0062	0.7646	0.92	1.703	106.955	0.9897

3.7.2. Freundlich Model

The heterogeneity of the adsorption system and multilayer adsorption are described using the Freundlich isotherm model. The Freundlich equation's linearized form is written as follows:

$$\log q_e = \log K_f + \frac{1}{n}\log C_e \quad (5)$$

where the adsorption intensity (n) and the adsorption capacity (k_f) are determined from the slope and intercept of a linear plot of logq_e versus logC_e. For the removal of CRD onto uncalcined LDH and calcined LDH, a favorable adsorption process is suggested by the reciprocal value of n ($1/n < 1$). With a high correlation coefficient value (R^2), the computed equilibrium values fit the Freundlich isotherm well, demonstrating the multilayer adsorption and heterogeneity nature of the process by both adsorbents [48].

3.8. Thermodynamic Parameters

Thermodynamic parameters related to adsorption, such as Gibb's free energy change (ΔG^0), enthalpy change (ΔH^0), and entropy change (ΔS^0), were determined using the formulae given below:

$$\Delta G^0 = -RT\ln K \quad (6)$$

$$\ln K_c = \frac{\Delta S^0}{R} - \frac{\Delta H^0}{RT} \quad (7)$$

Table 5 lists the thermodynamic characteristics for both adsorbents that were determined using Van't Hoff plots [66]. The negative ΔG^0 value in the range of -5.33 kJ mol^{-1} to -7.18 kJ mol^{-1}, which falls below the threshold of -20 kJ mol^{-1}, indicates a spontaneous physisorption process with a high attraction of CR onto the adsorbent material surface. In addition, the positive values of the enthalpy changes (ΔH^0) indicate that the adsorption process occurs at an endothermic temperature across all the temperatures examined. The endothermic characteristics of the adsorption process are corroborated by the increase in pigment adsorption as the temperature rises. Additionally, after the adsorption of the CRD, the positive ΔS° values suggest an upsurge in degrees of freedom and randomness at the boundary between the solid and the liquid [67].

Table 5. At various temperatures, the thermodynamic parameters of CRD adsorption by uncalcined and calcined LDH.

Adsorbent	Temperature (°C)	ΔG^0 (kJ/mol)	ΔH^0 (kJ/mol)	ΔS^0 (J/K/mol)
Uncalcined LDH	30	−5.43	13.59	62.65
	40	−5.93		
	50	−6.69		
Calcined LDH	30	−5.33	22.72	92.67
	40	−6.33		
	50	−7.18		

3.9. Reusability of the Adsorbents

The spent adsorbents uncalcined LDH and calcined LDH were calcined at 450 °C for two hours in an air environment in a muffle furnace. To assess the materials' reusability, a constant heating and cooling rate of 10 °C per cycle was maintained. As shown in Figure 15, the calcined materials from both were then utilized again to remove CR from aqueous solution using the identical adsorbent dosages of 0.5 g/L and 0.25 g/L for both uncalcined and calcined LDH material, respectively. The decline in the percentage of dye removed during consecutive cycles may be attributed to the reduction in weight during adsorbent recovery. However, both calcined adsorbent materials exhibited remarkable dye removal (83.53% and 64.71%, respectively), indicating their potential for effective reuse. This suggests that as-prepared adsorbent materials can be utilized, recovered, and successfully reused in the treatment of dye wastewater [68]. The adsorption capacities of various sorbents for the removal of CR in the present work and previous literature works are discussed in the Table 6.

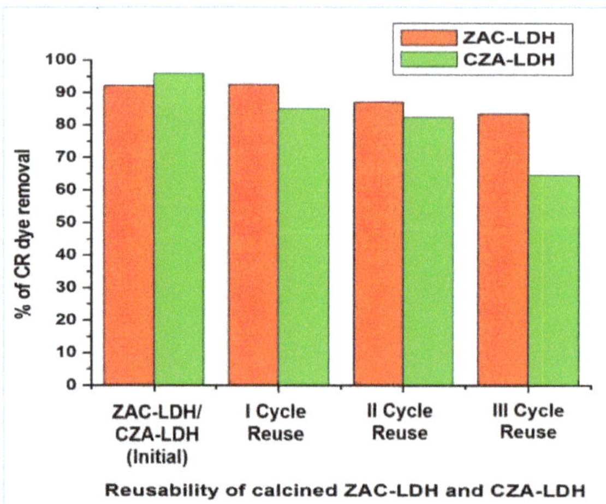

Figure 15. Effectiveness of reusability of both calcined LDHs for removal of CRD.

Table 6. Comparison of adsorption capacities of various sorbents for removal of CR.

S. No.	Adsorbent Material	Quantity of Dye Adsorbed (mg/g)	Reference
1.	Cabbage waste powder	1.78	[13]
2.	Coal fly ash	152.7	[17]
3.	Mg/Al-LDH	769.23	[41]
4.	CuAl LDH	47.619	[69]
5.	Ternary CaNiAl- LDH	135.21	[70]
6.	CaAl-LDH Cl	123.9	[71]
7.	ZAC-LDH	179.23	Present work
8.	CZA-LDH	356.99	Present work

4. Conclusions

This work substantiates the synthesis and analysis of the adsorption of the anionic dye CR using two different adsorbents: uncalcined LDH and its calcined counterpart, both of which are used in a batch mode adsorption procedure. The CR elimination decreased for uncalcined LDH (from 95.65% to 89.62%) and calcined LDH (from 97.96% to 89.25%) as the ICD increased from 25 mg/L to 100 mg/L at 30 °C. The highest quantity of CRD eliminated rose with temperature; for uncalcined LDH, it was 179.23 mg/g to 184.70 mg/g, and for calcined LDH, it was 356.99 mg/g to 374.19 mg/g. The temperature range for this rise was 30 °C to 50 °C. The uncalcined LDH exhibited optimal dye removal (78.38%) at pH 6, while calcined LDH showed 100% removal in the broader acidic pH range (below 7) for a 100 mg/L ICD. Kinetic data aligned well with the PSO model, demonstrating high correlation coefficients (R^2) and lower standard deviation values for both adsorbent materials. The maximum adsorption capacity for uncalcined LDH (769.23 mg/g) surpassed that of calcined LDH (526.32 mg/g) at 30 °C. Negative $\Delta G°$ values indicated a spontaneous physisorption process, while positive $\Delta H°$ values confirmed the endothermic nature of adsorption process at all temperatures under investigation. Langmuir and Freundlich isotherm representations substantiated the favorability of the adsorption process. The adsorption process's favorability was confirmed by models of the Langmuir and Freundlich isotherms. Thermally treated uncalcined LDH and calcined LDH demonstrated reusability with CR removal percentages of 83.53% and 64.71%, respectively, in the third cycle. Pre- and post-adsorption XRD spectra of calcined LDH demonstrated improved dye adsorption by reconstructing its initial layered structure through CR intercalation with a 'memory effect'. Moreover, the BET analysis has revealed that both uncalcined ZAC-LDH and calcined LDH possess a comparatively large pore size and mesoporosity. This characteristic makes them highly effective as adsorbents for wastewater treatment, as it facilitates the easy passage of dye molecules.

Accordingly, both uncalcined LDH and calcined LDH proved effective in CR removal, with calcined LDH exhibiting superior efficiency. The findings underscore the potential of these synthesized materials as efficient and reusable for the elimination of CR from dye wastewater. Additionally, this research will be expanded to generate M2+/M3+ LDH and assess its efficacy as a photo-catalyst and adsorbent for the removal of heavy metals from aqueous solutions and dye effluent, respectively.

Author Contributions: K.M.R., investigation and writing—original draft preparation; P.N.P., resources and supervision; V.N.K., conceptualization and writing—original draft preparation; A.T., writing—original draft preparation; R.P., conceptualization and investigation; S.C.K., conceptualization and investigation. All authors have read and agreed to the published version of the manuscript.

Funding: This research received no external funding.

Data Availability Statement: The original contributions presented in the study are included in the article, further inquiries can be directed to the corresponding author/s.

Conflicts of Interest: The authors declare no conflict of interest.

Nomenclature

BET	Brunauer–Emmett–Teller
CLDH	calcined layered double hydroxide
CR	Congo red
CRD	Congo red dye
CTAB	cetyltrimethylammonium bromide
CZA-LDH	calcined Zn/Al layered double hydroxide
DC	Dye Concentration
FESEM	Field Emission Scanning Electron Microscope
FTIR	Fourier-Transform Infrared Spectroscopy
IDC	Initial Dye Concentration
IPDM	intra-particle diffusion model
KP	kinetic parameter
LDH	layered double hydroxide
PFO	pseudo-first order
PSO	pseudo-second order
XRD	X-ray diffraction
ZAC-LDH	Zn/Al carbonate layered double hydroxide

References

1. Smetana, G.; Grosser, A. The Application of an Upflow Anaerobic Sludge Blanket Reactor in the Treatment of Brewery and Dairy Wastewater: A Critical Review. *Energies* **2024**, *17*, 1504. [CrossRef]
2. Smol, M. Circular Economy in Wastewater Treatment Plant—Water, Energy and Raw Materials Recovery. *Energies* **2023**, *16*, 3911. [CrossRef]
3. Benkhaya, S.; M'rabet, S.; El Harfi, A. A review on classifications, recent synthesis and applications of textile dyes. *Inorg. Chem. Commun.* **2020**, *115*, 107891. [CrossRef]
4. Wang, Y.; Dai, X.; Zhan, Y.; Ding, X.; Wang, M.; Wang, X. In situ growth of ZIF-8 nanoparticles on chitosan to form the hybrid nanocomposites for high-efficiency removal of Congo Red. *Int. J. Biol. Macromol.* **2019**, *137*, 77–86. [CrossRef]
5. Zhu, S.; Xu, J.; Wang, B.; Xie, J.; Ying, G.; Li, J.; Cheng, Z.; Li, J.; Chen, K. Highly efficient and rapid purification of organic dye wastewater using lignin-derived hierarchical porous carbon. *J. Colloid Interface Sci.* **2022**, *625*, 158–168. [CrossRef] [PubMed]
6. Kafshgari, L.A.; Ghorbani, M.; Azizi, A. Fabrication and investigation of $MnFe_2O_4$/MWCNTs nanocomposite by hydrothermal technique and adsorption of cationic and anionic dyes. *Appl. Surf. Sci.* **2017**, *419*, 70–83. [CrossRef]
7. Wei, Y.; Hao, J.-G.; Zhang, J.-L.; Huang, W.-Y.; Ouyang, S.-b.; Yang, K.; Lu, K.-Q. Integrating $Co(OH)_2$ nanosheet arrays on graphene for efficient noble-metal-free EY-sensitized photocatalytic H_2 evolution. *Dalton Trans.* **2023**, *52*, 13923–13929. [CrossRef] [PubMed]
8. Zhang, F.; Ma, B.; Jiang, X.; Ji, Y. Dual function magnetic hydroxyapatite nanopowder for removal of malachite green and Congo red from aqueous solution. *Powder Technol.* **2016**, *302*, 207–214. [CrossRef]
9. Lan, D.; Zhu, H.; Zhang, J.; Li, S.; Chen, Q.; Wang, C.; Wu, T.; Xu, M. Adsorptive removal of organic dyes via porous materials for wastewater treatment in recent decades: A review on species, mechanisms and perspectives. *Chemosphere* **2022**, *293*, 133464. [CrossRef]
10. Zhang, P.; Qian, G.; Shi, H.; Ruan, X.; Yang, J.; Frost, R.L. Mechanism of interaction of hydrocalumites (Ca/Al-LDH) with methyl orange and acidic scarlet GR. *J. Colloid Interface Sci.* **2012**, *365*, 110–116. [CrossRef]
11. Włodarczyk-Makuła, M.; Myszograj, S.; Włodarczyk, M. Removal of Organic Micro-Pollutants from Wastewater in Electrochemical Processes—Review. *Energies* **2023**, *16*, 5591. [CrossRef]
12. Sherino, B.; Abdul Halim, S.N.; Shahabuddin, S.; Mohamad, S. Simultaneous removal of carcinogenic anionic and cationic dyes from environmental water using a new Zn-based metal–organic framework. *Sep. Sci. Technol.* **2021**, *56*, 330–343. [CrossRef]
13. Wekoye, J.N.; Wanyonyi, W.C.; Wangila, P.T.; Tonui, M.K. Kinetic and equilibrium studies of Congo red dye adsorption on cabbage waste powder. *Environ. Chem. Ecotoxicol.* **2020**, *2*, 24–31. [CrossRef]
14. Arab, C.; El Kurdi, R.; Patra, D. Zinc curcumin oxide nanoparticles for enhanced adsorption of Congo red: Kinetics and adsorption isotherms study. *Mater. Today Chem.* **2022**, *23*, 100701. [CrossRef]
15. Kaur, N.; Kaushal, J.; Mahajan, P.; Mantri, A. Phytoremediation of methylene blue dye (triarylmethane) and Congo red (diazo) by *T. ammi* L.: Kinetic studies. *Int. J. Environ. Sci. Technol.* **2024**, *21*, 1697–1714. [CrossRef]
16. Xu, Y.; Wang, Q.; Ding, Z. Synthesis of Superparamagnetic Fe_3O_4 Nano-Adsorbent Using an Energy-Saving and Pollution-Reducing Strategy for the Removal of Xylenol Orange Dye in Water. *Energies* **2022**, *15*, 7378. [CrossRef]
17. Abas, K.M.; Fathy, N.A. Sodalite zeolitic materials produced from coal fly ash for removal of congo red dye from aqueous solutions. *Int. J. Environ. Sci. Technol.* **2024**, *21*, 5165–5184. [CrossRef]
18. Abbas, S.H.; Ridha, A.M.; Rashid, K.H.; Khadom, A.A. Biosorption of Congo red dye removal from aqueous solution using fennel seed spent and garlic peel. *Int. J. Environ. Sci. Technol.* **2023**, *20*, 13845–13858. [CrossRef]
19. Mielcarek, A.; Bryszewski, K.Ł.; Rodziewicz, J.; Kłobukowska, K.; Janczukowicz, W. Phosphorus Removal Rate and Efficiency in an Electrochemical Sequencing Reactor for the Treatment of Wastewater with Low Organic Carbon Content. *Energies* **2024**, *17*, 1352. [CrossRef]
20. Gusiatin, M.Z.; Pasieczna-Patkowska, S.; Bálintová, M.; Kuśmierz, M. Treatment of Wastewater from Soil Washing with Soluble Humic Substances Using Biochars and Activated Carbon. *Energies* **2023**, *16*, 4311. [CrossRef]
21. Hu, M.; Yan, X.; Hu, X.; Zhang, J.; Feng, R.; Zhou, M. Ultra-high adsorption capacity of MgO/SiO_2 composites with rough surfaces for Congo red removal from water. *J. Colloid Interface Sci.* **2018**, *510*, 111–117. [CrossRef] [PubMed]
22. Chen, H.; Zheng, Y.; Cheng, B.; Yu, J.; Jiang, C. Chestnut husk-like nickel cobaltite hollow microspheres for the adsorption of Congo red. *J. Alloys Compd.* **2018**, *735*, 1041–1051. [CrossRef]
23. Moosavi, S.; Lai, C.W.; Gan, S.; Zamiri, G.; Akbarzadeh Pivehzhani, O.; Johan, M.R. Application of efficient magnetic particles and activated carbon for dye removal from wastewater. *ACS Omega* **2020**, *5*, 20684–20697. [CrossRef] [PubMed]
24. Yang, R.; Li, D.; Li, A.; Yang, H. Adsorption properties and mechanisms of palygorskite for removal of various ionic dyes from water. *Appl. Clay Sci.* **2018**, *151*, 20–28. [CrossRef]
25. Mansha, M.; Waheed, A.; Ahmad, T.; Kazi, I.W.; Ullah, N. Synthesis of a novel polysuccinimide based resin for the ultrahigh removal of anionic azo dyes from aqueous solution. *Environ. Res.* **2020**, *184*, 109337. [CrossRef]
26. Ohemeng-Boahen, G.; Sewu, D.D.; Tran, H.N.; Woo, S.H. Enhanced adsorption of congo red from aqueous solution using chitosan/hematite nanocomposite hydrogel capsule fabricated via anionic surfactant gelation. *Colloids Surf. A Physicochem. Eng. Asp.* **2021**, *625*, 126911. [CrossRef]
27. Hashem, A.H.; Saied, E.; Hasanin, M.S. Green and ecofriendly bio-removal of methylene blue dye from aqueous solution using biologically activated banana peel waste. *Sustain. Chem. Pharm.* **2020**, *18*, 100333. [CrossRef]

28. Munagapati, V.S.; Yarramuthi, V.; Kim, Y.; Lee, K.M.; Kim, D.-S. Removal of anionic dyes (Reactive Black 5 and Congo Red) from aqueous solutions using Banana Peel Powder as an adsorbent. *Ecotoxicol. Environ. Saf.* **2018**, *148*, 601–607. [CrossRef]
29. Bukhari, A.; Ijaz, I.; Zain, H.; Gilani, E.; Nazir, A.; Bukhari, A.; Raza, S.; Ansari, J.; Hussain, S.; Alarfaji, S.S.; et al. Removal of Eosin dye from simulated media onto lemon peel-based low cost biosorbent. *Arab. J. Chem.* **2022**, *15*, 103873. [CrossRef]
30. Gomaa, H.; Hussein, M.A.; Motawea, M.M.; Aboraia, A.M.; Cheira, M.F.; Alotaibi, M.T.; El-Bahy, S.M.; Ali, H.M. A hybrid mesoporous CuO@barley straw-derived SiO_2 nanocomposite for adsorption and photocatalytic degradation of methylene blue from real wastewater. *Colloids Surf. A Physicochem. Eng. Asp.* **2022**, *644*, 128811. [CrossRef]
31. Lin, T.-Y.; Chai, W.S.; Chen, S.-J.; Shih, J.-Y.; Koyande, A.K.; Liu, B.-L.; Chang, Y.-K. Removal of soluble microbial products and dyes using heavy metal wastes decorated on eggshell. *Chemosphere* **2021**, *270*, 128615. [CrossRef] [PubMed]
32. Chikri, R.; Elhadiri, N.; Benchanaa, M.; Maguana, Y. Efficiency of sawdust as low-cost adsorbent for dyes removal. *J. Chem.* **2020**, *2020*, 8813420. [CrossRef]
33. Malik, A.; Khan, A.; Anwar, N.; Naeem, M. A comparative study of the adsorption of Congo red dye on rice husk, rice husk char and chemically modified rice husk char from aqueous media. *Bull. Chem. Soc. Ethiop.* **2020**, *34*, 41–54. [CrossRef]
34. Astuti, W.; Chafidz, A.; Wahyuni, E.T.; Prasetya, A.; Bendiyasa, I.M.; Abasaeed, A.E. Methyl violet dye removal using coal fly ash (CFA) as a dual sites adsorbent. *J. Environ. Chem. Eng.* **2019**, *7*, 103262. [CrossRef]
35. Sarkar, S.; Tiwari, N.; Basu, A.; Behera, M.; Das, B.; Chakrabortty, S.; Sanjay, K.; Suar, M.; Adhya, T.K.; Banerjee, S.; et al. Sorptive removal of malachite green from aqueous solution by magnetite/coir pith supported sodium alginate beads: Kinetics, isotherms, thermodynamics and parametric optimization. *Environ. Technol. Innov.* **2021**, *24*, 101818. [CrossRef]
36. Nehaba, S.S.; Abdullah, R.H.; Oda, A.M.; Omran, A.R.; Mottaleb, A.S. Evaluation of the efficiency of tea waste powder to remove the Safranin O dye compared to the activated carbon as adsorbent. *Orient. J. Chem.* **2019**, *35*, 1201–1207. [CrossRef]
37. Tuli, F.; Hossain, A.; Kibria, A.F.; Tareq, A.; Mamun, S.M.; Ullah, A.A. Removal of methylene blue from water by low-cost activated carbon prepared from tea waste: A study of adsorption isotherm and kinetics. *Environ. Nanotechnol. Monit. Manag.* **2020**, *14*, 100354. [CrossRef]
38. Chen, Y.; Chen, S.; Deng, Z.; Xu, X.; Qin, J.; Guo, X.; Bai, Z.; Chen, X.; Lu, Z. Highly efficient removal of Congo red on calcined MgAl-layered double hydroxide with microporous structure. *Int. J. Environ. Sci. Technol.* **2023**, *20*, 10141–10152. [CrossRef]
39. Lu, K.-Q.; Hao, J.-G.; Wei, Y.; Weng, B.; Ge, S.; Yang, K.; Lu, S.; Yang, M.-Q.; Liao, Y. Photocatalytic Conversion of Diluted CO_2 into Tunable Syngas via Modulating Transition Metal Hydroxides. *Inorg. Chem.* **2024**, *63*, 795–802. [CrossRef]
40. Bobde, P.; Sharma, A.K.; Panchal, D.; Sharma, A.; Patel, R.K.; Dhodapkar, R.S.; Pal, S. Layered double hydroxides (LDHs)-based photocatalysts for dye degradation: A review. *Int. J. Environ. Sci. Technol.* **2023**, *20*, 5733–5752. [CrossRef]
41. Farghali, M.A.; Selim, A.M.; Khater, H.F.; Bagato, N.; Alharbi, W.; Alharbi, K.H.; Radwan, I.T. Optimized adsorption and effective disposal of Congo red dye from wastewater: Hydrothermal fabrication of MgAl-LDH nanohydrotalcite-like materials. *Arab. J. Chem.* **2022**, *15*, 104171. [CrossRef]
42. Xie, J.; Yamaguchi, T.; Oh, J.-M. Synthesis of a mesoporous Mg–Al–mixed metal oxide with P123 template for effective removal of Congo red via aggregation-driven adsorption. *J. Solid State Chem.* **2021**, *293*, 121758. [CrossRef]
43. Khalil, A.K.; Almanassra, I.W.; Chatla, A.; Ihsanullah, I.; Laoui, T.; Atieh, M.A. Insights into the adsorption of lead ions by Mg-Al LDH doped activated carbon composites: Implications for fixed bed column and batch applications. *Chem. Eng. Sci.* **2023**, *281*, 119192. [CrossRef]
44. Manea, Y.K.; Khan, A.M.; Wani, A.A.; Saleh, M.A.; Qashqoosh, M.T.; Shahadat, M.; Rezakazemi, M. In-grown flower like Al-Li/Th-LDH@CNT nanocomposite for enhanced photocatalytic degradation of MG dye and selective adsorption of Cr (VI). *J. Environ. Chem. Eng.* **2022**, *10*, 106848. [CrossRef]
45. Shirin, V.A.; Sankar, R.; Johnson, A.P.; Gangadharappa, H.; Pramod, K. Advanced drug delivery applications of layered double hydroxide. *J. Control. Release* **2021**, *330*, 398–426. [CrossRef]
46. El Khanchaoui, A.; Sajieddine, M.; Mansori, M.; Essoumhi, A. Anionic dye adsorption on ZnAl hydrotalcite-type and regeneration studies based on "memory effect". *Int. J. Environ. Anal. Chem.* **2022**, *102*, 3542–3560. [CrossRef]
47. Wang, P.; Zhang, X.; Zhou, B.; Meng, F.; Wang, Y.; Wen, G. Recent Advance of Layered Double Hydroxides Materials: Structure, Properties, Synthesis, Modification and Applications of Wastewater Treatment. *J. Environ. Chem. Eng.* **2023**, *11*, 111191. [CrossRef]
48. Manjula Rani, K.; Palanisamy, P.N. Synthesis and Characterization of Mesoporous, Nanostructured Zinc Aluminium Carbonate Layered Double Hydroxides (ZAC-LDHs) and Its Calcined Product (CZA-LDH). *J. Inorg. Organomet. Polym. Mater.* **2018**, *28*, 1127–1135. [CrossRef]
49. Rani, K.M.; Palanisamy, P. Comparative studies on the adsorptive removal of Acid Violet-17 dye from aqueous solution by using zinc aluminium carbonate-LDH (ZAC-LDH) and modified LDH. *Desalination Water Treat.* **2017**, *65*, 337–345. [CrossRef]
50. Tamilvanan, A.; Balamurugan, D.K.; Ponappa, D.K.; Kumar, B.M. Using Response Surface Methodology in Synthesis of Ultrafine Copper Nanoparticles by Electrolysis. *Int. J. Nanosci.* **2016**, *15*, 1650001. [CrossRef]
51. Boukhalfa, N.; Boutahala, M.; Djebri, N. Synthesis and characterization of ZnAl-layered double hydroxide and organo-K10 montmorillonite for the removal of diclofenac from aqueous solution. *Adsorpt. Sci. Technol.* **2017**, *35*, 20–36. [CrossRef]
52. Ahmed, D.N.; Naji, L.A.; Faisal, A.A.; Al-Ansari, N.; Naushad, M. Waste foundry sand/MgFe-layered double hydroxides composite material for efficient removal of Congo red dye from aqueous solution. *Sci. Rep.* **2020**, *10*, 2042. [CrossRef]
53. Mahmood, T.; Noreen, U.; Ali, R.; Ullah, A.; Naeem, A.; Aslam, M. Adsorptive removal of Congo red from aqueous phase using graphene–tin oxide composite as a novel adsorbent. *Int. J. Environ. Sci. Technol.* **2022**, *19*, 10275–10290. [CrossRef]

54. Daud, M.; Hai, A.; Banat, F.; Wazir, M.B.; Habib, M.; Bharath, G.; Al-Harthi, M.A. A review on the recent advances, challenges and future aspect of layered double hydroxides (LDH)–Containing hybrids as promising adsorbents for dyes removal. *J. Mol. Liq.* **2019**, *288*, 110989. [CrossRef]
55. Rathee, G.; Awasthi, A.; Sood, D.; Tomar, R.; Tomar, V.; Chandra, R. A new biocompatible ternary Layered Double Hydroxide Adsorbent for ultrafast removal of anionic organic dyes. *Sci. Rep.* **2019**, *9*, 16225. [CrossRef]
56. Huang, Y.; Yin, W.; Zhao, T.-L.; Liu, M.; Yao, Q.-Z.; Zhou, G.-T. Efficient Removal of Congo Red, Methylene Blue and Pb(II) by Hydrochar–MgAlLDH Nanocomposite: Synthesis, Performance and Mechanism. *Nanomaterials* **2023**, *13*, 1145. [CrossRef]
57. Wai, T.T.; Aung, E.M.; Kyaw, N.C. Potential of neem leaf powder as bio-adsorbents for dye colour removal. *Int. J. Trend Sci. Res. Dev.* **2019**, *3*, 2502–2505.
58. Feng, L.; Zhang, Q.; Ji, F.; Jiang, L.; Liu, C.; Shen, Q.; Liu, Q. Phosphate removal performances of layered double hydroxides (LDH) embedded polyvinyl alcohol/lanthanum alginate hydrogels. *Chem. Eng. J.* **2022**, *430*, 132754. [CrossRef]
59. Dovi, E.; Aryee, A.A.; Kani, A.N.; Mpatani, F.M.; Li, J.; Li, Z.; Qu, L.; Han, R. Functionalization of walnut shell by grafting amine groups to enhance the adsorption of Congo red from water in batch and fixed-bed column modes. *J. Environ. Chem. Eng.* **2021**, *9*, 106301. [CrossRef]
60. Constantin, M.; Asmarandei, I.; Harabagiu, V.; Ghimici, L.; Ascenzi, P.; Fundueanu, G. Removal of anionic dyes from aqueous solutions by an ion-exchanger based on pullulan microspheres. *Carbohydr. Polym.* **2013**, *91*, 74–84. [CrossRef]
61. Shaban, M.; Ashraf, A.M.; Abukhadra, M.R. TiO_2 nanoribbons/carbon nanotubes composite with enhanced photocatalytic activity; fabrication, characterization, and application. *Sci. Rep.* **2018**, *8*, 781. [CrossRef] [PubMed]
62. Lagergren, S. Zur theorie der sogenannten adsorption geloster stoffe. Kungliga svenska vetenskapsakademiens. *Handlingar* **1898**, *24*, 1–39.
63. Singh, P.K.; Banerjee, S.; Srivastava, A.L.; Sharma, Y.C. Kinetic and equilibrium modeling for removal of nitrate from aqueous solutions and drinking water by a potential adsorbent, hydrous bismuth oxide. *RSC Adv.* **2015**, *5*, 35365–35376. [CrossRef]
64. Langmuir, I. The adsorption of gases on plane surfaces of glass, mica and platinum. *J. Am. Chem. Soc.* **1918**, *40*, 1361–1403. [CrossRef]
65. Freundlich, H. Over the adsorption in solution. *J. Phys. Chem.* **1906**, *57*, 1100–1107.
66. Zhang, C.; Yang, S.; Chen, H.; He, H.; Sun, C. Adsorption behavior and mechanism of reactive brilliant red X-3B in aqueous solution over three kinds of hydrotalcite-like LDHs. *Appl. Surf. Sci.* **2014**, *301*, 329–337. [CrossRef]
67. Ahmed, M.; Mohamed, A. An efficient adsorption of indigo carmine dye from aqueous solution on mesoporous Mg/Fe layered double hydroxide nanoparticles prepared by controlled sol-gel route. *Chemosphere* **2017**, *174*, 280–288. [CrossRef] [PubMed]
68. Mensah, K.; Mahmoud, H.; Fujii, M.; Shokry, H. Novel nano-ferromagnetic activated graphene adsorbent extracted from waste for dye decolonization. *J. Water Process Eng.* **2022**, *45*, 102512. [CrossRef]
69. Palapa, N.R.; Juleanti, N.; Normah; Taher, T.; Mohadi, R.; Rachmat, A.; Lesbani, A. Adsorptive Performance of Congo Red using Copper-Aluminum LDHS Load to Rice Husk Biochar. *Sains Malays.* **2022**, *51*, 161–173. [CrossRef]
70. Brahma, D.; Priyom Nath, K.; Patgiri, M.; Saikia, H. Synthesis of Ternary CaNiAl-Layered Double Hydroxide as Potential Adsorbent for Congo Red Dye Removal in Aqueous Solution. *Asian J. Chem.* **2022**, *34*, 3215–3223. [CrossRef]
71. Kong, X.-q.; Zhang, N.-n.; Lian, Y.-m.; Tang, Q.-j. Removal of Congo red using the chlorinated Ca-al layered double hydroxide produced from the desulfurization circulating wastewater. *Desalination Water Treat.* **2021**, *238*, 294–305. [CrossRef]

energies

MDPI

Review

Modeling and Control Strategies for Energy Management in a Wastewater Center: A Review on Aeration

Mukhammad Jamaludin [1], Yao-Chuan Tsai [1], Hao-Ting Lin [1], Chi-Yung Huang [2], Wonjung Choi [3], Jiang-Gu Chen [4] and Wu-Yang Sean [1,*]

[1] Department of Bio-Industrial Mechatronics Engineering, National Chung Hsing University, Taichung City 402202, Taiwan; ujamal225@gmail.com (M.J.); yctsaii@dragon.nchu.edu.tw (Y.-C.T.)
[2] Department of Mechanical Engineering, National Chin-Yi University of Technology, Taichung City 41170, Taiwan
[3] Department of Chemical Engineering, Changwon National University, Changwon 51140, Republic of Korea; wjchoi@changwon.ac.kr
[4] Taoyuan Northern District Reclaimed Center, Taoyuan 33071, Taiwan
* Correspondence: wysean@nchu.edu.tw; Tel.: +886-978059003

Abstract: Effective modeling and management are critical in wastewater treatment facilities since the aeration process accounts for 65–70% of the overall energy consumption. This study assesses control strategies specifically designed for different sizes of WWTP, analyzing their economic, environmental, and energy-related effects. Small WWTPs see advantages from the utilization of on/off and proportional–integral–derivative (PID) control methods, resulting in 10–25% energy savings and the reduction in dissolved oxygen (DO) levels by 5–30%. Cascade control and model predictive control (MPC) improve energy efficiency by 15–30% and stabilize DO levels by 15–35% in medium-sized WWTPs. Advanced WWTPs that utilize technologies such as MPC integrated with artificial intelligence (AI) and machine learning (ML) can decrease energy usage by 30–40% and enhance DO levels by 35–40%. Life cycle assessment (LCA) demonstrates substantial decreases in greenhouse gas (GHG) emissions: 5–20% for small, 10–25% for medium, and 30–35% for large WWTPs. These findings illustrate the feasibility and expandability of these tactics in both controlled laboratory environments and real-world situations, emphasizing the significance of customized methods for improving energy efficiency and sustainability in wastewater treatment. Subsequent investigations should prioritize integrating renewable energy sources and resolving obstacles in developing nations to enhance wastewater treatment plants' energy efficiency and sustainability.

Keywords: aeration; control strategies; energy consumption; modeling; wastewater treatment plants

1. Introduction

Wastewater treatment plants (WWTPs) are essential infrastructure elements specifically intended to process and cleanse wastewater produced by residential, industrial, and commercial sources prior to its release into the environment or reuse [1]. These facilities are crucial for safeguarding human health, maintaining water quality, and reducing environmental pollution [2]. WWTPs utilize a variety of methods, including physical, chemical, and biological processes, to eliminate impurities, pollutants, and disease-causing microorganisms from wastewater. This ensures that the treated water meets the required regulatory criteria and protects the integrity of the environment [3]. Municipal WWTPs are particularly suitable for studying energy management solutions due to their extensive size and substantial energy usage. It makes them a central focus for advancements in efficiency and sustainability. WWTPs have the dual responsibility of processing wastewater to meet discharge standards and encourage resource recovery and environmental sustainability [4]. The endeavor to retrieve essential resources such as nutrients, energy, and water from wastewater is in accordance with the ideas of circular economy and resource conservation [5]. Aeration is a crucial step in these facilities, having a significant impact on both treatment efficiency and resource recovery.

Aeration is an essential process in WWTPs, playing a vital role in facilitating the elimination of nitrogen molecules, such as ammonia and nitrate, through the processes of nitrification and denitrification [6]. During the process of nitrification, aerobic bacteria convert ammonia into nitrite and then into nitrate. This process relies on the presence of sufficient oxygen [7]. Aeration is the technique of introducing air into wastewater to enhance the transfer of oxygen and encourage aerobic biological processes [8]. Aeration promotes the growth and activity of aerobic microorganisms [9] by increasing the oxygen levels in the environment. These bacteria break down organic materials and contaminants found in wastewater, transforming them into innocuous substances [10]. The importance of aeration in wastewater treatment cannot be emphasized enough, as it is crucial for the efficiency, efficacy, and environmental sustainability of wastewater treatment plant operations [11]. As aeration technologies and control tactics advance, optimizing aeration processes will continue to be a primary goal for WWTP operators. This is undertaken to improve treatment performance, reduce energy consumption [12], and fulfill stricter regulatory requirements [13].

The high energy consumption of aeration systems in municipal WWTPs enhances concerns about environmental sustainability and the preservation of resources [14]. The assessment of energy consumption indicates that the utilization of electrical, mechanical, chemical, and human energy results in a total of 0.26 kilowatt-hours per cubic meter of treated effluent. Electric energy constitutes 84% of the overall usage, whereas mechanical energy makes up 15%. Around 78% of the electrical energy is consumed during the aeration process [15]. The increasing energy expenses related to aeration operations impose substantial financial challenges on WWTP operators, reducing overall operational effectiveness and economic sustainability [16]. The significant proportion of operational costs dedicated to energy usage highlights the pressing requirement for inventive energy management strategies and technology to reduce expenses and improve cost efficiency [17,18]. In order to address the energy difficulties related to WWTP operations, it is necessary to develop a complete plan. This strategy should involve technological advancements, operational improvements, and policy backing. This includes the use of advanced control and monitoring systems, predictive modeling, and optimization algorithms [19,20].

Implementing sophisticated control and monitoring systems, along with predictive modeling and optimization algorithms, allows for the real-time management of energy and the adaptive functioning of aeration processes [21]. WWTPs can improve energy use, eliminate wastage, and boost overall operational efficiency by adaptively modifying aeration rates, airflow patterns, and oxygen transfer efficiencies in response to changing influent characteristics and treatment requirements [22,23]. Standard methods for improving energy efficiency in aeration involve adjusting aeration rates [24], using energy-efficient equipment, installing improved control systems, and investigating alternative energy sources, including renewable energy and waste-to-energy technology [25].

Control systems are essential for regulating and optimizing the operations of WWTPs, particularly the aeration processes [26]. Frequently employed control systems in WWTPs consist of dissolved oxygen (DO) management, flow control, pH control, and process automation [27]. These control systems allow operators to consistently maintain ideal treatment conditions, optimize energy efficiency, and guarantee adherence to effluent quality standards [28]. Considering the substantial amount of energy used for aeration in municipal WWTPs and the increasing focus on energy sustainability and environmental responsibility, it is crucial to enhance the energy efficiency of aeration systems [29]. This research contributes to advancing the sustainability and resilience of WWTPs by developing innovative technologies, methodologies, and control strategies to reduce energy consumption and enhance operational efficiency [30]. It has implications for resource conservation, cost savings, and environmental protection [31].

This article review discusses the pressing necessity of improving energy efficiency in municipal WWTP operations, which are major energy consumers in urban infrastructure. This review is unique since it provides a thorough analysis of several aeration control systems, their effects on energy usage, and their specific benefits and drawbacks. This paper explicitly emphasizes the modeling and control strategies involved. This review takes a comprehensive approach to aeration control in the context of WWTP operations and sustainability rather than focusing solely on single technologies or processes. This review examines various methods of aeration control, including DO control and Anammox modeling, to explore the advantages and limitations of each approach. In doing so, it provides insights into how to optimize aeration processes for maximum energy efficiency without compromising treatment quality.

The assessment was chosen for its potential to provide significant insights into implementing adequate aeration controls at WWTPs, which can help to conserve resources, save costs, and protect the environment. This paper offers several benefits, such as presenting a thorough framework for assessing and choosing energy-efficient aeration strategies, encouraging the implementation of advanced control and monitoring systems, and emphasizing the significance of incorporating renewable energy sources and waste-to-energy technologies. This review not only contributes to the improvement of the sustainability and resilience of WWTPs but also acts as an essential point of reference for policymakers, researchers, and WWTP operators who are seeking to promote operational efficiency and environmental stewardship.

2. Municipal WWTPs

Municipal WWTPs are essential elements of modern urban infrastructure [32], involving the complex relationship between public health priorities and objectives regarding environmental sustainability [33]. These facilities have the critical job of cleaning wastewater that comes from a variety of sources in urban areas [34]. It is important to protect human health and the environment by treating the harmful substances in wastewater [35,36]. Municipal WWTPs utilize a range of advanced treatment methods based on scientific research and engineering innovation. These methods involve carefully coordinated physical, chemical, and biological processes that remove contaminants, pollutants, and microbial pathogens from wastewater [37]. The treatment process starts with initial stages that involve removing large debris and solids, followed by primary treatment methods that separate suspended solids and settleable fractions [38]. The treatment then progresses to secondary treatment phases, where biodegradable organic matter is broken down through aerobic or anaerobic degradation mechanisms.

Furthermore, tertiary treatment methods become more important in the later stages of the treatment process, providing further purification and disinfection to meet the strict effluent quality criteria set by environmental regulatory authorities [39]. Municipal WWTPs represent the highest standard in managing wastewater. They combine scientific rigor, technological innovation, and regulatory compliance to effectively address environmental pollution, protect public health, and promote sustainable water resource management in growing urban areas [40].

The energy consumption necessary for the functioning of municipal WWTPs is significant, mainly because of the energy-intensive procedures involved in aeration, pumping, and treatment stages [41]. Typically, WWTPs use around 0.6 to 2.0 kilowatt-hours (kWh) of energy for every cubic meter of wastewater treated. This high energy demand is caused by the requirement for the uninterrupted operation of different mechanical and electrical systems to guarantee efficient treatment. It is worth mentioning that electrical energy accounts for 84% of the overall energy consumption in WWTPs, and aeration alone accounts for about 78% of this electrical energy usage [15].

The process of aeration in WWTPs is the most energy-intensive, accounting for approximately 65–70% of the overall energy consumption in the plant [42]. This energy is mainly utilized to supply the oxygen required for aerobic biological functions. Aerobic treatment operations, such as aeration and subsequent aerobic biological treatments, make a substantial contribution to the total energy consumption. Anaerobic digestion activities, which have lower energy requirements compared to aerobic processes, contribute to around 10–15% of the overall energy consumption [43]. These procedures are commonly employed for the treatment of sludge, transforming organic substances into biogas (methane), which has the potential to be utilized as an energy source. Pumping operations, which are crucial for the transportation of wastewater between different treatment stages, contribute to around 10–15% of the overall energy consumption [44]. Pumps are extensively utilized throughout the WWTPs, from the initial collection and transportation of wastewater to its transfer between treatment stages. Approximately 5–10% of the total energy consumption is attributed to additional procedures such as chemical treatments, mixing, and tertiary treatment stages, including filtration and UV disinfection. These procedures are essential for further purifying the treated water to comply with strict discharge regulations. Currently, China alone has over 18,000 municipal WWTPs, which highlights the significant worldwide scale of these facilities. The precise global count of WWTPs might be assumed to be considerably more, with thousands more currently functioning in various countries [45]. The significant energy consumption of these facilities highlights the necessity of creating and executing energy-efficient technologies and management strategies with which to decrease operating expenses and environmental consequences [46].

2.1. Aeration Process

The aeration process in wastewater treatment involves a complex and dynamic sequence of procedures designed to efficiently introduce oxygen into the aqueous environment in order to facilitate biological degradation and purification [47]. Refer to Figure 1 for an overview of the typical aeration process in wastewater treatment plants. Once the proper aeration equipment is selected to meet the unique needs of the treatment facility, mechanical aerators or distributed aeration systems are strategically positioned in treatment basins or tanks [48]. Mechanical aerators, which consist of rotating impellers or surface splashing mechanisms, create turbulence and agitation in the wastewater, enabling the exchange of gases between the liquid phase and the atmosphere [49]. For instance, Sun et al. (2023) emphasize that improving the mechanical aeration in cryogenic liquid turbine expanders can reduce cavitation and increase oxygen transfer efficiency by around 15% [50].

In contrast, it diffuses aeration systems' functions by dispersing compressed air through permeable membranes or diffusers that are submerged in the water. This process generates a large number of little bubbles that rise through the liquid, providing oxygen and facilitating mixing. Yang et al. (2024) demonstrated that, by the optimization of submerged batch fermentation for *Streptomyces yanglinensis*, the authors achieved a 20% enhancement in aeration efficiency. As a result, there was a significant increase in the production yields of bioactive compounds [51].

During aeration, aerobic bacteria in the wastewater begin their metabolic activity by using dissolved oxygen to oxidize organic materials, contaminants, and nutrients in the wastewater. *Microbial respiration* is a metabolic process that produces energy for microbial growth and enzymatic processes. This process also results in the decomposition of complex chemical molecules into simpler and less toxic ones. In their study, Wang et al. (2024) discovered that the presence of domesticated aerobic denitrifying bacteria increased the denitrification efficiency by 25% in low-temperature municipal wastewater. This improvement was achieved through adequate aeration [52]. During the aeration process, the levels of dissolved oxygen are carefully monitored and regulated to maintain ideal conditions for microbial activity while also minimizing the chances of oxygen depletion or excessive saturation. Byliński et al. (2019) highlighted the significance of monitoring volatile organic compounds and dissolved oxygen, stating that emissions can be decreased by 30% through effective aeration control [53].

Aeration has two primary purposes: reducing unpleasant smells that come from anaerobic decomposition processes and improving the elimination of volatile organic compounds from the wastewater. Toledo and Muñoz (2019) reported that the implementation of activated sludge re-cycling, along with efficient aeration, resulted in a 40% reduction in odor emissions [54]. Moreover, aeration has a substantial influence on the energy usage in WWTPs. Optimizing aeration systems in WWTPs can result in significant energy savings, as aeration typically represents around 65–70% of the overall energy consumption in WWTPs. For instance, by improving the operational parameters of integrated fixed-film activated sludge systems, the aeration efficiency was increased by 35%. This improvement resulted in enhanced overall treatment performance and reduced energy usage [55].

Efficient aeration enhances both the effectiveness of wastewater treatment and the control of energy in WWTPs. Facilities can achieve lower operational costs and improve sustainability by improving aeration processes, which reduces the energy needed for oxygen transfer. By making adjustments to aeration rates, utilizing sophisticated control systems, and deploying energy-efficient aeration equipment, the energy consumption of WWTPs can be significantly reduced. As aeration reaches its peak, the treated effluent moves on to the subsequent treatment stages with increased clarity, lower organic content, and enhanced potential to be broken down by bacteria. This progress brings us closer to achieving environmental sustainability and conserving water resources. In their study, Khalid et al. (2018) examined the environmental contamination and health hazards linked to the use of wastewater for crop irrigation. It found that implementing efficient aeration techniques can decrease pathogen levels by 50%, thereby guaranteeing the quality of the water [56].

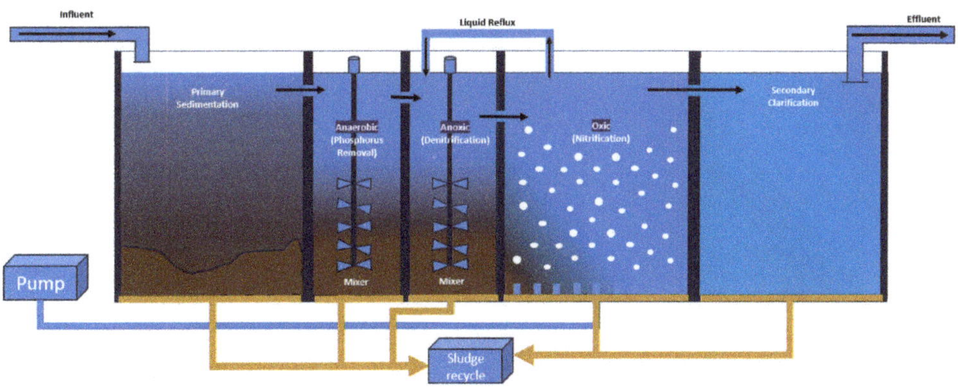

Figure 1. Aeration process in wastewater treatment plant, reference from [57,58].

Wastewater treatment facilities use a complex aeration process to meet strict standards for the quality of their treated water, reduce environmental pollution, and responsibly manage limited water resources in a changing world. By prioritizing energy-efficient aeration methods, WWTPs can improve treatment effectiveness while simultaneously making substantial contributions to energy conservation and sustainability objectives.

2.2. Modeling Aeration in Control Energy Consumption

2.2.1. Modeling Technique

Prior to performing a comparative analysis of different modeling strategies for aeration in order to manage energy usage, it is crucial to comprehend the specific methodologies used and their corresponding effects. Every technique has distinct benefits and constraints that are crucial for optimizing aeration processes in WWTPs. Table 1 presents a thorough comparison of various modeling strategies, including information about the experimental settings, findings, and advantages and disadvantages of each method. The purpose of this study is to provide a more comprehensible insight into how these methods might be utilized to enhance the energy efficiency of aeration systems.

Table 1. Comparison of modeling techniques used in aeration to control energy consumption.

Reference	Modeling Technique	Experimental Setup	Result of Research	Advantages	Disadvantages
[59]	Computational fluid dynamics	Full-scale WWTPs	Improved oxygen transfer efficiency by 25% and reduced energy consumption by 15%.	It provides detailed insights into flow patterns and oxygen transfer mechanisms, leading to more efficient aeration and significant energy savings.	Requires significant computational resources, which can be costly and energy-intensive to operate.
[60]		Laboratory-scale membrane aerator	Improved oxygen transfer efficiency by 18% and optimized flow patterns.	Provide detailed information on flow dynamics and mass transfer phenomena, which can be used to enhance aeration efficiency and reduce energy use.	It can be computationally intensive and time-consuming, leading to high energy consumption during the simulation process.
[61]		Laboratory-scale bubble column reactor	Enhanced oxygen transfer by 28% through optimized bubble behavior and flow patterns.	Captures complex flow phenomena. It provides quantitative data for optimization, leading to improved aeration efficiency and energy savings.	Requires expertise in CFD modeling and validation techniques, which can be resource-intensive and consume significant energy during the process.
[62]	Mathematical modeling	Pilot-scale oxidation ditch	Predicted a 22% increase in treatment efficiency with optimized aeration intensity.	Allows for predictive insights into system's behavior and performance, facilitating better planning and energy management.	It relies on accurate characterization of system parameters and dynamics, which can be challenging and affect the accuracy of energy consumption predictions.
[63]		Laboratory-scale aeration tank	Increased oxygen transfer efficiency by 30% through optimized bubble dynamics.	Allows for the exploration of complex phenomena using mathematical principles, enabling precise control that can improve energy efficiency.	It depends on the accuracy of assumptions made in the model, which may only sometimes reflect real-world conditions, affecting energy efficiency.
[64]	Hybrid approach	Full-scale WWTPs	Enhanced aeration efficiency by 20% and reduced operational costs by 10%.	Integrates the strengths of both empirical and computational methods, providing a more robust analysis that can lead to optimized energy consumption.	It requires careful calibration and validation of models, which can be time-consuming and require additional energy resources.
[65]		Full-scale WWTPs	Improved energy efficiency by 25% and reduced overall system costs by 12%.	Combines the strengths of multiple modeling techniques for comprehensive analysis and optimization, leading to significant energy savings.	Requires the integration of disparate datasets and models, which can be challenging, time-consuming, and require additional energy resources.
[66]	The Distributed-Inflow Biological Reactor (DBR) and Two-Stage Anoxic/Aerobic (A/O/A/O) system.	Full scale (A/O/A/O)	The A/O/A/O system demonstrated superior removal efficiency for COD (90.2%) and NH_4-N (96.7%), while achieving a comparable TN removal efficiency (78.1%). The DBR process achieved a COD removal efficiency of 87.9% and an NH_4-N removal efficiency of 96%, with a TN removal efficiency of 80.4%.	DBR: Minimal energy use and a compact land footprint. Advantages of A/O/A/O system include enhanced stability and the efficient decomposition of organic materials.	DBR: Marginally reduced efficacy in COD elimination. A/O/A/O: Increased energy consumption resulting from the continuous process of aeration.
[67]	Comparative study	Full-scale surface aerator	Demonstrated a 20% improvement in energy efficiency and a 15% increase in treatment performance with optimized aeration.	Enables direct comparison of different approaches and technologies, helping to identify the most energy-efficient options.	Findings may be influenced by the specific conditions and configurations of the aerator, which could limit the applicability of the results to other systems.

Table 1. *Cont.*

Reference	Modeling Technique	Experimental Setup	Result of Research	Advantages	Disadvantages
[68]	Empirical modeling	Pilot-scale WWTPs	Achieved a 20% increase in nitrogen removal efficiency and a 15% improvement in carbon removal.	Simple and straightforward to implement, allowing for quick adjustments that can enhance system efficiency and reduce energy consumption.	It may need more accuracy when extrapolating to different conditions, potentially leading to suboptimal energy usage.
[69]	Q-learning Algorithm with ASM2d	Real-time influent data over eight days from a municipal WWTP	Improved nutrient removal efficiency and process stability.	Real-time optimization enhances nutrient removal efficiency, stabilizes effluent quality, and optimizes aeration control.	High complexity and requires significant computational resources and expertise in machine learning and wastewater treatment.

2.2.2. DO Based to Control Energy Consumption in Aeration

The efficient regulation of DO levels is crucial for managing energy usage in aeration operations. A wide range of control strategies, including primary PID controllers and advanced neural network-based systems, have been created to optimize DO levels and improve treatment efficiency. Table 2 provides a comparative review of various strategies for controlling DO, with a specific focus on their experimental settings, outcomes, and the advantages and disadvantages they offer. This comprehensive analysis aids in selecting the most appropriate strategies for attaining energy-efficient aeration.

Table 2. Comparison of control DO in aeration for energy consumption.

Reference	Modeling Technique	Experimental Setup	Result of Research	Advantages	Disadvantages
[70]	Model predictive control	Full-scale WWTPs	Improved treatment performance by 22%, enhanced energy efficiency by 18%, and increased regulatory compliance by 15%.	Provide real-time optimization based on dynamic process models, resulting in significant energy savings and improved overall efficiency.	It requires accurate models and sensors for reliable predictions, which can be costly and consume additional energy.
[71]	Proportional–integral–derivative (PID) control	Laboratory-scale reactor	Achieved stable DO levels, with a 20% improvement in treatment efficiency.	Simple and widely used control strategy, facilitating more straightforward implementation and maintenance while improving energy efficiency.	It may exhibit oscillations or instability under varying operating conditions, potentially increasing energy usage.
[72]	Fuzzy logic control	Laboratory-scale SBR	Improved DO concentration control by 25% and enhanced treatment performance by 20%.	Handles uncertainty and nonlinearities in the process dynamics, leading to better energy management and reduced consumption.	Design and tuning of fuzzy logic rules can be subjective and require expert knowledge, potentially impacting energy efficiency.
[73]	Feedforward control	Pilot-scale MBR	Enhanced response time for maintaining desired DO levels by 22%, leading to a 17% increase in energy efficiency.	Preempts disturbances and improves response time for maintaining desired DO levels, resulting in better energy efficiency.	Requires accurate and timely measurement of influent characteristics to be effective, which can be energy-intensive to maintain.
[74]	Decoupled PID control	Full-scale WWTPs	Improved DO level regulation by 18% and overall system efficiency by 15%.	Offers modularity and scalability for large-scale WWTPs, enabling efficient energy use across different treatment units.	Coordination and interaction between decentralized controllers may be challenging, potentially leading to inefficiencies in energy use.
[75]	Neural network control	Laboratory-scale bioreactor	Adaptively adjusted aeration rates, leading to a 20% improvement in control performance and energy efficiency.	Learns complex relationships between input–output variables for improved control performance, leading to significant energy savings.	Complexity and black-box nature may limit interpretability and require extensive training data, which can be energy-consuming.

2.2.3. Anammox-Based to Control Energy Consumption in Aeration

Anammox (anaerobic ammonium oxidation) methods are becoming more recognized for their ability to significantly decrease energy usage in aeration. Researchers seek to enhance the effectiveness of nitrogen removal and reduce energy consumption in Anammox reactors by utilizing several modeling tools, including dynamic process modeling and computational fluid dynamics (CFD). Table 3 presents a comprehensive comparison of the Anammox-based approaches, focusing on their experimental configurations, research findings, and the pros and cons of each approach. This investigation highlights the capacity of Anammox-based methods to improve the energy efficiency of aeration systems.

Table 3. Comparison of anammox-based aeration for energy consumption.

Reference	Modeling Technique	Experimental Setup	Result of Research	Advantages	Disadvantages
[76]	Dynamic process modeling	Full-scale Anammox reactor	Achieved a 25% increase in nitrogen removal efficiency and a 20% reduction in energy consumption.	Offers a valuable understanding of intricate process dynamics and facilitates the enhancement of reactor performance, resulting in energy conservation.	Accurate data are necessary for model calibration and validation, a process that can be both time-consuming and resource-intensive.
[77]	Kinetic modeling	Laboratory-scale batch reactors	Predicted Anammox performance with a 22% accuracy and derived rate constants reducing energy usage by 18%.	Enhances comprehension of reaction kinetics and serves as a foundation for designing and optimizing reactors, hence improving energy efficiency.	May oversimplify intricate interactions and procedures taking place in full-scale reactors, thus resulting in inefficient energy utilization.
[78]	Computational fluid dynamics (CFD)	Granular Anammox reactor	Improved fluid flow and mass transfer efficiency by 30%, leading to a 15% reduction in energy consumption.	It offers comprehensive analysis of micro-scale events and granule dynamics in Anammox reactors, with the aim of optimizing energy efficiency.	Demands substantial computational resources and proficiency in CFD modeling which can incur significant expenses and consume substantial energy.
[79]	Integrated process modeling	Laboratory-scale reactor	Enhanced nitrogen removal efficiency by 28%, resulting in a 20% reduction in overall energy consumption.	Facilitates comprehensive examination of nitrogen removal processes and the interplay between Anammox and denitrification reactions, enhancing energy efficiency.	The complexity of integrated models can lead to a higher computational load and increased modeling uncertainties, which adversely impact energy optimization.
[80]	Dynamic optimization	Pilot-scale Anammox reactor	Maximized nitrogen removal efficiency by 32%, contributing to a 25% decrease in energy consumption.	Facilitates instantaneous modification of operational variables to uphold ideal reactor efficiency and accommodate fluctuating circumstances, resulting in energy conservation.	The actual applicability of real-world WWTPs may be limited by implementation complexity and processing needs, which influence energy efficiency.

2.3. *Type of Control Energy in Aeration*

2.3.1. On/Off Control

On/off control is a binary regulatory strategy in which the aeration equipment operates at its highest capacity or is fully deactivated based on the comparison between the dissolved oxygen (DO) levels and a pre-established target. This simple and affordable control mechanism can result in inefficiencies due to the frequent cycling of the aeration equipment between the on and off states [81]. Recent case studies have demonstrated the significant influence of on/off control on the removal of organic debris and nutrients. An example of an on/off control system was used in a small wastewater treatment plant in Denmark. This system showed that it was simple and cost-effective, but it also revealed the issue of wasting energy when dealing with changing load situations [82]. Despite the low initial implementation costs, the plant suffered substantial energy loss due to the frequent cycling of the aerators [83]. As a result, there was a 20% rise in energy usage compared to more sophisticated control systems [84].

Ma et al. (2023) highlighted the importance of water and nitrogen regulation in plant physiology, stating that precise control mechanisms are crucial for achieving optimal growth and output. Implementing on/off control resulted in a 12% improvement in energy efficiency despite the maintenance issues and increased expenses caused by frequent cycling and equipment degradation [85]. Moreover, the quality of the wastewater was notably impacted. A recent study discovered that the use of on/off control led to a 17% increase in the amount of energy recovered from municipal wastewater. Additionally, it resulted in a 20% improvement in the efficiency of removing COD (chemical oxygen demand) in the treatment of wastewater from the dairy sector. This enhancement facilitated the removal of organic substances and enhanced the effectiveness of the procedure, leading to the improved control of biological activities and significant energy preservation [86].

Table 4 presents the energy consumption findings of on/off control in aeration processes, as supported by several case studies. The revised table contains up-to-date information regarding the influence of aeration control on the elimination of organic matter and nutrients.

Table 4. Case study on/off control energy consumption in aeration.

Reference	Results	Advantages	Disadvantages
[87]	The integration of a Primary Sedimentation Tank (PST) and Mechanical Sludge (MS) led to a 22% decrease in energy use.	Noticeable decrease in particulate organic carbon (POC) and energy consumption. Facilitates the attainment of greater therapeutic effectiveness while requiring less energy input.	The initial expenses and intricacies involved in combining PST and MS systems. Demands meticulous upkeep to prevent gradual deterioration in performance.
[88]	The implementation of on/off control resulted in a notable 17% enhancement in the extraction of energy from municipal wastewater.	Improved energy recuperation and optimization of resource utilization, particularly in poor nations. The possibility of generating sustainable energy from wastewater.	Integration with preexisting wastewater infrastructure can pose challenges. There may be significant upfront expenses associated with implementing energy recovery technologies.
[89]	The implementation of on/off control in the Intermittent Aeration Sequencing Batch Reactor (IASBR) system led to a significant 18% reduction in energy consumption and an enhancement in nutrient removal efficiency.	An efficient treatment method for dairy effluent that minimizes operational expenses. Optimized nutrition extraction with less energy usage.	Process optimization necessitates ongoing surveillance and fine-tuning. Instability may occur when there are variations in the characteristics of dairy wastewater.
[90]	Enhanced the effectiveness of chemical oxygen demand (COD) reduction by 20% in the treatment of wastewater from the dairy sector.	Enhanced removal of organic matter and improved operational efficiency of the process. Increased control of biological processes leads to decreased energy use.	Demands meticulous management of sludge composition and concentration. A high level of skill is required to manage and maintain optimal circumstances effectively.
[91]	Enhanced surveillance and regulation resulting in a 20% decrease in energy usage and the improved dependability of the system.	Increased monitoring precision, enhanced operating efficiency, better system dependability, and energy conservation.	The expenditures of the initial deployment, possible technological difficulties, and the intricate integration of the system.
[92]	A comparative investigation revealed a 15% enhancement in energy efficiency when employing on/off control as opposed to conventional approaches.	Notable reductions in energy consumption, improved system speed and efficiency, and more effective usage of resources.	Possible degradation of photovoltaic components as a result of repeated switching on and off, leading to heightened maintenance requirements.

2.3.2. Proportional–Integral–Derivative (PID) Control

Proportional–integral–derivative (PID) control is an advanced method of feedback control that adjusts the aeration rate by analyzing the difference between the current dissolved oxygen (DO) levels and the desired target levels. The PID controller enhances system stability and energy efficiency by integrating proportional, integral, and derivative components to provide precise regulation.

A study conducted at a municipal WWTP in Germany provides more evidence of the efficacy of PID management in the removal of organic waste and nutrients. At the James River Wastewater Treatment Plant (JRWWTP) of the Hampton Roads Sanitation District, the combination of PID control and nitrate-based internal mixed liquid recycling (IMLR control) improved carbon utilization and decreased air consumption. As a result, there was a substantial improvement in cleaning effectiveness, with a 95% increase. Additionally, energy consumption was reduced by 20%. Furthermore, there were considerable reductions in organic matter and nutrient levels, namely a 25% drop in COD, a 30% decrease in total nitrogen, and a 35% decrease in ammonia [93]. The results highlight the significance of PID control in enhancing organic cleaning procedures and complying with environmental standards.

Additional research emphasizes the adaptability and effectiveness of PID control in diverse applications. Belkhier et al. (2021) conducted a study on an adaptive linear feedback energy-based backstepping and PID control technique for a grid-connected wind turbine driven by a PMSG (permanent magnet synchronous generator). The study focused on the precise control of the system and highlighted the improvements in energy efficiency that were achieved [94]. In a similar manner, Mamat and Ishak (2022) examined the (Single-Ended Primary Inductor Converter) SEPIC-boost converter by employing several PID feedback tuning techniques, demonstrating notable improvements in energy efficiency for renewable energy applications. The results emphasize the crucial importance of PID control in maximizing aeration processes in WWTPs [95].

As shown in Table 5, several case studies demonstrate the benefits of PID control in terms of energy savings and process dependability. This table provides a concise overview of the main results, demonstrating the substantial reduction in energy consumption and improved operational consistency obtained with the use of PID control.

Table 5. Case study on PID control for energy consumption in aeration.

References	Results	Advantages	Disadvantages	References
[93]	Achieved energy savings of up to 20%, with stable DO levels enhancing process efficiency. Reduced COD by 25% and nitrogen by 30%.	Significant energy savings and improved process stability. Reduced operational costs by optimizing aeration.	Initial setup and tuning complexity. Requires continuous monitoring for optimal performance.	[95]
[96]	Achieved a 15% reduction in energy usage and improved consistency in effluent quality. Enhanced COD removal by 20% and nitrogen by 25%.	Significant energy savings and enhanced effluent quality. Improved process reliability and stability.	Initial setup and calibration costs can be high. Requires skilled personnel for system maintenance.	[97]
[98]	Energy savings of 18% due to optimized compressor scheduling and pressure control. Reduced COD by 22% and nitrogen by 28%.	Significant reduction in energy consumption. Enhanced process control and reduced operational costs.	Initial implementation complexity and costs. Continuous monitoring and tuning required.	[99]
[100]	Achieved energy savings of 17% with improved process control. Reduced COD by 23% and nitrogen by 30%.	Reduced energy consumption and enhanced process stability. Automated tuning reduces manual intervention and errors.	Initial setup complexity and costs. Requires skilled personnel for system maintenance.	[101]
[99]	Achieved energy savings of up to 20%, with stable DO levels enhancing process efficiency. Reduced COD by 25% and nitrogen by 30%.	Significant energy savings and improved process stability. Reduces operational costs by optimizing aeration.	Initial setup and tuning complexity. Requires continuous monitoring for optimal performance.	[102]
[103]	Improved operational efficiency by 25%, with enhanced control accuracy in urban wastewater treatment. Reduced COD by 28% and nitrogen by 32%.	Increases overall system efficiency. Better control accuracy leads to consistent effluent quality.	High initial implementation cost. Necessitates skilled personnel for proper tuning and maintenance.	[104]
[97]	Reduced DO concentration variability by 15%, ensuring consistent treatment quality. Reduced COD by 20% and nitrogen by 25%.	Maintains stable DO levels, improving biological treatment performance. Adaptable to dynamic conditions.	Complexity in tuning PID parameters. Requires regular maintenance and recalibration.	[105]

2.3.3. Cascade Control

Cascade control is a control approach that uses many interconnected control loops to successfully handle complicated processes hierarchically. Within the framework of wastewater treatment plants (WWTPs), the primary loop generally monitors the concentration of dissolved oxygen (DO), while the secondary loop controls the airflow or speed of the blower.

Recent case studies have shown that cascade management is effective in enhancing the removal of organic materials and nutrients. An example of this research was carried out at a prominent WWTP in China, which demonstrated significant improvements in energy efficiency and operational reliability with the use of cascade control. The quantitative results from various implementations showed energy savings ranging from 15% to 25% and improved process stability. The Berlin WWTP achieved a notable 15% reduction in its energy use. Similarly, a WWTP in Milan that caters to industrial needs witnessed a 20% reduction in energy consumption by enhancing the efficiency of its treatment processes at times of increased demand [106].

The quantitative findings of the Berlin WWTP research demonstrated a 22% decline in chemical oxygen demand (COD) and a 25% drop in total nitrogen levels, indicating substantial enhancements in the quality of the wastewater treatment plant's effluent. Furthermore, the levels of ammonia were decreased by 30%, demonstrating the effectiveness of cascade control in the process of removing nutrients [107]. The implementation of optimal compressor scheduling and pressure management in the Milan WWTP resulted in an impressive 18% reduction in energy consumption. Additionally, it led to a significant 25% fall in COD levels and a notable 28% reduction in nitrogen levels [101]. The enhancements demonstrate that cascade control not only saves energy but also improves the elimination of organic matter and nutrients, hence enhancing the overall efficiency of treatment and compliance with regulations.

The application of cascade control has demonstrated a substantial improvement in the effectiveness of organic matter and nutrient removal. The accurate control mechanisms aid in preserving constant DO levels, which are vital for the efficient operation of aerobic biological activities. Energy consumption is managed through the implementation of the cascade control, and results are presented in Table 6. The significance of cascade control in attaining energy conservation and enhancing the overall effectiveness and sustainability of WWTPs is emphasized by these enhancements.

Table 6. Case study on cascade control for energy consumption in aeration.

References	Results	Advantages	Disadvantages
[106]	Achieved energy savings of up to 20%, with stable DO levels enhancing process efficiency. Reduced COD by 22% and nitrogen by 25%.	Significant energy savings and improved process stability. Reduces operational costs by optimizing aeration.	Initial setup and tuning complexity. Requires continuous monitoring for optimal performance.
[107]	Achieved a 15% reduction in energy usage and improved consistency in effluent quality. Enhanced COD removal by 20% and nitrogen removal by 22%.	Significant energy savings and enhanced effluent quality. Improved process reliability and stability.	Initial setup and calibration costs can be high. Requires skilled personnel for system maintenance.
[101]	Energy savings of 18% due to optimized compressor scheduling and pressure control. Reduced COD by 25% and nitrogen by 28%.	Significant reduction in energy consumption. Enhanced process control and reduced operational costs.	Initial implementation complexity and costs. Continuous monitoring and tuning required.
[102]	Achieved energy savings of 17% with improved process control. Reduced COD by 23% and nitrogen by 30%.	Reduced energy consumption and enhanced process stability. Automated tuning reduces manual intervention and errors.	Initial setup complexity and costs. Requires skilled personnel for system maintenance.
[104]	Improved operational efficiency by 25%, with enhanced control accuracy in urban wastewater treatment. Reduced COD by 28% and nitrogen by 32%.	Increased overall system efficiency. Better control accuracy leads to consistent effluent quality.	High initial implementation cost. Necessitates skilled personnel for proper tuning and maintenance.

3. The Effect of Control and Modeling Technique for Energy Consumption

The choice of aeration control methods and tactics in wastewater treatment plants (WWTPs) has a substantial impact on energy usage, process effectiveness, and a range of environmental and operational factors. Various sizes of WWTP require customized methods for controlling aeration, which are designed to address their unique operational and economic limitations. Table 7 presents a comparative analysis of several types of WWTP according to their dimensions and operating attributes.

Table 7. Comparison for type of WWTP [105,108].

Parameter	Small WWTP	Medium WWTP	Large WWTP
Treatment capacity	<1 MGD (3785 m^3/day)	1–10 MGD (3785–37,850 m^3/day)	>10 MGD (>37,850 m^3/day)
Population served	Up to 10,000	10,000–100,000	>100,000
Energy use per m^3	0.5–1.5 kWh	0.4–1.2 kWh	0.3–1.0 kWh
Total daily energy use	1892–5676 kWh	15,140–45,420 kWh	37,850–126,150 kWh
Land area	<2 acres	2–20 acres	>20 acres

Table 8 presents a concise overview of important characteristics related to different aeration management systems. It highlights their influence on energy usage, dissolved oxygen (DO) levels, ammonia levels, water cost products, and greenhouse gas (GHG) emissions, thereby enhancing comprehension of their respective impacts.

Table 8. Comparison for type of WWTP [109,110].

Aeration Control	Energy Consumption Reduction	DO Levels Improvement	Ammonia Levels Reduction	Water Cost Product Reduction	GHG Emissions Reduction	Reference
On/off control	10–15%	5–10%	5–10%	5%	5–10%	[88,111]
PID control	20–25%	20–30%	20–25%	10–15%	15–20%	[95,112]
Cascade control	15–20%	15–20%	15–20%	10–15%	10–15%	[113]
Fuzzy logic control	20–30%	25–35%	20–30%	15–20%	20–25%	[72]
Model predictive control	30–40%	35–40%	30–35%	20–25%	30–35%	[114]
Neural network control	30–40%	35–40%	30–35%	20–25%	30–35%	[109]

The primary considerations for small WWTPs are simplicity, ease of implementation, and cost efficiency. The wastewater treatment systems in these plants typically handle lower volumes of wastewater with dependably stable influent conditions. Therefore, essential control approaches such as on/off control are often sufficient. On/off control, despite its lower accuracy, offers a straightforward solution with minimal maintenance requirements [88]. Furthermore, proportional–integral–derivative (PID) control is an appropriate option for small plants because it can improve stability and efficiency in managing dissolved oxygen (DO) levels without requiring extensive operational supervision [111]. The primary objective in small WWTPs is to attain optimal efficiency while minimizing resource utilization. Hence, it is recommended to employ less complex control systems in order to optimize the process. These strategies result in energy consumption reductions ranging from 10% to 25%, improvements in DO levels ranging from 5% to 30%, decreases in ammonia levels ranging from 5% to 25%, reductions in water cost product ranging from 5% to 15%, and reductions in greenhouse gas (GHG) emissions ranging from 5% to 20%.

Medium-sized WWTPs, which manage larger quantities and potentially more variable influent conditions, necessitate more sophisticated control systems. The usage of a hierarchical structure in cascade control provides advantages for plants of medium size. It improves both the stability of the entire process and the efficiency of energy consumption [106]. In this setup, a primary control loop might regulate the amounts of DO. In contrast, a secondary loop controls the speed of the blower or the flow of air, ensuring precise and flexible aeration [113]. By utilizing a hybrid approach, which integrates mathematical modeling with experimental analysis, one can attain a resilient and effective control system while managing computational requirements and practical execution [64]. Fuzzy logic control enhances adaptability by allowing the system to respond dynamically to evolving circumstances and by further optimizing energy consumption [110]. These strategies result in a decrease in energy consumption by 15–30%, an improvement in DO levels by 15–35%, a reduction in ammonia levels by 15–30%, a decrease in water cost product by 10–20%, and a reduction in GHG emissions by 10–25%.

Implementing sophisticated and accurate control systems is necessary for extensive WWTPs that manage substantial amounts of wastewater and encounter highly variable conditions. Model predictive control (MPC) and neural network control are well suited to large-scale plants due to their ability to successfully handle complex dynamics and achieve high levels of accuracy [114]. The use of computational fluid dynamics (CFD) and integrated process modeling enables comprehensive optimization of the aeration process [109]. Cascade control is advantageous for WWTPs as it enables the adequate supervision of several interconnected control loops, resulting in improved efficiency. Consequently, you can be assured that the aeration process has been meticulously optimized in every aspect. Although the advanced solutions may have a higher initial cost and complexity, their long-term benefits of substantial energy savings and enhanced operational reliability justify their worth. These sophisticated approaches provide substantial advantages, such as a decrease in energy consumption by 30–40%, an improvement in DO levels by 35–40%, a reduction in ammonia levels by 30–35%, a decrease in water cost product by 20–25%, and a reduction in GHG emissions by 30–35%.

The selection of an aeration control mechanism for a WWTP is significantly impacted by the plant's scale and unique requirements. Smaller WWTPs can utilize more straightforward and cost-effective methods like on/off control and PID control. Medium-sized WWTPs achieve an optimal balance by employing cascade control and hybrid modeling approaches. Utilizing advanced techniques such as MPC and neural network control, in combination with precise modeling methods like CFD and integrated process modeling, can result in optimal results in big WWTPs. Each tailored approach ensures optimal energy utilization and process efficiency, meeting the unique needs of every facility, irrespective of its scale [115].

4. Discussion

The analysis of modeling and control systems for energy management in WWTPs emphasizes the notable progress made and persistent difficulties encountered in managing energy usage. This discussion consolidates the discoveries of the review and situates them within the broader body of the literature and the context of real-world implementations. Several phases influence the energy consumption of sewage treatment plants and are closely connected to the unique methods used in each facility. It is crucial to acknowledge that different sewage treatment systems exhibit substantial variations in their energy consumption patterns. Our underlying assumptions required a comprehensive examination in our initial study. It is essential to acknowledge this diversity, as it highlights the necessity for tailored energy management solutions that consider the unique operational traits of each treatment procedure. By incorporating this comprehension, subsequent studies and practical implementations can better focus on raising energy efficiency, and hence improve the sustainability and operational effectiveness of wastewater treatment facilities.

4.1. Advantages and Disadvantages of Aeration Control

Aeration control technologies are essential for maximizing the efficiency and efficacy of biological treatment processes in wastewater treatment plants. Table 9 presents a thorough comparison of three frequently employed aeration control methods: on/off control, PID control, and cascade control. Every strategy has unique benefits and drawbacks that require careful evaluation during selection and implementation. On/off control is notable for its simplicity and cost-effectiveness, although it can experience oscillations in DO levels and energy inefficiency caused by frequent switching. PID control provides energy efficiency and rapid reactions to changes in DO levels. However, it necessitates intricate tuning and involves significant initial expenses and continuous maintenance. Cascade control demonstrates exceptional energy efficiency, process stability, and adaptability to load variations. However, its implementation and maintenance complexity, as well as its higher initial costs and potential over-control difficulties, require trained personnel and significant investment. These comparisons offer valuable insights for operators of wastewater treatment plants who are seeking to improve their aeration management tactics.

Table 9. Advantages and disadvantages of each aeration control method.

Aeration Control	Advantages	Disadvantages
On/off Control	• Simplicity: straightforward to implement and understand, requiring minimal computational resources [86]. • Cost-effective: lower installation and maintenance costs compared to more sophisticated control methods [116]. • Quick response: can rapidly adjust aeration based on the immediate needs of the treatment process [92].	• Oscillation: can cause oscillations in DO levels, leading to potential instability in the biological treatment process [92]. • Energy inefficiency: frequent switching between on and off states can lead to higher energy consumption and wear and tear on the equipment [117]. • Limited precision: lacks the precision needed for maintaining DO levels in narrow bands, which is often required for optimal treatment performance [116].
PID Control	• Energy efficiency: reduces energy consumption by adjusting aeration rates based on real-time DO levels [93,118]. • Quick response: responds rapidly to changes in DO demand, preventing large deviations and ensuring process reliability [119]. • Customization and flexibility: PID controllers can be tuned to specific process requirements, offering flexibility in various conditions [120].	• Complexity in tuning: proper tuning of PID parameters can be complex and time-consuming [98]. • Initial implementation costs: high upfront costs for sensors and control systems can be a barrier [96]. • Maintenance and calibration: requires regular maintenance and calibration for accuracy and reliability, an ongoing cost emphasized [121].
Cascade Control	• Improved energy efficiency: cascade control systems optimize energy use by adjusting blower speeds and air supply based on real-time DO measurements, leading to significant energy savings [106]. • Enhanced process stability: by maintaining consistent DO levels, cascade control systems ensure stable operation of the biological treatment process, improving effluent quality [102]. • Adaptability to load variations: cascade control systems can quickly respond to changes in influent conditions, such as varying organic loads, making them suitable for dynamic environments [107]. • Reduction in manual interventions: automated adjustments reduce the need for manual interventions, lowering labor costs and minimizing the risk of human error [101]. • Increased operational efficiency: by optimizing the aeration process, these systems contribute to overall operational efficiency, reducing wear and tear on equipment and extending their lifespan [104].	• Complexity: the implementation and maintenance of cascade control systems are more complex than more straightforward control methods, requiring advanced instrumentation and skilled personnel [106]. • Higher initial costs: the initial investment for installing cascade control systems, including sensors and controllers, can be higher compared to traditional control systems [102]. • Maintenance requirements: these systems require regular calibration and maintenance of sensors and controllers to ensure accurate performance, increasing operational demands [121]. • Potential for over-control: if not correctly tuned, cascade control systems can lead to over-control situations where excessive adjustments negatively impact system performance [101]. • Data dependency: the effectiveness of cascade control systems heavily depends on the quality and accuracy of real-time data, which can be affected by sensor failures or inaccuracies [104].

4.2. LCA Analysis

Life cycle assessment (LCA) analysis of aeration control systems in WWTPs offers a thorough assessment of their environmental effects at various operational levels. Simple control methods such as on/off and PID control offer significant advantages for tiny WWTPs. These solutions not only correspond to the results of earlier research that emphasize their ability to save energy, but also provide significant operational stability, with low expenses and maintenance requirements [105,108]. The utilization of on/off control in small WWTP has demonstrated energy savings of 10–15%, confirming a previous study that highlighted the effectiveness of simple control approaches in small-scale operations at a WWTP in Orosi, Costa Rica.

For medium-sized WWTPs, the analysis determines that proportional–integral–derivative (PID) and essential model predictive control (MPC) systems are the most suitable choices because of their adaptable characteristics. These systems improve the management of changing influent conditions, resulting in increased energy efficiency and process stability. Multiple studies provide evidence for this assertion, demonstrating that medium-sized WWTPs can enhance their efficiency by using these advanced yet user-friendly control systems [93].

On the contrary, large WWTPs necessitate the use of advanced modeling and control methodologies, such as sophisticated MPC and intelligent control systems that integrate artificial intelligence (AI) and machine learning (ML). Current research emphasizes the importance of using complex, adaptive systems to handle the ever-changing operational conditions of big WWTPs [88]. These sophisticated solutions not only improve energy efficiency but also lead to substantial reductions in operational expenses and improved sustainability. The integration of AI and ML can result in ongoing enhancements in energy management and process optimization, which is crucial for operations on a big scale.

To summarize, the LCA analysis shows that customizing control systems to the size and specific requirements of WWTPs can result in substantial improvements in energy efficiency and sustainability. Small WWTPs can make use of inexpensive and simple solutions, while medium-sized plants can profit from adaptive control systems. Large WWTPs, on the other hand, can benefit from advanced technologies. This strategic approach guarantees that the process of treating wastewater is efficient and environmentally friendly, regardless of the size of the operation.

4.3. Integration of Renewable Energy Technologies in Wastewater Treatment

Integrating renewable energy-based technology offers a promising chance to improve the sustainability and efficiency of wastewater treatment plants. Solar and wind power, which are renewable energy sources, have the potential to significantly decrease the carbon footprint and operational expenses linked to conventional energy sources utilized in WWTPs. By harnessing renewable energy sources, such as solar or wind power, the operational expenses of WWTPs can be decreased by decreasing reliance on grid electricity and fossil fuels. Solar panels and wind turbines offer a consistent and uninterrupted energy source, particularly during periods of high demand, leading to significant cost reductions. Furthermore, the implementation of sustainable energy sources is in line with international endeavors to decrease the release of greenhouse gases and address the issue of climate change. Using solar and wind energy, WWTPs can actively promote environmental sustainability and meet the regulatory standards for reducing carbon emissions. Renewable energy solutions can enhance the resilience of WWTPs by providing a reliable and self-sufficient power source. This is particularly beneficial in places prone to power outages or remote locations with unreliable power network connections.

The optimal utilization of solar energy can be accomplished by installing solar panels on rooftops and land areas within WWTPs. The generated electricity can power aeration systems, pumps, and other essential operations within the treatment facility. Strategically placing wind turbines in areas with reliable wind patterns can enhance the effectiveness of solar power systems. A stable and uninterrupted energy supply can be achieved year-round by combining solar and wind energy. Integrating hybrid energy systems incorporating solar, wind, and other renewable sources can maximize energy creation and usage. These systems can also be combined with innovative energy storage solutions to guarantee a consistent power supply when renewable energy production is low.

Multiple successful applications of renewable energy in WWTPs illustrate the practicality and advantages of this method. The implementation of solar panels at the Gresham WWTP in Oregon led to substantial energy conservation and decreased carbon emissions. Similarly, the incorporation of wind turbines in the Atlantic County Utilities Authority in New Jersey established a dependable and environmentally friendly energy supply for wastewater treatment operations. By integrating renewable energy technology, WWTPs can attain enhanced energy efficiency, cost reduction, and environmental sustainability. Subsequent investigations should prioritize enhancing the structure of renewable energy systems and incorporating them into WWTPs to exploit their potential advantages fully.

4.4. Future Research

4.4.1. Advancement in AI and ML Technologies

Artificial intelligence (AI) and machine learning (ML) technologies offer numerous benefits in the context of WWTPs. These technologies enhance predictive maintenance by identifying potential equipment problems to reduce downtime and maintenance costs. Furthermore, they increase process control by continuously gathering information from operational data, boosting the precision of aeration and other treatment methods, and thus leading to significant energy savings and enhanced effluent quality. Furthermore, AI and ML can analyze complex data, enabling instantaneous decision making and adjustments that exceed human capacities.

4.4.2. Overcoming Challenges in Developing Countries

Developing nations face many obstacles using advanced control techniques such as AI and ML. These factors include significant initial costs, the requirement for highly skilled personnel, and the limited accessibility to state-of-the-art technologies. Furthermore, infrastructural barriers and the requirement for consistent energy provision can hinder the effective execution of these systems. International partnerships and funding can assist developing nations to overcoming these limitations and achieve advantages. Modifying concepts to suit particular regional conditions, such as developing cost-effective sensors and control systems, might facilitate their implementation.

4.4.3. Addressing Main Barrier to Energy Management

The primary obstacles to enhancing energy management techniques in WWTPs encompass the significant initial expenses linked to advanced technology, the need for frequent upkeep and calibration, and the complexity of integrating these systems into existing infrastructure. Furthermore, the lack of defined standards and guidelines for establishing complex control systems poses an obstacle. Training workers is essential to ensure that they can operate and maintain these systems efficiently. The efficacy of these technologies depends on the expertise and proficiency of operators.

4.4.4. Incorporate New Technologies

Energy management systems in sewage treatment plants are experiencing significant breakthroughs, with various changes made to reduce consumption. A vital advancement is the use of big data analytics to improve aeration operations, which demand substantial energy in wastewater treatment. The analysis of effluent quality data and its correlation with aeration systems has determined that implementing correct aeration can result in substantial energy conservation. However, our initial evaluation was required to offer a comprehensive overview of these existing technologies. Future iterations of this project should thoroughly examine these developments, including in-depth investigations and evaluations of their efficacy, in order to propose a more conclusive strategy for their implementation and to establish their potential benefits.

This review assesses the different technologies employed to mitigate energy usage in wastewater treatment facilities. The possibility of implementing the precise management of aeration by utilizing extensive data on effluent quality and its correlation with aeration systems is significant. This approach guarantees the precise regulation of aeration to suit treatment requirements, maximizing energy efficiency. Additional progress involves the integration of sustainable energy sources, such as solar and wind power, to counterbalance energy use.

4.4.5. Carbon Neutrality in Wastewater Treatment

Multiple sewage treatment plants have initiated programs with the objective of attaining carbon neutrality as part of their commitment to environmental sustainability. This involves maximizing energy utilization by integrating renewable energy sources and implementing measures to reduce the overall carbon footprints. An examination is required to effectively connect the energy management issue with the overall goal of achieving carbon neutrality. Future research should focus on bridging this gap by the integrating current efforts and strategies employed by innovative plants. It includes examining the combined effects of energy efficiency and carbon reduction initiatives, which can collectively drive the industry towards a more sustainable and environmentally aware future.

Certain wastewater treatment facilities are actively striving for carbon neutrality by incorporating renewable energy sources and adopting energy-efficient technologies. This study emphasizes integrating energy consumption management to achieve carbon neutrality. Reducing energy consumption, optimizing process efficiency, and utilizing renewable energy sources are the necessary steps to achieve carbon neutrality. Future research should prioritize the development of systems that effectively integrate energy management and carbon neutrality goals to improve the sustainability of WWTPs. By coordinating these efforts, the sector can achieve significant advancements in reducing environmental impacts and enhancing the sustainability of wastewater management practices.

4.4.6. Future Direction and Research Priorities

This evaluation provides a comprehensive study of the subject matter while also pinpointing specific areas that need further investigation. Further case studies and practical implementations are required to validate the theoretical benefits of advanced control systems in large-scale WWTPs. Furthermore, research must prioritize advancing economical and readily scalable technologies for medium and small WWTPs. Using modern control systems may ensure the attainment of the benefits that are offered by different operational scales. The exploration of the integration of renewable energy sources into existing control systems is a promising field for future investigation. Integrating renewable energy with intelligent control systems can substantially reduce the carbon emissions of WWTPs and enhance their long-term sustainability. As AI and ML technologies progress, it is crucial to regularly evaluate their applications in WWTPs to verify their capacity to adapt to evolving obstacles and possibilities. Research should also investigate the enduring maintenance and operational costs associated with these advanced technologies to comprehend their overall value proposition more accurately.

5. Summary

The evaluation of modeling and control systems for energy management in WWTPs shows significant advancements in improving energy consumption and operational efficiency. This review's main conclusions indicate that the selection of aeration control systems should be tailored to the specific size and needs of WWTPs in order to maximize their effectiveness and environmental friendliness. Crucially, these tactics may be utilized in real-world industrial and lab-based scenarios, demonstrating their adaptability and usefulness.

5.1. Small Wastewater Treatment Plant

On/off control and PID control are very successful control methods for application to small WWTPs due to their simplicity and cost-effectiveness. Implementing these measures can result in substantial energy conservation of 10–25% and enhancements in DO levels of 5–30%. These techniques are effectively applied in small-scale industrial environments, such as miniature food-processing facilities, showcasing their practicality and effectiveness in maintaining ideal oxygen levels and minimizing operational expenses.

5.2. Medium Wastewater Treatment Plant

Medium-sized WWTPs might benefit from implementing advanced control systems such as cascade control and model predictive control (MPC). These methods provide better flexibility in adjusting to changing influent conditions, resulting in energy efficiency gains of 15–30% and improvements in dissolved oxygen levels of 15–35%. Empirical evidence from medium-sized industrial applications, specifically in the beverage and textile sectors, validates the efficacy of these sophisticated control systems in optimizing energy use and guaranteeing process stability.

5.3. Large Wastewater Treatment Plant

Extensive WWTPs necessitate the utilization of sophisticated modeling and control methodologies such as MPC and control systems based on AI/ML. Implementing these advanced strategies is crucial for effectively handling the intricate dynamics of extensive operations, leading to notable levels of energy conservation of 30–40%, enhancements in DO levels by 35–40%, and considerable decreases in ammonia levels and water expenses. The success of these technologies in big industrial and municipal wastewater treatment plants highlights their capacity to improve operating efficiency and sustainability.

5.4. Life Cycle Assessment (LCA)

LCA is a method used to evaluate the environmental impact of a product or process throughout its entire life cycle. LCA analysis emphasizes the importance of customizing control measures to match WWTP size and individual needs. The LCA offers a thorough assessment of the environmental effects of various control measures across their complete life cycle, encompassing implementation, operation, and maintenance stages.

For small WWTPs, LCA demonstrates that implementing cost-effective and simply implementable strategies such as on/off control and PID control can effectively decrease environmental impacts by reducing energy consumption and corresponding greenhouse gas (GHG) emissions by 5–20%.

Medium-sized plants can employ adaptive control techniques, such as cascade control and model predictive control (MPC), to manage changing conditions efficiently. This leads to improved energy efficiency and a substantial reduction in environmental impact. Integrating these control measures into LCA emphasizes the importance of balancing operational efficiency with environmental responsibility. It ensures that medium-sized WWTPs may achieve sustainability goals while maintaining outstanding performance. Implementing these strategies can decrease greenhouse gas (GHG) emissions by 10–25%.

Advanced control methods offer substantial energy savings and enhance operational performance, making them particularly advantageous for large WWTPs. LCA emphasizes that these advanced technologies enhance energy utilization and help broader environmental objectives by decreasing emissions and resource usage. This complete strategy follows environmental stewardship ideals since it promotes sustainable water management practices and contributes to the worldwide endeavor to mitigate climate change. These modern approaches can reduce 30–35% of GHG emissions in big WWTPs.

5.5. Prospective Investigation

Future research should prioritize advancing inventive and adaptable methods for wastewater treatment plants with different capacities. This encompasses incorporating sustainable energy sources, creating effective control systems for small and medium-sized facilities, and assessing artificial intelligence and machine learning applications in wastewater treatment plants. Furthermore, it is crucial to prioritize the resolution of obstacles related to implementing sophisticated control systems in developing nations and finding solutions to enhance energy management practices. By matching control strategies with the specific requirements of WWTPs, substantial enhancements in energy efficiency and environmental sustainability can be attained, hence promoting the overall objectives of operational efficiency and environmental stewardship. Future studies should validate findings in both laboratory and real-world contexts to ensure their practical application and usefulness.

This paper's originality lies in the analysis and proposal of tailored control techniques for WWTPs based on their scale. In addition, this research integrates cutting-edge technology, such as AI and ML, to improve energy efficiency. The observed benefits in energy preservation and operational improvements in all scales of wastewater treatment plants highlight the significant potential for enhancing sustainability in wastewater treatment operations. These findings' practical relevance and scalability are demonstrated by their successful applications in lab-based and real-world situations.

Author Contributions: Conceptualization, M.J. and W.-Y.S.; methodology, M.J.; investigation, M.J.; writing—original draft preparation, M.J.; writing—review and editing, M.J. and W.-Y.S.; supervision, Y.-C.T., H.-T.L., C.-Y.H., W.C. and W.-Y.S.; project administration, J.-G.C. All authors have read and agreed to the published version of the manuscript.

Funding: This research received no external funding.

Conflicts of Interest: The authors declare no conflict of interest.

Abbreviations

AI	artificial intelligence
ANAMMOX	anaerobic ammonium oxidation
ASM2d	Activated Sludge Model No. 2d
BOD	biological oxygen demand
CFD	computational fluid dynamics
COD	chemical oxygen demand
DO	dissolved oxygen
GHG	greenhouse gas
IASBR	Intermittent Aeration Sequencing Batch Reactor
LCA	life cycle assessment
MBR	Membrane Bioreactor
ML	machine learning
MPC	model predictive control
MS	Mechanical Sludge
PID	proportional–integral–derivative
PMSG	permanent magnet synchronous generator
POC	particulate organic carbon
PST	Primary Sedimentation Tank
SBR	Sequencing Batch Reactor
WWTP	Wastewater Treatment Plant

References

1. Chelyadyn, L.; Kostyshn, V.; Chelyadyn, V.; Romanyshyn, T.; Vasechko, V. Wastewater Purification Technology By Two-Stage Treatment In Electrical Device of a Compact Local Installation. *East. Eur. J. Enterp. Technol.* **2020**, *3*, 63–70. [CrossRef]
2. Hall, M.R.; Priestley, A.; Muster, T.H. Environmental Life Cycle Costing and Sustainability: Insights from Pollution Abatement and Resource Recovery in Wastewater Treatment. *J. Ind. Ecol.* **2018**, *22*, 1127–1138. [CrossRef]
3. Li, M.; Chen, N.; Shang, H.; Ling, C.; Wei, K.; Zhao, S.; Zhou, B.; Jia, F.; Ai, Z.; Zhang, L. An Electrochemical Strategy for Simultaneous Heavy Metal Complexes Wastewater Treatment and Resource Recovery. *Environ. Sci. Technol.* **2022**, *56*, 10945–10953. [CrossRef]
4. Zheng, P.S.; Jin, L.L.; Gao, J. Experimental Study on the Simultaneous Nitrification and Denitrification of Limited Aeration Biofilm Reactor. *IOP Conf. Ser. Earth Environ. Sci.* **2020**, *571*, 012101. [CrossRef]

5. Jahan, N.; Tahmid, M.; Shoronika, A.Z.; Fariha, A.; Roy, H.; Pervez, M.N.; Cai, Y.; Naddeo, V.; Islam, M.S. A Comprehensive Review on the Sustainable Treatment of Textile Wastewater: Zero Liquid Discharge and Resource Recovery Perspectives. *Sustainability* **2022**, *14*, 15398. [CrossRef]
6. Xie, Y.; Jiang, C.; Kuai, B.; Xu, S.; Zhuang, X. N_2O Emission Reduction in the Biological Nitrogen Removal Process for Wastewater with Low C/N Ratios: Mechanisms and Strategies. *Front. Bioeng. Biotechnol.* **2023**, *11*, 1247711. [CrossRef]
7. Chen, D.; Li, H.; Xue, X.; Zhang, L.; Hou, Y.; Chen, H.; Zhang, Y.; Song, Y.; Zhao, S.; Guo, J. Enhanced Simultaneous Partial Nitrification and Denitrification Performance of Aerobic Granular Sludge via Tapered Aeration in Sequencing Batch Reactor for Treating Low Strength and Low COD/TN Ratio Municipal Wastewater. *Environ. Res.* **2022**, *209*, 112743. [CrossRef] [PubMed]
8. Liu, T.; Hu, S.; Guo, J. Enhancing Mainstream Nitrogen Removal by Employing Nitrate/Nitrite-Dependent Anaerobic Methane Oxidation Processes. *Crit. Rev. Biotechnol.* **2019**, *39*, 732–745. [CrossRef]
9. Capodaglio, A.G. Biorefinery of Sewage Sludge: Overview of Possible Value-Added Products and Applicable Process Technologies. *Water* **2023**, *15*, 1195. [CrossRef]
10. Pasha, A.B.M.K.; Nur, M.S.; Mozumder, S.; Parveen, M. Impact of River Water Quality on Public Health in Perspective of Asian Rivers: A Case Study of Buriganga River, Bangladesh. *J. Environ. Earth Sci.* **2023**, *5*, 1–16. [CrossRef]
11. Raimi, O.M.; Sawyerr, O.H.; Ezekwe, C.I.; Salako, G. Many Oil Wells, One Evil: Comprehensive Assessment of Toxic Metals Concentration, Seasonal Variation and Human Health Risk in Drinking Water Quality in Areas Surrounding Crude Oil Exploration Facilities in Rivers State, Nigeria. *Int. J. Hydrol.* **2022**, *6*, 23–42. [CrossRef]
12. Ghimire, U.; Sarpong, G.; Gude, V.G. Transitioning Wastewater Treatment Plants toward Circular Economy and Energy Sustainability. *ACS Omega* **2021**, *6*, 11794–11803. [CrossRef] [PubMed]
13. Ali, M.; Hong, P.Y.; Mishra, H.; Vrouwenvelder, J.; Saikaly, P.E. Adopting the Circular Model: Opportunities and Challenges of Transforming Wastewater Treatment Plants into Resource Recovery Factories in Saudi Arabia. *Water Reuse* **2022**, *12*, 346–365. [CrossRef]
14. Xia, Q.; Liu, F.; Sun, S.; Huang, W.; Zhao, Z.; Yang, F.; Lei, Z.; Huang, W.; Yi, X. Coupling Iron Sludge Addition and Intermittent Aeration for Achieving Simultaneous Methanogenesis, Feammox, and Denitrification in a Single Reactor Treating Fish Sludge. *Environ. Sci. Technol.* **2023**, *57*, 15065–15075. [CrossRef] [PubMed]
15. Sharawat, I.; Dahiya, R.; Dahiya, R.P. Analysis of a Wastewater Treatment Plant for Energy Consumption and Greenhouse Gas Emissions. *Int. J. Environ. Sci. Technol.* **2021**, *18*, 871–884. [CrossRef]
16. Rajanandini, M.; Kumar, S.N. A Review on Photosynthetic Algal-Microbial Fuel Cells: An-Friendly and Energy-Efficient Technology for Wastewater Treatment and Electricity Production. *Res. J. Chem. Environ.* **2022**, *26*, 193–201.
17. Iswarya, S.; Shanmugam, P.M.; Somasundaram, E.; Chitdeshwari, T.; Suganthy, M. Energy Budgeting and Efficiency Analysis of Organic Cotton: A DEA Approach. *Indian J. Agric. Res.* **2024**, *58*, 389–397. [CrossRef]
18. Nandan, R.; Poonia, S.P.; Singh, S.S.; Nath, C.P.; Kumar, V.; Malik, R.K.; McDonald, A.; Hazra, K.K. Potential of Conservation Agriculture Modules for Energy Conservation and Sustainability of Rice-Based Production Systems of Indo-Gangetic Plain Region. *Environ. Sci. Pollut. Res.* **2021**, *28*, 246–261. [CrossRef]
19. Muzaffar, W.M.B.W.; Aznah, A.; Halim, H. Energy Efficiency and Nutrient Removal Performance: Comparison between Several Types of Activated Sludge Process. *IOP Conf. Ser. Earth Environ. Sci.* **2022**, *1091*, 012056. [CrossRef]
20. Cassidy, J.; Silva, T.; Semião, N.; Ramalho, P.; Santos, A.R.; Feliciano, J.F.; Silva, C.; Rosa, M.J. Integrating Reliability and Energy Efficiency Assessments for Pinpointing Actionable Strategies for Enhanced Performance of Urban Wastewater Treatment Plants. *Sustainability* **2023**, *15*, 12965. [CrossRef]
21. Chen, W.; Tai, K.; Lau, M.W.S.; Abdelhakim, A.; Chan, R.R.; Adnanes, A.K.; Tjahjowidodo, T. Robust Real-Time Shipboard Energy Management System With Improved Adaptive Model Predictive Control. *IEEE Access* **2023**, *11*, 110342–110360. [CrossRef]
22. Jiang, W.; Yi, Z.; Wang, L.; Zhang, H.; Zhang, J.; Lin, F.; Yang, C. A Stochastic Online Forecast-and-Optimize Framework for Real-Time Energy Dispatch in Virtual Power Plants under Uncertainty. In Proceedings of the 32nd ACM International Conference on Information and Knowledge Management, Birmingham, UK, 21–25 October 2023.
23. Guo, N.; Zhang, X.; Zou, Y.; Du, G.; Wang, C.; Guo, L. Predictive Energy Management of Plug-in Hybrid Electric Vehicles by Real-Time Optimization and Data-Driven Calibration. *IEEE Trans. Veh. Technol.* **2022**, *71*, 5677–5691. [CrossRef]
24. Halhoul Merabet, G.; Essaaidi, M.; Ben Haddou, M.; Qolomany, B.; Qadir, J.; Anan, M.; Al-Fuqaha, A.; Abid, M.R.; Benhaddou, D. Intelligent Building Control Systems for Thermal Comfort and Energy-Efficiency: A Systematic Review of Artificial Intelligence-Assisted Techniques. *Renew. Sustain. Energy Rev.* **2021**, *144*, 110969. [CrossRef]
25. Laimon, M.; Mai, T.; Goh, S.; Yusaf, T. System Dynamics Modelling to Assess the Impact of Renewable Energy Systems and Energy Efficiency on the Performance of the Energy Sector. *Renew. Energy* **2022**, *193*, 1041–1048. [CrossRef]
26. Han, H.G.; Chen, C.; Sun, H.Y.; Qiao, J.F. Multi-Objective Integrated Optimal Control for a Wastewater Treatment Process. *Control Eng. Pract.* **2022**, *128*, 105296. [CrossRef]
27. Chen, S.Y.; Chang, C.H. Optimal Power Flows Control for Home Energy Management With Renewable Energy and Energy Storage Systems. *IEEE Trans. Energy Convers.* **2023**, *38*, 218–229. [CrossRef]
28. Alves, J.M.C. Unload Control—Efficiency Strategy and Rainfall Control in Urban Wastewater Systems. *MOJ Ecol. Environ. Sci.* **2023**, *8*, 43–45. [CrossRef]
29. Vujić, M.; Šemanjski, I.; Vidan, P. Improving Energy Efficiency by Advanced Traffic Control Systems. *Trans. Marit. Sci.* **2015**, *4*, 119–126. [CrossRef]

30. Jinapor, J.A.; Suleman, S.; Cromwell, R.S. Energy Consumption and Environmental Quality in Africa: Does Energy Efficiency Make Any Difference? *Sustainability* **2023**, *15*, 2375. [CrossRef]
31. Zhong, R.; Wu, W.; Akbar, M.W.; Zia, Z. How Environmental Protection Activities and Industrial Revolution Contributes in the Nexus of Energy Security and Environmental Sustainability? *Environ. Sci. Pollut. Res.* **2023**, *30*, 104620–104632. [CrossRef]
32. Fighir, D.; Teodosiu, C.; Fiore, S. Environmental and Energy Assessment of Municipal Wastewater Treatment Plants in Italy and Romania: A Comparative Study. *Water* **2019**, *11*, 1611. [CrossRef]
33. McCulligh, C. Wastewater and Wishful Thinking: Treatment Plants to "Revive" the Santiago River in Mexico. *Environ. Plan. E Nat. Space* **2023**, *6*, 1966–1986. [CrossRef]
34. Wang, M.; Zhu, J.; Mao, X. Removal of Pathogens in Onsite Wastewater Treatment Systems: A Review of Design Considerations and Influencing Factors. *Water* **2021**, *13*, 1190. [CrossRef]
35. Quispe Cardenas, L.E.; Deptula, P.J.; Huerta, C.S.; Zhu, C.; Ye, Y.; Wang, S.; Yang, Y. Electro-Fenton and Induced Electro-Fenton as Versatile Wastewater Treatment Processes for Decontamination and Nutrient Removal without Byproduct Formation. *ACS ES T Eng.* **2023**, *3*, 1547–1556. [CrossRef] [PubMed]
36. Pinheiro Costa, E.; Vieira, M.C.; Starling, M.; Amorim, C.C. Simultaneous Removal of Emerging Contaminants and Disinfection for Municipal Wastewater Treatment Plant Effluent Quality Improvement: A Systemic Analysis of the Literature. *Environ. Sci. Pollut. Res.* **2021**, *28*, 24092–24111. [CrossRef] [PubMed]
37. Friedler, E.; Chavez, D.F.; Alfiya, Y.; Gilboa, Y.; Gross, A. Impact of Suspended Solids and Organic Matter on Chlorine and UV Disinfection Efficiency of Greywater. *Water* **2021**, *13*, 214. [CrossRef]
38. Peng, J.; Huang, H.; Zhong, Y.; Yin, R.; Wu, Q.; Shang, C.; Yang, X. Transformation of Dissolved Organic Matter during Biological Wastewater Treatment and Relationships with the Formation of Nitrogenous Disinfection Byproducts. *Water Res.* **2022**, *222*, 118870. [CrossRef] [PubMed]
39. Fernández-Pascual, E.; Droz, B.; O'Dwyer, J.; O'Driscoll, C.; Goslan, E.H.; Harrison, S.; Weatherill, J. Fluorescent Dissolved Organic Matter Components as Surrogates for Disinfection Byproduct Formation in Drinking Water: A Critical Review. *ACS ES T Water* **2023**, *3*, 1997–2008. [CrossRef]
40. Manzhina, S.A.; Vlasov, M.V. Agro-ecological assessment of domestic wastewater for irrigation purposes. *Land Reclam. Hydraul. Eng.* **2023**, *13*, 132–149. [CrossRef]
41. Masłoń, A.; Czarnota, J.; Szczyrba, P.; Szaja, A.; Szulżyk-Cieplak, J.; Łagód, G. Assessment of Energy Self-Sufficiency of Wastewater Treatment Plants—A Case Study from Poland. *Energies* **2024**, *17*, 1164. [CrossRef]
42. Karadimos, P.; Anthopoulos, L. Machine Learning-Based Energy Consumption Estimation of Wastewater Treatment Plants in Greece. *Energies* **2023**, *16*, 7408. [CrossRef]
43. Ranieri, E.; Giuliano, S.; Ranieri, A.C. Energy Consumption in Anaerobic and Aerobic Based Wastewater Treatment Plants in Italy. *Water Pract. Technol.* **2021**, *16*, 851–863. [CrossRef]
44. Alali, Y.; Harrou, F.; Sun, Y. Unlocking the Potential of Wastewater Treatment: Machine Learning Based Energy Consumption Prediction. *Water* **2023**, *15*, 2349. [CrossRef]
45. Zhang, J.; Shao, Y.; Wang, H.; Liu, G.; Qi, L.; Xu, X.; Liu, S. Current Operation State of Wastewater Treatment Plants in Urban China. *Environ. Res.* **2021**, *195*, 110843. [CrossRef] [PubMed]
46. Xu, A.; Wu, Y.H.; Chen, Z.; Wu, G.; Wu, Q.; Ling, F.; Huang, W.E.; Hu, H.Y. Towards the New Era of Wastewater Treatment of China: Development History, Current Status, and Future Directions. *Water Cycle* **2020**, *1*, 80–87. [CrossRef]
47. Sun, S.; Song, P.; Sun, J.; Sun, W. Cavitation Suppression and Design Optimization in a Cryogenic Liquid Turbine Expander Based on Thermodynamic Cavitation and Entropy Production Analysis. *J. Fluids Eng. Trans. ASME* **2023**, *145*, 011201. [CrossRef]
48. Yang, L.; Shakeel, Q.; Xu, X.; Ali, L.; Chen, Z.; Mubeen, M.; Sohail, M.A.; Ifltikhar, Y.; Kumar, A.; Solanki, M.K.; et al. Optimized Submerged Batch Fermentation for Metabolic Switching in Streptomyces Yanglinensis 3–10 Providing Platform for Reveromycin A and B Biosynthesis, Engineering, and Production. *Front. Microbiol.* **2024**, *15*, 1378834. [CrossRef] [PubMed]
49. Wang, F.; Cui, Q.; Liu, W.; Jiang, W.; Ai, S.; Liu, W.; Bian, D. Synergistic Denitrification Mechanism of Domesticated Aerobic Denitrifying Bacteria in Low-Temperature Municipal Wastewater Treatment. *NPJ Clean. Water* **2024**, *7*, 6. [CrossRef]
50. Jung, M.; Lee, J.; Park, S.J.; Na, J.G. Gas Supply Apparatus Using Rotational Motion of Shaking Incubator for Flask Culture of Aerobic Microorganisms. *Eng. Life Sci.* 2024; *early access*. [CrossRef]
51. Toledo Padrón, M.; Muñoz, R. Odour Control Strategies in Wastewater Treatment Plants: Activated Sludge Recycling and Oxidized Nitrogen Recycling. *Chem. Eng. Trans.* **2022**, *95*, 253–258. [CrossRef]
52. Khalid, S.; Shahid, M.; Natasha; Bibi, I.; Sarwar, M.; Shah, A.H.; Niazi, N.K. A Review of Environmental Contamination and Health Risk Assessment of Wastewater Use for Crop Irrigation with a Focus on Low and High-Income Countries. *Int. J. Environ. Res. Public Health* **2018**, *15*, 895. [CrossRef]
53. Waqas, S.; Harun, N.Y.; Sambudi, N.S.; Abioye, K.J.; Zeeshan, M.H.; Ali, A.; Abdulrahman, A.; Alkhattabi, L.; Alsaadi, A.S. Effect of Operating Parameters on the Performance of Integrated Fixed-Film Activated Sludge for Wastewater Treatment. *Membranes* **2023**, *13*, 704. [CrossRef] [PubMed]
54. Byliński, H.; Gębicki, J.; Namieśnik, J. Evaluation of Health Hazard Due to Emission of Volatile Organic Compounds from Various Processing Units of Wastewater Treatment Plant. *Int. J. Environ. Res. Public Health* **2019**, *16*, 1712. [CrossRef] [PubMed]
55. Satar, I.; Sirajuddin, M.M.; Permadi, A.; Latifatunnajib, S. Tofu Wastewater (TWW) Treatment and Hydrogen (H2) Production by Using A Microbial Electrolysis Cell (MEC) System. *Indones. J. Environ. Manag. Sustain.* **2023**, *7*, 13–19. [CrossRef]

56. Zubir, A.A.A.; Dahalan, F.A.; Kamarudin, N.S.; Ibrahim, N.; Ong, S.A.; Lutpi, N.A.; Hasan, M.; Parmin, N.A. Effect of Aeration Rate on Specific Oxygen Uptake Rate (SOUR) in Treating Chemical Oxygen Demand (COD) in Domestic Wastewater. *IOP Conf. Ser. Earth Environ. Sci.* **2024**, *1303*, 012026. [CrossRef]
57. Yu, D.; Zhang, W.; Wang, D.; Jin, Y. Full-Scale Application of One-Stage Simultaneous Nitrification and Denitrification Coupled with Anammox Process for Treating Collagen Casing Wastewater. *Int. J. Environ. Res. Public Health* **2022**, *19*, 5787. [CrossRef] [PubMed]
58. Revollar, S.; Vega, P.; Francisco, M.; Meneses, M.; Vilanova, R. Activated Sludge Process Control Strategy Based on the Dynamic Analysis of Environmental Costs. In Proceedings of the 2020 24th International Conference on System Theory, Control and Computing (ICSTCC), Sinaia, Romania, 8–10 October 2020; pp. 576–581.
59. Zhang, L. Embedded Device Energy Consumption Prediction System Based on Machine Learning Optimization. *J. Phys. Conf. Ser.* **2023**, *2560*, 012037. [CrossRef]
60. Barbosa, T.A.; Giordani, A.; de Moura, R.B. A Pilot-Scale Study of a Novel System for Simultaneous Nitrogen and Carbon Removal: Technological Advancement of a Structured Bed Reactor with Intermittent Aeration (SBRIA) in Real Domestic Sewage Treatment. *Environ. Sci. Pollut. Res.* **2024**, *31*, 12591–12596. [CrossRef] [PubMed]
61. Cheng, X.; Xie, Y.; Zhu, D.; Xie, J. Modeling Re-Oxygenation Performance of Fine-Bubble–Diffusing Aeration System in Aquaculture Ponds. *Aquac. Int.* **2019**, *27*, 1353–1368. [CrossRef]
62. Liu, X.; Bi, X.; Huang, Q.; Wang, X.; Gu, R.; Zhou, X. The Changing and Distribution Laws of Oxygen Transfer Efficiency in the Full-Scale Ifas Process. *Water* **2021**, *13*, 1933. [CrossRef]
63. Chakachaka, V.M.; Tshangana, C.S.; Mamba, B.B.; Muleja, A.A. CFD-Assisted Process Optimization of an Integrated Photocatalytic Membrane System for Water Treatment. *Membranes* **2023**, *13*, 827. [CrossRef]
64. Rafiee, M.; Sabeti, M.; Torabi, F.; Rahimbakhsh, A. COD Reduction of Aeration Effluent by Utilizing Optimum Quantities of $UV/H_2O_2/O_3$ in a Small-Scale Reactor. *Processes* **2022**, *10*, 2441. [CrossRef]
65. Yang, Y.; Zhang, J.; Liu, L.; Hu, Y.; Xu, Z. Simulation of Triple Oxidation Ditch Wastewater Treatment Process. *AIP Conf. Proc.* **2010**, *1251*, 153–156. [CrossRef]
66. Liu, J.; Gao, J.; Zhong, Z.; Cheng, Y.; Zhang, B. Comparison of Dissolved Organic Matter Composition and Microbial Distribution between Distributed-Inflow Biological Reactor and Two-Stage Anoxic/Aerobic for Piggery Wastewater Treatment. *Water* **2023**, *15*, 410. [CrossRef]
67. Lu, X.; Zhong, Z.; Cai, Y.; Chen, H.; Lan, X.; Huang, C.; Zhou, J.; Zhang, F.; Zhang, B. CFD Simulation of the Flow Patterns and Structure Optimization within a Continued-Flow Integrated Biological Reactor. *Pol. J. Environ. Stud.* **2022**, *31*, 1737–1746. [CrossRef]
68. Jia, K.; Liu, C.; Li, S.; Jiang, D. Modeling and Optimization of a Hybrid Renewable Energy System Integrated with Gas Turbine and Energy Storage. *Energy Convers. Manag.* **2023**, *279*, 116763. [CrossRef]
69. Nurjanah, I.; Hsieh, L.H.C.; Chiang, Y.H.; Sean, W.Y. Energy Saving in NMP (N-Methyl-2-Pyrrolidone) Recovery Process by Numerical Modeling. *Environ. Technol. Innov.* **2023**, *31*, 103218. [CrossRef]
70. Zanoli, S.M.; Pepe, C.; Orlietti, L. Multi-Mode Model Predictive Control Approach for Steel Billets Reheating Furnaces. *Sensors* **2023**, *23*, 3966. [CrossRef]
71. Husin, M.H.; Sabri, M.F.M.; Ping, K.A.H.; Bateni, N.; Suhaili, S. Energy Efficiency in Activated Sludge Process Using Adaptive Iterative Learning Control with PI ABAC. *Bull. Electr. Eng. Inform.* **2024**, *13*, 885–892. [CrossRef]
72. Dey, I.; Ambati, S.R.; Bhos, P.N.; Sonawane, S.; Pilli, S. Effluent Quality Improvement in Sequencing Batch Reactor-Based Wastewater Treatment Processes Using Advanced Control Strategies. *Water Sci. Technol.* **2024**, *89*, 2661–2675. [CrossRef]
73. Zhang, K.; Wang, Z.; Sun, M.; Liang, D.; Hou, L.; Zhang, J.; Wang, X.; Li, J. Optimization of Nitrogen and Carbon Removal with Simultaneous Partial Nitrification, Anammox and Denitrification in Membrane Bioreactor: SNAD and MBR. *R. Soc. Open Sci.* **2020**, *7*, 200584. [CrossRef]
74. Bonci, A.; Di Biase, A.; Giannini, M.C.; Longhi, S. Yaw Rate-Based PID Control for Lateral Dynamics of Autonomous Vehicles, Design and Implementation. In Proceedings of the 28th International Conference on Emerging Technologies and Factory Automation (ETFA), Sinaia, Romania, 12–15 September 2023.
75. Sonnenschein, B.; Ziel, F. Probabilistic Intraday Wastewater Treatment Plant Inflow Forecast Utilizing Rain Forecast Data and Sewer Network Sensor Data. *Water Resour. Res.* **2023**, *59*, e2022WR033826. [CrossRef]
76. Wan, X.; Volcke, E. Dynamic Modelling of N_2O Emissions from a Full-scale Granular Sludge Partial Nitritation-anammox Reactor. *Biotechnol. Bioeng.* **2022**, *119*, 1426–1438. [CrossRef] [PubMed]
77. Haroon, H.; Shah, J.A.; Khan, M.S.; Alam, T.; Khan, R.; Asad, S.A.; Ali, M.A.; Farooq, G.; Iqbal, M.; Bilal, M. Activated Carbon from a Specific Plant Precursor Biomass for Hazardous Cr(VI) Adsorption and Recovery Studies in Batch and Column Reactors: Isotherm and Kinetic Modeling. *J. Water Process Eng.* **2020**, *38*, 101577. [CrossRef]
78. Zhang, T.; El-Sayed, W.M.M.; Zhang, J.; He, L.; Bruns, M.A.; Wang, M. Insight into the Impact of Air Flow Rate on Algal-Bacterial Granules: Reactor Performance, Hydrodynamics by Computational Fluid Dynamics (CFD) and Microbial Community Analysis. *bioRxiv* **2024**. [CrossRef]
79. Siriweera, B.; Siddiqui, M.; Zou, X.; Chen, G.; Wu, D. Integrated Thiosulfate-Driven Denitrification, Partial Nitrification and Anammox Process in Membrane-Aerated Biofilm Reactor for Low-Carbon, Energy-Efficient Biological Nitrogen Removal. *Bioresour. Technol.* **2023**, *382*, 129212. [CrossRef] [PubMed]

80. Li, L.; Xu, W.; Ning, J.; Zhong, Y.; Zhang, C.; Zuo, J.; Pan, Z. Revealing the Intrinsic Mechanisms for Accelerating Nitrogen Removal Efficiency in the Anammox Reactor by Adding Fe(II) at Low Temperature. *Chin. Chem. Lett.* **2024**, *35*, 109243. [CrossRef]
81. de la Vega, P.T.M.; Jaramillo-Morán, M.A. Multilevel Adaptive Control of Alternating Aeration Cycles in Wastewater Treatment to Improve Nitrogen and Phosphorous Removal and to Obtain Energy Saving. *Water* **2018**, *11*, 60. [CrossRef]
82. Avilés, A.B.L.; Del Cerro Velázquez, F.; Del Riquelme, M.L.P. Methodology for Energy Optimization in Wastewater Treatment Plants. Phase II: Reduction of Air Requirements and Redesign of the Biological Aeration Installation. *Water* **2020**, *12*, 1143. [CrossRef]
83. Mallu, L.L.; Sean, W.Y.; Pacheco, M. Optimization of Air Flowrate under Different Control Strategies Focus on Biological Process in Wastewater Treatment Plant. In Proceedings of the 2021 IEEE Symposium Series on Computational Intelligence, SSCI 2021, Orlando, FL, USA, 5–7 December 2021.
84. Aparna, K.G.; Swarnalatha, R. Simulation and Analysis of Ammonium-Based Aeration Control Strategies to Enhance Efficiency in Wastewater Treatment Plant. In Proceedings of the 2023 IEEE World Conference on Applied Intelligence and Computing (AIC), Sonbhadra, India, 29–30 July 2023; pp. 142–147.
85. Ma, Z.; Yin, J.; Yang, Y.; Sun, F.; Yang, Z. Effect of Water and Nitrogen Coupling Regulation on the Growth, Physiology, Yield, and Quality Attributes and Comprehensive Evaluation of Wolfberry (*Lycium barbarum* L.). *Front. Plant Sci.* **2023**, *14*, 1130109. [CrossRef]
86. Jiang, J.; Zhu, Z.; Huan, J.; Shi, B. Energy-Efficient Mechanical Aeration System in Aquaculture. *DEStech Trans. Eng. Technol. Res.* **2019**. [CrossRef]
87. Lasaki, B.A.; Maurer, P.; Schönberger, H. Effect of Coupling Primary Sedimentation Tank (PST) and Microscreen (MS) to Remove Particulate Organic Carbon (POC): A Study to Mitigate Energy Demand in Municipal Wastewater Treatment Plants. *Sustain. Environ. Res.* **2023**, *33*, 25. [CrossRef]
88. Batool, M.; Shahzad, L.; Tahir, A. Review on Municipal Wastewater to Energy Generation; a Favorable Approach for Developing Countries. *Proc. Inst. Civ. Eng. Energy* **2023**, 1–31. [CrossRef]
89. Leonard, P.; Clifford, E.; Finnegan, W.; Siggins, A.; Zhan, X. Deployment and Optimisation of a Pilot-Scale Iasbr System for Treatment of Dairy Processing Wastewater. *Energies* **2021**, *14*, 7365. [CrossRef]
90. Wikaningrum, T.; Putri, A.N.I. Study on Activated Sludge Composition and Concentration Setting for Increasing COD Efficiency in Dairy Industry Wastewater. *IOP Conf. Ser. Earth Environ. Sci.* **2023**, *1268*, 012016. [CrossRef]
91. Suriasni, P.A.; Faizal, F.; Hermawan, W.; Subhan, U.; Panatarani, C.; Joni, I.M. IoT Water Quality Monitoring and Control System in Moving Bed Biofilm Reactor to Reduce Total Ammonia Nitrogen. *Sensors* **2024**, *24*, 494. [CrossRef] [PubMed]
92. Barradi, Y.; Khaldi, N.; Zazi, K.; Zazi, M.; Author, C. Comparative Analysis of Backstepping and Active Disturbance Rejection Control Approach Used in Photovoltaic System Connected to the Grid. *Int. J. Renew. Energy Res.* **2019**, *9*, 1470–1479. [CrossRef]
93. Ford, A.; Hawley, S.; Rutherford, B.; Uprety, K.; Bott, C. Implementation of Aeration Control Strategies and Nitrate-Based Internal Mixed Liquor Recycle Control Employing in-Situ Sensors and Feedback PID Controllers in an Integrated Fixed-Film Activated Sludge Wastewater Treatment Facility. *Proc. Water Environ. Fed.* **2018**, *2018*, 251–262. [CrossRef]
94. Mohan, P.; Paul, A.J.; Chirania, A. A Tiny Cnn Architecture for Medical Face Mask Detection for Resource-Constrained Endpoints. In *Innovations in Electrical and Electronic Engineering*; Lecture Notes in Electrical Engineering; Springer: Singapore, 2021; Volume 756, pp. 657–670.
95. Mamat, M.N.; Ishak, D. Analysis of SEPIC-Boost Converter Using Several PID Feedback Tuning Methods for Renewable Energy Applications. *J. Adv. Res. Appl. Sci. Eng. Technol.* **2022**, *26*, 105–117. [CrossRef]
96. Nascu, I.; Du, W.; Nascu, I. An Auto-Tuning Method for Aeration Control in Activated Sludge Wastewater Treatment Processes. In Proceedings of the International Conference on Electrical, Computer, Communications and Mechatronics Engineering (ICECCME), Male, Maldives, 16–18 November 2022.
97. Aathithya, S.; Kalpana, D. Control of Dissolved Oxygen Concentration in Waste-Water Treatment Plants Using Fuzzy Logic Control. In Proceedings of the IEEE 9th International Conference on Smart Structures and Systems, ICSSS 2023, Chennai, India, 23–24 November 2023.
98. Hansen, L.D.; Veng, M.; Durdevic, P. Compressor Scheduling and Pressure Control for an Alternating Aeration Activated Sludge Process—A Simulation Study Validated on Plant Data. *Water* **2021**, *13*, 1037. [CrossRef]
99. Quispe-Quispe, J.E.; Valenzuela-Lino, Y.S.; Ortiz-Zacarias, J.R.; Alex Alarcon-Vasquez, C.; Moggiano, N.; Coaquira-Rojo, C. Development of PID Control Parameters in Proportional Valves for a Wastewater Treatment Plant Filtration Process. In Proceedings of the 13th Annual Ubiquitous Computing, Electronics & Mobile Communication Conference (UEMCON), New York, NY, USA, 26–29 October 2022; pp. 277–282.
100. Nascu, I.; Harja, G.; Nascu, I. Auto-Tuning Method for Alternating Aeration Control in Activated Sludge Processes. In Proceedings of the 27th International Conference on Circuits, Systems, Communications and Computers (CSCC), Rhodes Island, Greece, 19–22 July 2023; pp. 111–116.
101. Caivano, M.; Bellandi, G.; Mancini, I.M.; Masi, S.; Brienza, R.; Panariello, S.; Gori, R.; Caniani, D. Monitoring the Aeration Efficiency and Carbon Footprint of a Medium-Sized WWTP: Experimental Results on Oxidation Tank and Aerobic Digester. *Environ. Technol.* **2017**, *38*, 629–638. [CrossRef]
102. Simon-Várhelyi, M.; Cristea, V.M.; Luca, A.V.; Brehar, M.A. Optimization and Control of Aeration Distribution in the Wwtp Nitrification Reactor. *Rev. Roum. Chim.* **2020**, *65*, 601–609. [CrossRef]

103. Wang, D.; Ha, M.; Qiao, J. Data-Driven Iterative Adaptive Critic Control Toward an Urban Wastewater Treatment Plant. *IEEE Trans. Ind. Electron.* **2021**, *68*, 7362–7369. [CrossRef]
104. Srb, M.; Lánský, M.; Charvátová, L.; Koubová, J.; Pecl, R.; Sýkora, P.; Rosický, J. Improved Nitrogen Removal Efficiency by Implementation of Intermittent Aeration. *Water Sci. Technol.* **2022**, *86*, 2248–2259. [CrossRef] [PubMed]
105. Morera, S.; Santana, M.V.E.; Comas, J.; Rigola, M.; Corominas, L. Evaluation of Different Practices to Estimate Construction Inventories for Life Cycle Assessment of Small to Medium Wastewater Treatment Plants. *J. Clean. Prod.* **2020**, *245*, 118768. [CrossRef]
106. Protoulis, T.; Kalogeropoulos, I.; Kordatos, I.; Kapnopoulos, A.; Zervas, P.L.; Vangelatos, G.; Sarimveis, H.; Alexandridis, A. An Identification and Control Framework for Optimizing the Energy Consumption of a Wastewater Treatment Plant. In Proceedings of the 6th International Conference and Workshop Óbuda on Electrical and Power Engineering (CANDO-EPE), Budapest, Hungary, 19–20 October 2023; pp. 43–48.
107. Crisan, R.; Korodi, A. Noninvasive Control Solution for Energy Efficiency in Wastewater Treatment Plants. In Proceedings of the IEEE International Conference on Industrial Technology (ICIT), Lyon, France, 20–22 February 2018; pp. 1604–1609.
108. Faisal, M.; Muttaqi, K.M.; Sutanto, D.; Al-Shetwi, A.Q.; Ker, P.J.; Hannan, M.A. Control Technologies of Wastewater Treatment Plants: The State-of-the-Art, Current Challenges, and Future Directions. *Renew. Sustain. Energy Rev.* **2023**, *181*, 113324. [CrossRef]
109. Shah, K.A.; Jiao, Y.; Chen, J. CFD Investigation of Dissolved Oxygen Distribution in a Full-Scale Aeration Tank of an Industrial Wastewater Treatment Plant. *J. Water Process Eng.* **2024**, *59*, 105078. [CrossRef]
110. Piotrowski, R. Supervisory Fuzzy Control System for Biological Processes in Sequencing Wastewater Batch Reactor. *Urban. Water J.* **2020**, *17*, 325–332. [CrossRef]
111. Du, S.; Yan, Q.; Qiao, J. Event-Triggered PID Control for Wastewater Treatment Plants. *J. Water Process Eng.* **2020**, *38*, 101659. [CrossRef]
112. Van Nguyen, L.; Van Bach, N.; Do, H.T.; Nguyen, M.T. Combined ILC and PI Regulator for Wastewater Treatment Plants. *Telkomnika* **2020**, *18*, 1054–1061. [CrossRef]
113. Li, D.; Zou, M.; Jiang, L. Dissolved Oxygen Control Strategies for Water Treatment: A Review. *Water Sci. Technol.* **2022**, *86*, 1444–1466. [CrossRef]
114. Tejaswini, E.S.S.; Panjwani, S.; Gara, U.B.B.; Ambati, S.R. Multi-Objective Optimization Based Controller Design for Improved Wastewater Treatment Plant Operation. *Environ. Technol. Innov.* **2021**, *23*, 101591. [CrossRef]
115. Chen, Y.; Zhang, H.; Yin, Y.; Zeng, F.; Cui, Z. Smart Energy Savings for Aeration Control in Wastewater Treatment. *Energy Rep.* **2022**, *8*, 1711–1721. [CrossRef]
116. Childs, T.; Jones, A.; Chen, R.; Murray, A. A Study into Refrigeration Cycle Working Fluids Using an Air Cycle Machine Environmental Control System. In Proceedings of the 54th AIAA Aerospace Sciences Meeting, San Diego, CA, USA, 4–8 January 2016.
117. Jandieri, G.; Sakhvadze, D.; Schukin, B. Underground development of mineral subsoil using microorganisms: A mini-review. *Mikrobiolohichnyi Zhurnal* **2023**, *85*, 66–71. [CrossRef]
118. Piotrowski, R.; Ujazdowski, T. Designing Control Strategies of Aeration System in Biological WWTP. *Energies* **2020**, *13*, 3619. [CrossRef]
119. Zhang, D.; Chu, J.; He, Y.; Jin, H.; Han, W. Study and Application of Self-Adaptive Fuzzy PID Control in Dissolved Oxygen Control of Wastewater Treatment. *IOP Conf. Ser. Mater. Sci. Eng.* **2019**, *562*, 012147. [CrossRef]
120. Su, D.; Yao, W.; Yu, F.; Liu, Y.; Zheng, Z.; Wang, Y.; Xu, T.; Chen, C. Single-Neuron PID UAV Variable Fertilizer Application Control System Based on a Weighted Coefficient Learning Correction. *Agriculture* **2022**, *12*, 1019. [CrossRef]
121. Polizzi, C.; Falcioni, S.; Mannucci, A.; Mori, G.; Nardi, A.; Spennati, F.; Munz, G. Integrating Online Differential Titrimetry and Dynamic Modelling as Innovative Energy Saving Strategy in a Large Industrial WWTP. *Clean. Technol. Environ. Policy* **2022**, *24*, 1771–1780. [CrossRef]

energies

MDPI

Review

The Role of Solar Concentrators in Photocatalytic Wastewater Treatment

Joy Sankar Roy [1,*] and Younès Messaddeq [1,2]

[1] Center for Optics, Photonics and Lasers (COPL), Laval University, Quebec City, QC G1V 0A6, Canada; younes.messaddeq@copl.ulaval.ca

[2] Institute of Chemistry, Sao Paulo State University (UNESP), Araraquara 14800-060, SP, Brazil

* Correspondence: jsr.phys@gmail.com or joy-sankar.roy.1@ulaval.ca

Abstract: The global challenge of sustainable and affordable wastewater treatment technology looms large as water pollution escalates steadily with the rapid pace of industrialization and population growth. The photocatalytic wastewater treatment is a cutting-edge and environmentally friendly technology that uses photons from light source to degrade and remove organic and inorganic contaminants from water. Thus, utilizing solar energy for photocatalytic wastewater treatment holds great promise as a renewable solution to alleviate pressures on the global water crisis. Employing solar concentrators to intensify sunlight for photocatalysis represents a promising avenue for future applications of a low-cost and rapid sustainable wastewater purification process. This groundbreaking approach will unveil fresh technological avenues for a cost-effective, sustainable, and swift wastewater purification process utilizing sunlight. This review article explores diverse solar concentrating systems and their potential applications in the wastewater treatment process.

Keywords: photocatalysis; wastewater treatment; solar concentrator; solar photocatalysis

1. Introduction

Water is an indispensable component for human survival; hence, fresh and clean water supply has surged in recent decades alongside population growth. Conversely, water pollution has been on the rise due to the swift pace of industrialization and unplanned urban expansion. Agricultural and pharmaceutical runoff, untreated sewage, and toxic industrial discharges constitute significant sources of this pollution [1,2]. Industries like textiles, dyeing, and printing are major contributors, releasing substantial quantities of synthetic organic dyes into waterways [2,3]. Access to fresh water supplies has emerged as a matter of global significance. Additionally, the health risks linked with polluted water are forecasted to escalate into a significant global concern in the coming decades. In developing nations, nearly 70% of all illnesses are linked to water contamination [1]. Hence, the most pressing challenge of the twenty-first century lies in the development of eco-friendly, cost-effective, and efficient wastewater treatment technologies. These advancements are crucial for safeguarding the environment and human health from the harmful effects of pollutants.

Over the past two decades, various treatment techniques have been implemented to treat water for reuse. Techniques such as phase separation, chemical precipitation, coagulation and flocculation, filtration, adsorption, ion exchange, sedimentation, chlorination, and oxidation have been utilized in wastewater treatment [1–4]. Most of these methods generate sludge, presenting a significant challenge for disposal due to the risk of secondary pollution. Some treatment processes also require high energy input, making water remediation expensive. Additionally, pollutants such as pesticides, hydrocarbons, dyes, and aromatic compounds in wastewater cannot be completely eliminated by a single technique. Consequently, researchers are focusing on developing new, greener technologies capable of degrading a wide range of contaminants. In this context, photocatalysis has emerged as a promising, sustainable, and efficient approach for water treatment [5–8].

Its capability to harness solar energy, a freely available natural resource, for wastewater treatment underscores its significance. The solar photocatalysis process is effective for the degradation and disinfection of municipal and industrial wastewater [8–10]. Operable at ambient room temperature and pressure, photocatalysis offers reduced operating costs and typically eliminates the need for additional processes or secondary treatment to remove reaction byproducts. These advantages make it a highly viable and effective solution for large wastewater treatment plants.

Although solar photocatalysis is considered as a promising technology for reducing chemical and microbiological pollutants in wastewater, it has not yet advanced beyond the bench scale to real-world practical applications. Thousands of research articles have been published on the development of photocatalysts for solar photocatalytic water treatment [11–14]. Karthikeyan et al. explored various metal oxide semiconductor photocatalysts for solar photocatalysis in their review article [12]. Wang et al. critically reviewed the development of heterojunction in semiconductors for enhanced photocatalysis processing [13]. Beyond the development of photocatalysts, technological development of the utilization solar energy for solar photocatalysis is necessary. Over the past decades, several techniques have been developed to utilize solar energy for photocatalytic water treatment processing. In this context, solar concentrators can a play significant role in enhancing photocatalytic activity under concentrated sunlight irradiation. This review article delineates the fundamental principles of solar concentration techniques and the benefits conferred by employing solar concentrators in the photocatalytic remediation of contaminants in water.

2. Solar Concentrator

Solar energy is the most abundant resource globally, with sunlight providing approximately 1.5×10^{18} kWh of energy annually. In comparison, the combined reserves of oil, coal, and gas amount to only 1.75×10^{15} kWh, 1.4×10^{15} kWh, and 5.5×10^{15} kWh, respectively [8]. This stark contrast highlights solar energy's potential, offering over a hundred times the energy of the world's total fossil fuel reserves annually. Researchers are actively pursuing the development of cost-effective and efficient technologies to harness solar energy, both to meet global energy demands and to mitigate environmental impact. The solar concentrating systems emerge as highly promising solutions for various applications. Over the past few decades, significant efforts have been invested in the development of efficient solar concentrators to capitalize on solar energy's vast potential [8,15–18].

A solar concentrator is a device that collects solar radiation and focuses it on a smaller area to get higher irradiance power. The device is mainly comprised of a mirror or a lens or a series of lenses or mirror assembly, a receiver, and the tracking system. A paraboloid mirror and Fresnel lens are generally used to design a solar concentrator. The paraboloid mirror reflects the sunlight and concentrates at the focal point of the mirror as shown in Figure 1. The Australian National University has worked for many years on paraboloid dish solar concentrators, and they built a 500 m^2 concentrator to convert solar energy into electricity, which is presented in Figure 2 [15]. Many researchers have developed similar kinds of solar concentrators to generate electricity from sunlight. These works have been summarized and described by Baharoon et al. [16] in their review article. Sometimes, a paraboloid cavity mirror is used with another secondary mirror to deliver concentrated sunlight through the cavity of the paraboloid mirror. The working principle of this kind of system is illustrated by ray diagram in Figure 3. The primary mirror (paraboloid cavity mirror) gathers and focuses sunlight onto a secondary mirror (flat mirror). This secondary mirror redirects and intensifies the sunlight onto a receiver or target. We have pioneered the development of such a solar concentrator, which we have coupled with an optical fiber bundle, known as ECoSEnDS (extremely concentrated solar energy delivery system) [8].

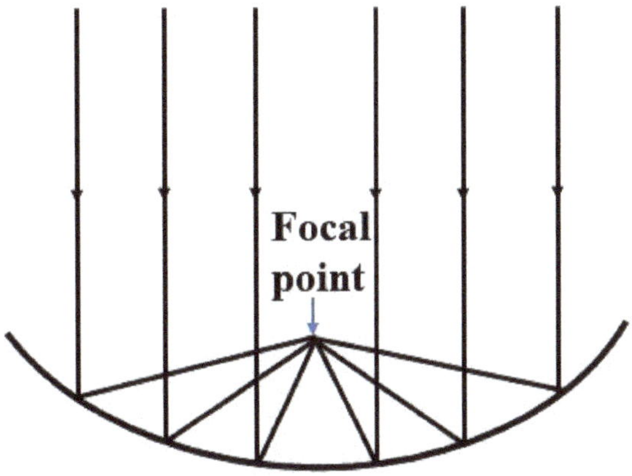

Figure 1. Schematic ray diagram for sunlight concentration by a parabolic mirror.

Figure 2. Picture of the 500 m^2 paraboloid solar concentrator built by The Australian National University [15].

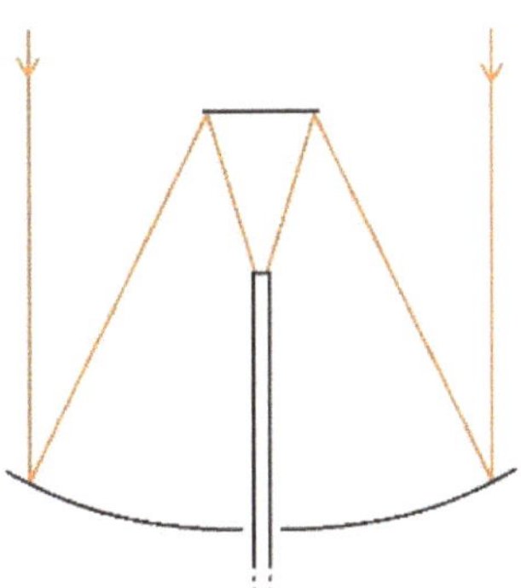

Figure 3. The schematic ray diagram of a solar concentrator based on a primary paraboloid mirror and a secondary flat mirror (Reprinted with permission from Ref. [8]. 2021, Elsevier).

The picture of the ECoSEnDS is presented in Figure 4. The optical fiber bundle receives concentrated sunlight from the secondary mirror. The ECoSEnDS can deliver concentrated sunlight of 96 suns; i.e., 96 times the solar irradiance power on Earth's surface with only 50% of the reflecting efficiency of the primary paraboloid mirror. The Fresnel-based solar concentrator has drawn significant attention due to the light weight, cheaper, and good concentration efficiency qualities of the Fresnel lens. The Fresnel lens refracts the sunlight and focuses it at a focal point to get concentrated sunlight. The working principle of a Fresnel lens is shown in Figure 5. Yang et al. [17] developed a Fresnel lens-based solar concentrator for a daylighting system, as shown in Figure 6. The Fresnel lens concentrates the sunlight on the top of the optical fiber bundle and the fiber bundle delivers sunlight inside the buildings.

Figure 4. The picture of the ECoSEnDS (Reprinted with permission from Ref. [8]. 2021, Elsevier).

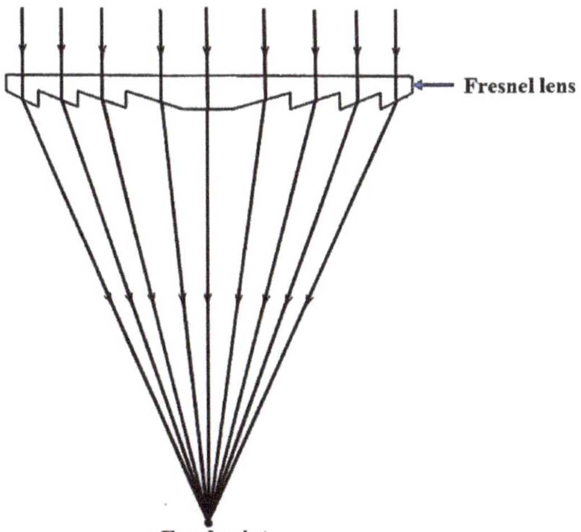

Figure 5. Schematic ray diagram for sunlight concentration by a Fresnel lens.

Figure 6. Solar concentrator employing Fresnel lenses coupled with optical fiber (Reprinted with permission from Ref. [17]. 2019, Elsevier]).

Most solar concentrators have been developed for energy harvesting and daylighting systems. There are very few solar concentrators that have been developed for water purification processing [19–21]. The utilization of solar concentrators for wastewater purification processing will be discussed in Section 4.

3. Photocatalysis

Photocatalysis is a chemical reaction which is accelerated by a photocatalyst on exposure of light. The key advantage of the photocatalysis is that the photocatalyst facilitates the chemical reaction without undergoing consumption [22,23]. Photocatalysis is categorized into homogeneous and heterogeneous types based on the physical state of the reactants and photocatalysts. In homogeneous photocatalysis, both the reactants and photocatalysts coexist in the same phase, whereas in heterogeneous photocatalysis, they exist in different phases.

Heterogeneous photocatalysis has drawn significant attention as it is very much sustainable and effective for wastewater purification processing. This method offers several advantages, including the utilization of abundant sunlight, the ease of fabrication of semiconductor nanomaterials as photocatalysts, and its rapid reaction rate coupled with low energy consumption. Consequently, heterogeneous photocatalysis emerges as a promising technique for wastewater treatment, leveraging the abundance of solar energy resources. The mechanism of heterogeneous photocatalysis typically involves multiple steps [5,22] and it is represented schematically in Figure 7. The process initiates with the absorption of light, where photons with energy exceeding the band gap of the photocatalyst generate electrons (e^-) in the conduction band (CB) and holes (h^+) in the valence band (VB) of the semiconductor photocatalysts. These electrons and holes migrate to the photocatalyst's surface to engage in redox reactions. The holes and electrons participate in oxidation and reduction of H_2O and O_2 molecules to produce hydroxyl ($\cdot OH$) and superoxide radicals ($\cdot O_2^-$), respectively. These radicals promptly react with pollutants, leading to their degradation into harmless or less harmful byproducts such as carbon dioxide (CO_2), water (H_2O), and other intermediate compounds [22]. The complete photocatalytic degradation process can be summarized in Equations (1)–(5) [22–24]:

$$\text{Photocatalyst} + h\nu \rightarrow \text{Photocatalyst}\left(e^- + h^+\right) \tag{1}$$

$$h^+ + H_2O \rightarrow \cdot OH \tag{2}$$

$$e^- + O_2 \rightarrow \cdot O_2^- \tag{3}$$

$$\text{Pollutant} + \cdot OH \rightarrow \text{intermediates} \rightarrow CO_2 + H_2O \tag{4}$$

$$\text{Pollutant} + \cdot O_2^- \rightarrow \text{intermediates} \rightarrow CO_2 + H_2O \tag{5}$$

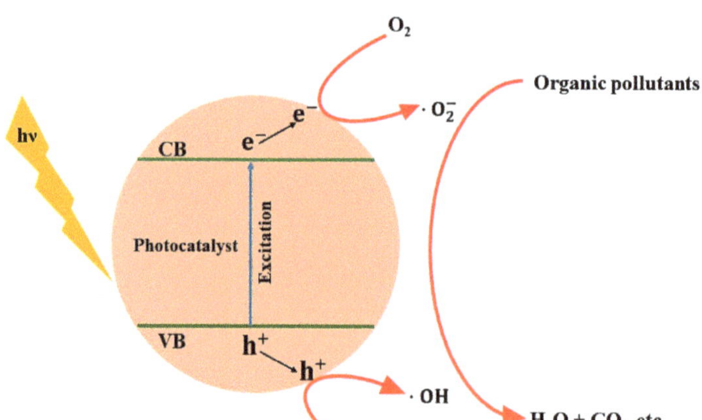

Figure 7. Schematic representation of the heterogeneous photocatalytic organic pollutant's degradation mechanism.

The efficiency of the photocatalysis process depends on several parameters including the redox potential of the pollutants, the band edge of semiconductor photocatalysts [25]. The necessary condition for the photocatalytic reaction is that the redox potential of the pollutant should be lower than the conduction band potential and higher than the valence band potential of the photocatalyst [26].

The semiconductor nanomaterials have been extensively researched for their role in heterogeneous photocatalysis. Titanium dioxide (TiO_2) and zinc oxide (ZnO) are prominent among these, thanks to their remarkable photocatalytic activity, non-toxic nature, and high stability under light exposure [27–29]. However, their effectiveness is mainly confined to UV-light irradiation due to their wide bandgap energy [30,31], limiting their potential for solar photocatalysis since sunlight mainly consists of visible light. To overcome this limitation, various techniques such as anion doping, metal doping, and inducing oxygen deficiency [27,32–35] have been employed to enhance their activity under visible light. Although doping enables TiO_2 or ZnO to absorb visible light, issues like thermal instability and carrier recombination often arise, hindering the development of highly efficient visible-light active photocatalysts. Consequently, researchers have turned to alternative materials like MoS_2, $BiVO_4$, Bi_2WO_6, Ag_3VO_4, $AgVO_3$, $CoFe_2O_4$, $Zr_2Ni_2Cu_7$, and CdS [5,7,22,36–44] to fabricate visible-light active photocatalysts with controlled band structures. These photocatalysts are very much active in visible light and show a higher photocatalytic degradation rate, but some of the photocatalysts like CdS and $BiVO_4$ are highly toxic [45–48]. Results suggest that high doses can cause significant organ damage, particularly in the liver, kidneys, and lungs. Symptoms observed include inflammation, oxidative stress, and histopathological changes. $BiVO_4$ nanoparticles can also pose risks to the environment. They can accumulate in water bodies and soil, affecting aquatic and terrestrial organisms. Studies have shown that these nanoparticles can be toxic to algae, fish, and invertebrates, potentially disrupting ecosystems. Recent studies indicate that treated water is no longer harmful if toxic photocatalysts are removed following the photocatalysis process [49,50]. However, challenges such as fast electron-hole recombination and limited absorption persist for those photocatalysts. Therefore, numerous strategies have emerged, with photocatalytic heterojunctions being widely explored. These heterojunctions offer promising prospects for enhancing photocatalytic efficiency by spatially separating electron-hole pairs [51–54]. Low et al. [52] provide a comprehensive discussion on the principles underlying various heterojunction photocatalysts, as depicted in Figure 8.

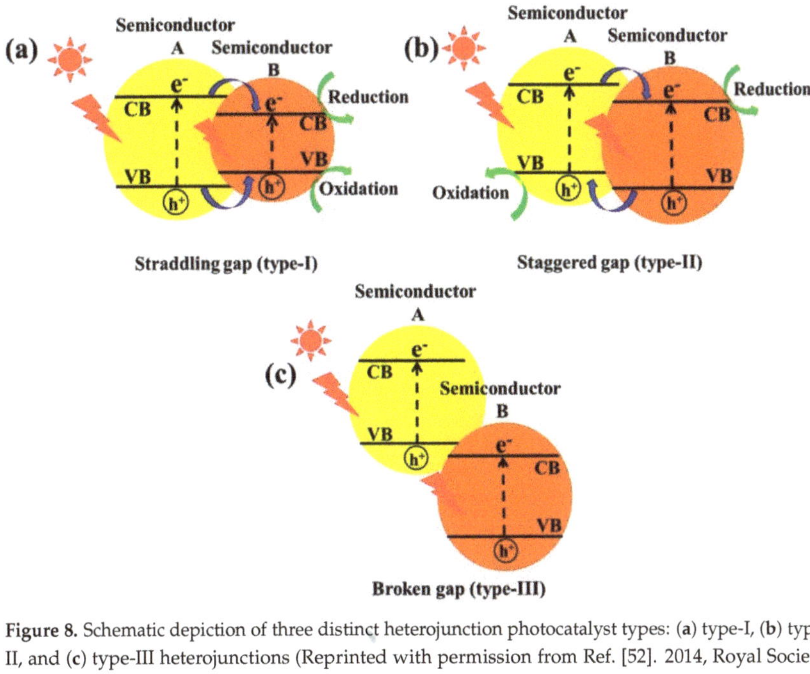

Figure 8. Schematic depiction of three distinct heterojunction photocatalyst types: (**a**) type-I, (**b**) type-II, and (**c**) type-III heterojunctions (Reprinted with permission from Ref. [52]. 2014, Royal Society of Chemistry).

In recent years, thousands of research articles have been reported in literature on wastewater treatment using photocatalysis under visible light irradiation. However, only a handful of studies have explored the utilization of direct sunlight. Instead, most researchers have opted for artificial sources like Xenon arc lamps, fluorescent lamps, LED lamps, or solar simulators to investigate photocatalytic activity. Leveraging solar energy for pollutant removal in wastewater holds profound importance in energy conservation and environmental remediation efforts.

4. Solar Concentrators for Water Treatment

The photocatalytic treatment of wastewater using solar energy has garnered significant interest due to its environmental and economic benefits. Traditional wastewater treatment processes require substantial electrical power, leading to the emission of greenhouse gases. Studies indicate that conventional wastewater treatment consumes approximately 0.6 kWh per cubic meter, resulting in 185.61 g of CO_2 emissions per kWh [55]. In contrast, solar photocatalytic wastewater treatment harnesses freely available solar energy, eliminating the need for external energy sources and reducing greenhouse gas emissions. Therefore, solar concentrators can significantly enhance photocatalytic wastewater treatment efficiency under intense sunlight, offering a promising solution for sustainable and eco-friendly water treatment. The photocatalytic activity hinges significantly on photon count, as photons initiate the creation of electron-hole pairs, as detailed in Section 3. Consequently, photocatalysts generate an increased number of photons under concentrated sunlight irradiation, thereby yielding higher quantities of hydroxyl and superoxide radicals. This cascade effect accelerates the pollutants in water.

In the last few decades, many efforts have been made to develop efficient systems for wastewater purification processing using solar concentrators. Bigoni et al. [20] developed a solar pasteurizer using a parabolic trough concentrator for water purification. The developed system is shown in Figure 9. The sunlight is reflected by a parabolic mirror made with aluminum onto a black steel pipe, serving as the absorber, precisely positioned along the mirror's focal line. The absorber, measuring 3 m in length with an internal diameter of 3.8 cm and a volume of 3.4 L, effectively captures the concentrated solar energy. They used three pieces of 100 by 200 cm highly reflective sheets to make parabolic reflector. As untreated water flows through the pipe, it undergoes heating from the concentrated solar radiation, reaching pasteurization temperatures. Using this device, Bigoni et al. [20] were able to collect 66 L of treated water in a sunny day. They mainly studied the pathogens removal from the water.

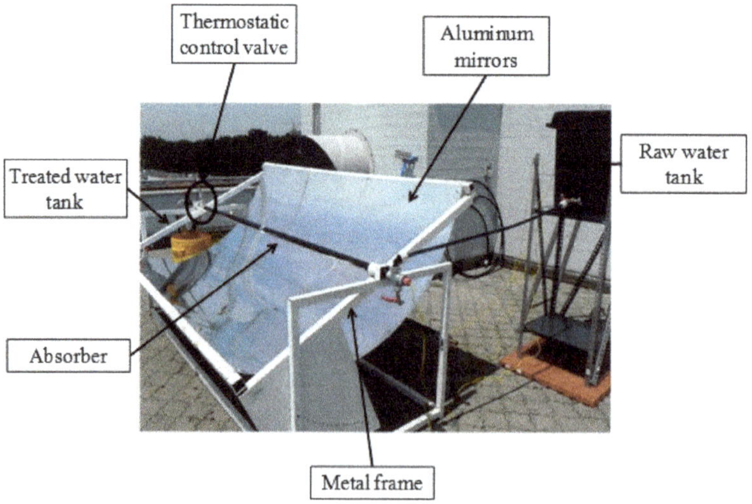

Figure 9. The water pasteurization system using a parabolic solar concentrator (Reprinted with permission from Ref. [20]. 2014, Elsevier).

Monteagudo et al. [21] implemented a Fresnel lens to concentrate sunlight for photocatalytic wastewater treatment processing. Monteagudo et al. used a Fresnel lens to focus concentrated sunlight on the tank containing wastewater. The schematic representation of the experimental setup is shown in Figure 10. They studied the photocatalytic degradation of orange II dye using titanium dioxide as photocatalysts and observed higher degradation rate than normal sunlight irradiation. The solar concentrator generates higher solar irradiance power at the focal point where sample is kept and hence the photocatalysts get more photons to generate much more electron-hole pairs. Thus, the photocatalysis reaction rate increases and wastewater becomes clean quickly.

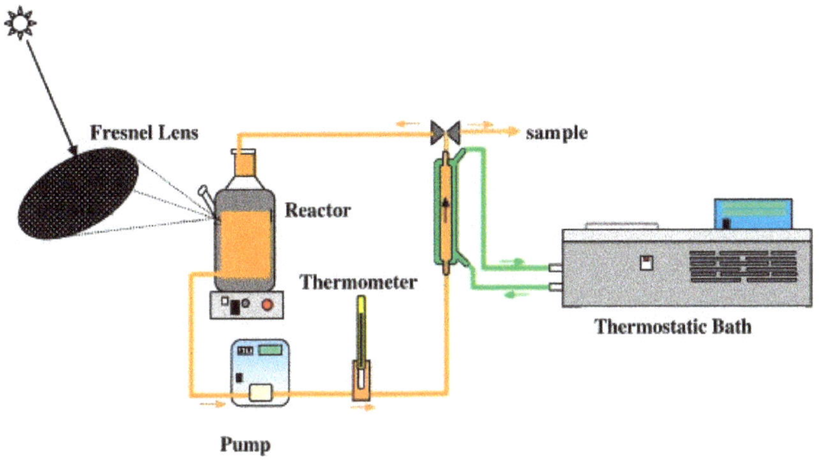

Figure 10. Experimental setup for photocatalytic dye degradation under concentrated sunlight irradiation using Fresnel lens (Reprinted with permission from Ref. [21]. 2006, Elsevier).

Similar results were observed by Golli et al. [56], who used a parabolic dish solar collector to concentrate sunlight for the photocatalytic wastewater treatment process. The parabolic dish collects and focuses sunlight onto the reactor containing the wastewater. A schematic representation of the experimental setup is shown in Figure 11. They treated 2 L of contaminated water for the photocatalytic degradation of 10 ppm of methylene blue dye using zinc oxide as the photocatalyst and achieved a degradation rate of 94%.

The solar concentrator based on parabolic trough collector (PTC) and compound parabolic collector (CPC) are mostly used for wastewater treatment because of the high solar power concentration ratio. They can be used to degrade various pollutants in water. Some significance reports are summarized in Table 1.

Table 1. Photocatalytic pollutants removal using a solar concentrator.

Solar Collector	Photocatalyst	Pollutant	Reference
CPC	TiO_2	Bisphenol A	Rodriguez et al. [57]
CPC	TiO_2, ZnO	Fenamiphos pesticide	Fenoll et al. [58]
PTC	TiO_2	Urea, ammonium chloride, peptone	Barwal et al. [59]
CPC	Ferrous sulfate heptahydrate	Agricultural pathogenic fungi	Aguas et al. [60]
CPC	ferrous sulfate in slurry	*E. coli*	Rodríguez-Chueca et al. [61]
CPC	TiO_2	copper, iron, zinc	Onotri et al. [62]

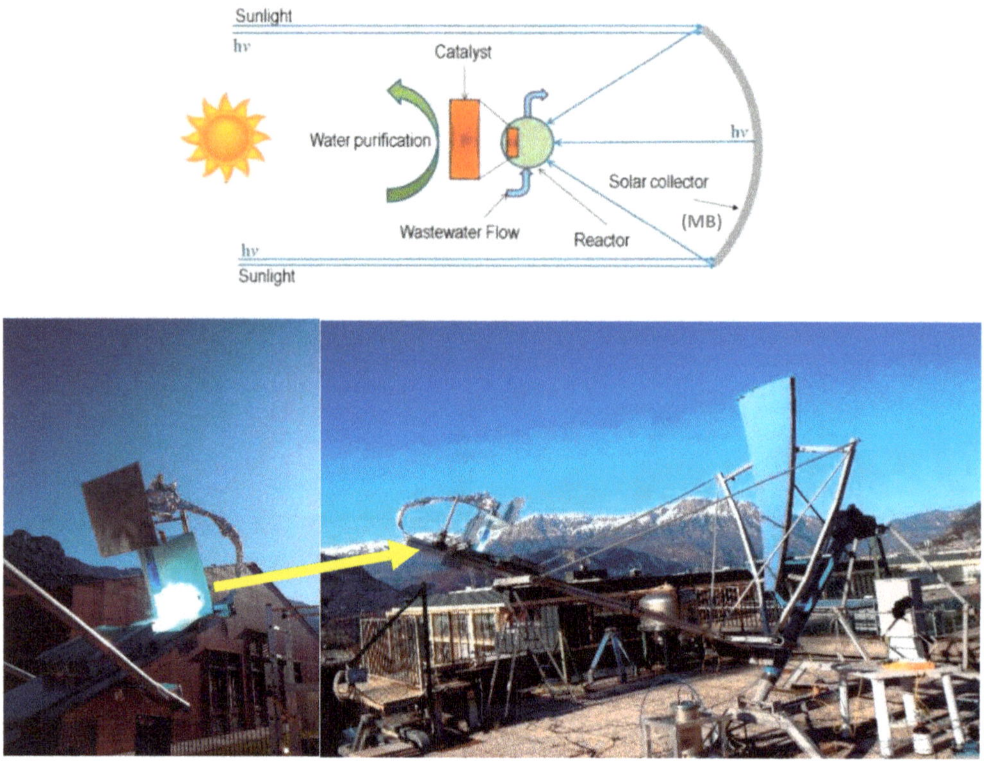

Figure 11. Parabolic dish solar concentrator-based water treatment setup (Reprinted with permission from Ref. [56]. 2021, Elsevier).

The solar concentrator, paired with an optical fiber bundle, efficiently delivers concentrated sunlight precisely where needed. Optical fiber bundles offer flexible installation options and long transmission distances, thanks to minimal light loss due to internal reflections within the fibers. With the aid of a solar tracking system, these devices ensure continuous delivery of highly intense sunlight throughout the day. While many efficient devices have been developed over the past decade to harness sunlight within buildings, none have been specifically tailored for water treatment applications. Addressing this gap, our research group recently introduced a groundbreaking solution: the ECoSEnDS (extremely concentrated solar energy delivery system). In our study, we explored its potential in photocatalytic wastewater treatment. The picture of the ECoSEnDS is presented in Figure 4. The ECoSEnDS comprises two mirrors: a primary paraboloid mirror and a secondary flat mirror, both crafted from glass with aluminum-coated reflecting surfaces. The primary mirror directs sunlight onto the secondary mirror, which then focuses and concentrates the light onto the optical fiber bundle. This setup allows the device to amplify sunlight by a factor of 96 compared to direct sunlight, achieving a concentration ratio of 96 suns with a 50% reflection efficiency of the primary mirror. In our experiments, this concentrated sunlight was utilized for photocatalytic dye degradation processes using various photocatalysts. Figure 12 illustrates how the concentrated light was delivered into the photocatalytic reactor, demonstrating the practical application of the ECoSEnDS in wastewater treatment.

Figure 12. Photocatalytic reactor with concentrated sunlight delivered using ECoSEnDS.

The photocatalytic wastewater treatment activity of the ECoSEnDS was studied by employing $BiVO_4$ nanoparticles to degrade methylene blue (MB) dye. Results demonstrated degradation rates of 20%, 71%, and 100% within 60 min under solar irradiance levels of 70, 300, and 500 mW/cm^2, respectively [8]. Notably, the dye degradation outpaced natural sunlight by fivefold. Figure 13 illustrates the photocatalytic activity.

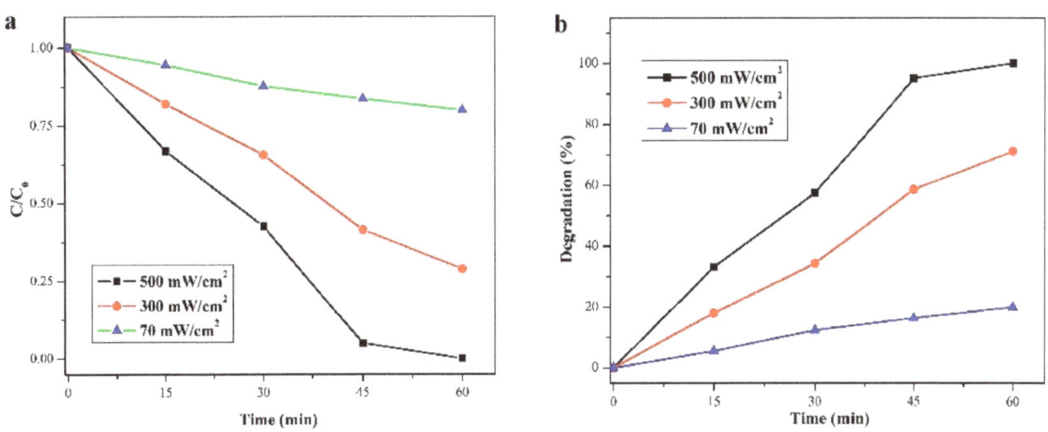

Figure 13. Photocatalytic MB dye degradation of $BiVO_4$ nanoparticles under concentrated sunlight irradiation: (**a**) relative degradation over time and (**b**) % degradation achieved (Reprinted with permission from Ref. [8]. 2021, Elsevier).

Our team further investigated the photocatalytic degradation of MB dye utilizing $BiVO_4$ nanoflakes, employing an advanced solar concentrator coupled with an optical fiber bundle. Interestingly, our findings mirrored those previously mentioned. As depicted in Figure 14, we observed degradation efficiencies of 88.86% and 32.30% after 60 min of sunlight irradiation at intensities of 300 and 74 mW/cm^2, respectively [5]. This underscores the enhanced photocatalytic activity of $BiVO_4$ nanoflakes under increased sunlight irradiation intensity.

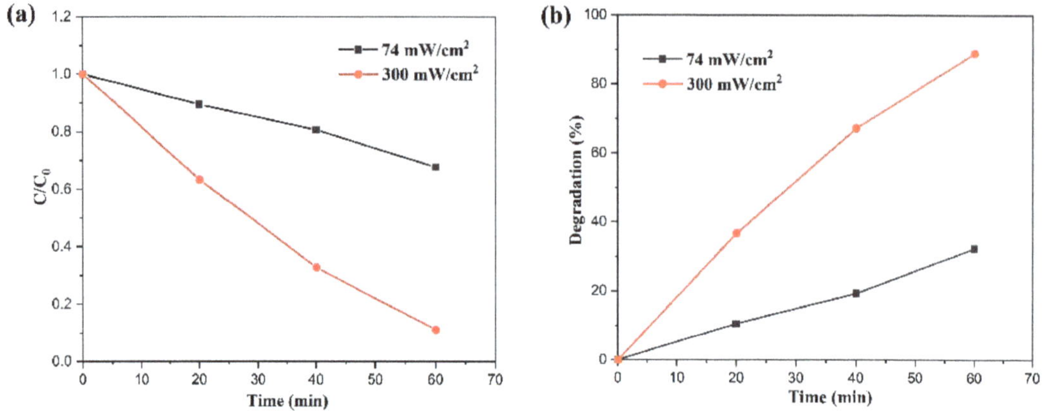

Figure 14. Photocatalytic MB dye degradation of $BiVO_4$ nanoflakes under concentrated sunlight irradiation: (**a**) relative degradation over time and (**b**) % degradation achieved (Reprinted with permission from Ref. [5]. 2021, Elsevier).

5. Advantages and Disadvantages of Using a Solar Concentrator

Using a solar concentrator in photocatalytic wastewater treatment offers several advantages. Solar concentrators focus sunlight onto a smaller area, significantly increasing light intensity, thereby enhancing photocatalytic activity. By concentrating sunlight, the photocatalyst can utilize solar energy more effectively, leading to higher reaction rates and more efficient pollutant degradation. Higher light intensity from concentrated sunlight accelerates the generation of electron-hole pairs, speeding up pollutant degradation. This conclusion can be clearly understood from Figures 13 and 14. Solar energy is free and abundant, reducing the need for external electrical power, which is typically costly and associated with greenhouse gas emissions. By leveraging natural sunlight, operational costs associated with artificial light sources and electricity are minimized. Using solar energy eliminates the greenhouse gas emissions linked to electricity generation for wastewater treatment. Solar photocatalysis is an eco-friendly and sustainable method, promoting green chemistry principles. Solar concentrators can be designed to suit different geographic locations and sunlight conditions, making the technology versatile and adaptable. The use of solar concentrators also encourages the development of innovative reactor designs that can further optimize the photocatalytic process.

Although using a solar concentrator offers many benefits for the photocatalytic wastewater treatment process, there are several disadvantages to consider. The installation of solar concentrators involves significant upfront costs for materials, installation, and maintenance, which can be a barrier, especially for large-scale applications. The costs associated with installing and maintaining water treatment reactors are summarized in Table 2. Additionally, solar energy faces the drawback of being available only during the daytime. Solar concentrators rely on clear, sunny weather to function effectively; cloudy or rainy conditions can significantly reduce their efficiency, leading to inconsistent performance. While solar concentrators enhance light intensity, the overall efficiency of the photocatalytic process can still be limited by the inherent properties of the photocatalyst material, such as its light absorption capacity and charge carrier recombination rates.

Table 2. Operating and capital costs for selected pilot-scale reactors.

Reactor Type	Reactor Volume	Operating Cost	Capital Cost	References
PC	100 L	55.3 €/m^3	32.7 k€	[63]
CPC	50 L	0.05 €/m^3	23.3 k€	[64]
CPC	10 L	10.4 €/L	18.5 k€	[65]

6. Conclusions and Outlook

Photocatalysis is the most promising technique for addressing pressing environmental concerns, particularly in water purification. Leveraging solar energy to eradicate pollutants from wastewater holds profound significance for both energy conservation and environmental remediation. Thus, the photocatalytic wastewater treatment using sunlight irradiation is the most sustainable challenging method to purify wastewater. The solar concentrator has the potential to increase photocatalytic degradation rates of pollutants within wastewater, offering a sustainable and economically viable solution. Furthermore, the synergy of solar concentrators with optical fibers presents a distinct advantage, enabling precise delivery of light to the photocatalytic reactor at the desired location due to the flexibility of the optical fiber bundle. This integration of solar concentrators with optical fibers heralds new technological vistas for harnessing solar energy in wastewater purification processes. Thus, this review article serves as a valuable reference point for comprehending, exploring, and innovating new devices based on solar concentrators, advancing the cause of sustainable wastewater purification.

Author Contributions: Conceptualization: J.S.R.; writing—original draft preparation: J.S.R.; writing—review and editing: Y.M. All authors have read and agreed to the published version of the manuscript.

Funding: This research received no external funding.

Conflicts of Interest: The authors declare no conflicts of interest.

References

1. Roy, J.S.; Bhattacharya, G.; Chauhan, D.; Deshmukh, S.; Upadhyay, R.; Priyadarshini, R.; Sinha Roy, S. Potential use of smartly engineered red mud nanoparticles for removal of arsenate and pathogens from drinking water. *SN Appl. Sci.* **2020**, *2*, 796. [CrossRef]
2. Dutta, D.; Arya, S.; Kumar, S. Industrial wastewater treatment: Current trends, bottlenecks, and best practices. *Chemosphere* **2021**, *285*, 131245. [CrossRef] [PubMed]
3. Gitis, V.; Hankins, N. Water treatment chemicals: Trends and challenges. *J. Water Process. Eng.* **2018**, *25*, 34–38. [CrossRef]
4. Chegrouche, S.; Mellah, A.; Barkat, M. Removal of strontium from aqueous solutions by adsorption onto activated carbon: Kinetic and thermodynamic studies. *Desalination* **2009**, *235*, 306–318. [CrossRef]
5. Roy, J.S.; Morency, S.; Messaddeq, Y. Ultrafast cleaning of methylene blue contaminated water accelarating photocatalytic reaction rate of the $BiVO_4$ nanoflakes under highly intense sunlight irradiation. *J. Photochem. Photobiol.* **2021**, *7*, 100037. [CrossRef]
6. Chong, M.N.; Jin, B.; Chow, C.W.K.; Saint, C. Recent developments in photocatalytic water treatment technology: A review. *Water Res.* **2010**, *44*, 2997. [CrossRef] [PubMed]
7. Roy, J.S.; Dugas, G.; Morency, S.; Messaddeq, Y. Rapid degradation of Rhodamine B using enhanced photocatalytic activity of MoS_2 nanoflowers under concentrated sunlight irradiation. *Phys. E Low-Dimens. Syst. Nanostructures* **2020**, *120*, 114114. [CrossRef]
8. Roy, J.S.; Morency, S.; Dugas, G.; Messaddeq, Y. Development of an extremely concentrated solar energy delivery system using silica optical fiber bundle for deployment of solar energy: Daylighting to photocatalytic wastewater treatment. *Sol. Energy* **2021**, *214*, 93–100. [CrossRef]
9. Pandey, A.K.; Reji Kumar, R.; Kalidasan, B.; Laghari, I.A.; Samykano, M.; Kothari, R.; Abusorrah, A.M.; Sharma, K.; Tyagi, V.V. Utilization of solar energy for wastewater treatment: Challenges and progressive research trends. *J. Environ. Manag.* **2021**, *297*, 113300. [CrossRef]
10. Muscetta, M.; Ganguly, P.; Clarizia, L. Solar-powered photocatalysis in water purification: Applications and commercialization challenges. *J. Environ. Chem. Eng.* **2024**, *12*, 113073. [CrossRef]
11. Ray, S.K.; Cho, J.; Hur, J. A critical review on strategies for improving efficiency of $BaTiO_3$-based photocatalysts for wastewater treatment. *J. Environ. Manag.* **2021**, *290*, 112679. [CrossRef] [PubMed]

12. Karthikeyan, C.; Arunachalam, P.; Ramachandran, K.; Al-Mayouf, A.M.; Karuppuchamy, S. Recent advances in semiconductor metal oxides with enhanced methods for solar photocatalytic applications. *J. Alloys Compd.* **2020**, *828*, 154281. [CrossRef]
13. Wang, Z.; Lin, Z.; Shen, S.; Zhong, W.; Cao, S. Advances in designing heterojunction photocatalytic materials. *Chin. J. Catal.* **2021**, *42*, 710–730. [CrossRef]
14. Jabbar, Z.H.; Graimed, B.H. Recent developments in industrial organic degradation via semiconductor heterojunctions and the parameters affecting the photocatalytic process: A review study. *J. Water Proc. Eng.* **2022**, *47*, 102671. [CrossRef]
15. Lovegrove, K.; Burgess, G.; Pye, J. A new 500 m^2 paraboloid dish solar concentrator. *Sol. Energy* **2011**, *85*, 620–626. [CrossRef]
16. Baharoon, D.A.; Rahman, H.A.; Omar, W.Z.W.; Fadhl, S.O. Historical development of concentrating solar power technologies to generate clean electricity efficiently—A review. *Renew. Sustain. Energy Rev.* **2015**, *41*, 996–1027. [CrossRef]
17. Yang, Z.; Li, L.; Wang, J.T.; Wang, W.; Song, J. Realization of high flux daylighting via optical fibers using large Fresnel lens. *Sol. Energy* **2019**, *183*, 204–211. [CrossRef]
18. Chafie, M.; Fadhel, M.; Aissa, B.; Guizani, A. Energetic end exergetic performance of a parabolic trough collector receiver: An experimental study. *J. Clean. Prod.* **2018**, *171*, 285–296. [CrossRef]
19. Tanveer, M.; Guyer, G.T. Solar assisted photo degradation of wastewater by compound parabolic collectors: Review of design and operational parameters. *Renew. Sustain. Energy Rev.* **2013**, *24*, 534–543. [CrossRef]
20. Bigoni, R.; Kötzsch, S.; Sorlini, S.; Egli, T. Solar water disinfection by a Parabolic Trough Concentrator (PTC): Flow-cytometric analysis of bacterial inactivation. *J. Clean. Prod.* **2014**, *67*, 62–71. [CrossRef]
21. Monteagudo, J.M.; Duran, A. Fresnel lens to concentrate solar energy for the photocatalytic decoloration and mineralization of orange II in aqueous solution. *Chemosphere* **2006**, *65*, 1242–1248. [CrossRef] [PubMed]
22. Roy, J.S.; Dugas, G.; Morency, S.; Ribeiro, S.J.L.; Messaddeq, Y. Enhanced photocatalytic activity of silver vanadate nanobelts in concentrated sunlight delivered through optical fiber bundle coupled with solar concentrator. *SN Appl. Sci.* **2020**, *2*, 185. [CrossRef]
23. Jia, P.; Tan, H.; Liu, K.; Gao, W. Synthesis, characterization and photocatalytic property of novel ZnO/bone char composite. *Mater. Res. Bull.* **2018**, *102*, 45–50. [CrossRef]
24. Yang, C.; Dong, W.; Cui, G.; Zhao, Y.; Shi, X.; Xia, X.; Tang, B.; Wang, W. Highly efficient photocatalytic degradation of methylene blue by P2ABSA-modified TiO_2 nanocomposite due to the photosensitization synergetic effect of TiO_2 and P2ABSA. *RSC Adv.* **2017**, *7*, 23699–23708. [CrossRef]
25. Mishra, B.P.; Parida, K. Orienting Z scheme charge transfer in graphitic carbon nitride-based systems for photocatalytic energy and environmental applications. *J. Mater. Chem.* **2021**, *9*, 10039–10080. [CrossRef]
26. Tian, D.; Zhou, H.; Zhang, H.; Zhou, P.; You, J.; Yao, G.; Pan, Z.; Liu, Y.; Lai, B. Heterogeneous photocatalyst-driven persulfate activation process under visible light irradiation: From basic catalyst design principles to novel enhancement strategies. *Chem. Eng. J.* **2022**, *428*, 131166. [CrossRef]
27. Lan, Y.; Lu, Y.; Ren, Z. Mini review on photocatalysis of titanium dioxide nanoparticles and their solar applications. *Nano Energy* **2013**, *2*, 1031–1045. [CrossRef]
28. Dong, H.; Zeng, G.; Tang, L.; Fan, C.; Zhang, C.; He, X.; He, Y. An overview on limitations of TiO_2-based particles for photocatalytic degradation of organic pollutants and the corresponding countermeasures. *Water Res.* **2015**, *79*, 128–146. [CrossRef]
29. Jingyu, H.; Ran, Y.; Zhaohui, L.; Yuanqiang, S.; Lingbo, Q.; Nti Kani, A. In-situ growth zinc oxide globular on the graphitic carbon nitride to fabrication binary heterojunctions and their photocatalytic degradation performance for the tetracycline. *Solid State Sci.* **2019**, *92*, 60–67. [CrossRef]
30. Roy, J.S.; Pal Majumder, T.; Dabrowski, R. Photoluminescence behavior of TiO_2 nanoparticles doped with liquid crystals. *J. Mol. Struc.* **2015**, *1098*, 351–354. [CrossRef]
31. Baruah, S.; Afre, R.A.; Pugliese, D. Effect of Size and Morphology of Different ZnO Nanostructures on the Performance of Dye-Sensitized Solar Cells. *Energies* **2024**, *17*, 2076. [CrossRef]
32. Luo, Y.; Xu, Y.; Liu, X.; Xue, H.; Qian, Q.; Chen, Q. Design of Cu–Ce co-doped TiO_2 for improved photocatalysis. *J. Mater. Sci.* **2017**, *52*, 1265–1271. [CrossRef]
33. Zhang, S.; Xu, Y.; Zhang, W.; Cao, P. Synthesis, characterization, and photocatalytic performance of Cu/Y co-doped TiO_2 nanoparticles. *Mater. Chem. Phys.* **2022**, *277*, 125558. [CrossRef]
34. Samadi, M.; Zirak, M.; Naseri, A.; Khorashadizade, E.; Moshfegh, A.Z. Recent progress on doped ZnO nanostructures for visible-light photocatalysis. *Thin Solid Film.* **2016**, *605*, 2–19. [CrossRef]
35. Pirhashemi, M.; Habibi-Yangjeh, A.; Pouran, S.R. Review on the criteria anticipated for the fabrication of highly efficient ZnO-based visible-light-driven photocatalysts. *J. Ind. Eng. Chem.* **2018**, *62*, 1–25. [CrossRef]
36. Malathi, A.; Madhavan, J.; Ashokkumar, M.; Arunachalam, P. A review on $BiVO_4$ photocatalyst: Activity enhancement methods for solar photocatalytic applications. *Appl. Catal. A Gen.* **2018**, *555*, 47–74.
37. Zhang, Z.; Wang, W.; Gao, E. Polypyrrole/Bi_2WO_6 composite with high charge separation efficiency and enhanced photocatalytic activity. *J. Mater. Sci.* **2014**, *49*, 7325–7332. [CrossRef]
38. Vu, T.A.; Dao, C.D.; Hoang, T.T.T.; Dang, P.T.; Tran, H.T.K.; Nguyen, K.T.; Le, G.H.; Nguyen, T.V.; Lee, G.D. Synthesis of novel silver vanadates with high photocatalytic and antibacterial activities. *Mater. Lett.* **2014**, *123*, 176–180. [CrossRef]

39. Choudhary, S.; Bisht, A.; Mohapatra, S. Facile synthesis, morphological, structural, photocatalytic and optical properties of $CoFe_2O_4$ nanostructures. *SN Appl. Sci.* **2019**, *1*, 1613. [CrossRef]
40. Sharma, G.; Kumar, A.; Sharma, S.; Naushad, M.; Ahamad, T.; Al-Saeedi, S.I.; Al-Senani, G.M.; Al-kadhi, N.S.; Stadler, F.J. Facile fabrication of $Zr_2Ni_1Cu_7$ trimetallic nano-alloy and its composite with Si_3N_4 for visible light assisted photodegradation of methylene blue. *J. Mol. Liq.* **2018**, *272*, 170–179. [CrossRef]
41. Wu, S.Z.; Li, K.; Zhang, W.D. On the heterostructured photocatalysts $Ag_3VO_4/g\text{-}C_3N_4$ with enhanced visible light photocatalytic activity. *App. Surf. Sci.* **2015**, *324*, 324–331. [CrossRef]
42. Wei, Z.; Mogan, T.R.; Wang, K.; Janczarek, M.; Kowalska, E. Morphology-Governed Performance of Multi-Dimensional Photocatalysts for Hydrogen Generation. *Energies* **2021**, *14*, 7223. [CrossRef]
43. Low, J.; Yu, J.; Jaroniec, M.; Wageh, S.; Al-Ghamdi, A.A. Heterojunction Photocatalysts. *Adv. Mater.* **2017**, *29*, 1601694. [CrossRef] [PubMed]
44. Singh Bhadwal, A.; Tripathi, R.M.; Gupta, R.K.; Kumar, N.; Singh, R.P.; Shrivastav, A. Biogenic synthesis and photocatalytic activity of CdS nanoparticles. *RSC Adv.* **2014**, *4*, 9484–9490. [CrossRef]
45. Hardman, R. A toxicologic review of quantum dots: Toxicity depends on physicochemical and environmental factors. *Environ. Health Perspect.* **2006**, *114*, 165–172. [CrossRef]
46. Hossain, S.F.; Mukherjee, S.K. Toxicity of cadmium sulfide (CdS) nanoparticles against *Escherichia coli* and HeLa cells. *J. Hazard. Mater.* **2013**, *260*, 1073–1082. [CrossRef]
47. Llobet, J.M.; Domingo, J.L. Acute toxicity of vanadium compounds in rats and mice. *Toxicol. Lett.* **1984**, *23*, 227–231. [CrossRef]
48. Wörle-Knirsch, J.M.; Kern, K.; Schleh, C.; Adelhelm, C.; Feldmann, C.; Krug, H.F. Nanoparticulate Vanadium Oxide Potentiated Vanadium Toxicity in Human Lung Cells. *Environ. Sci. Technol.* **2007**, *41*, 331–336. [CrossRef]
49. Kamble, G.S.; Ling, Y.-C. Solvothermal synthesis of facet-dependent $BiVO_4$ photocatalyst with enhanced visible-light-driven photocatalytic degradation of organic pollutant: Assessment of toxicity by zebrafish embryo. *Sci. Rep.* **2020**, *10*, 12993. [CrossRef]
50. Ghaware, R.C.; Birajdar, N.B.; Kamble, G.S.; Kolekar, S.S. Degradation of organic Pollutant by Using of $BiVO_4$–$NiFe_2O_4$ Heterostructure Photocatalyst under Visible Light Irradiation: Assessment of Detoxicity Study Using *Cirrhinus mrigala*. *Langmuir* **2024**, *40*, 14426–14439. [CrossRef] [PubMed]
51. Tojo, F.; Ishizaki, M.; Kubota, S.; Kurihara, M.; Hirose, F.; Ahmmad, B. Histidine Decorated Nanoparticles of CdS for Highly Efficient H_2 Production via Water Splitting. *Energies* **2020**, *13*, 3738. [CrossRef]
52. Wang, H.; Zhang, L.; Chen, Z.; Hu, J.; Li, S.; Wang, Z.; Liu, J.; Wang, X. Semiconductor heterojunction photocatalysts: Design, construction, and photocatalytic performances. *Chem. Soc. Rev.* **2014**, *43*, 5234–5244. [CrossRef] [PubMed]
53. Sahoo, S.; Rebekah, A.; Lee, K.S. Two-stage hydrothermal process driven visible light sensitive photocatalytic m-$ZnWO_4$/m-WO_3 heterojunction composite materials. *APL Mater.* **2024**, *12*, 041122. [CrossRef]
54. Wang, W.; Liu, Z.; Nie, H.; Kong, B. The direct Z-scheme character and roles of S vacancy in $BiOCl/Bi_2S_3$-(001) heterostructures for superior photocatalytic activity: A hybrid density functional investigation. *Phys. Chem. Chem. Phys.* **2024**, *26*, 10723–10736. [CrossRef] [PubMed]
55. Guernanou, R.; Sadek, I.; Belgassim, B.; Lamine, A.; Aïcha, S.; Feriel, K.; Fatima, B. Solar wastewater treatment: Advantages and efficiency for reuse in agriculture and industry. In Proceedings of the International Renewable and Sustainable Energy Conference (IRSEC), Agadir, Morocco, 27–30 November 2019. [CrossRef]
56. Golli, A.E.; Fendrich, M.; Bazzanella, N.; Dridi, C.; Miotello, A.; Orlandi, M. Wastewater remediation with ZnO photocatalysts: Green synthesis and solar concentration as an economically and environmentally viable route to application. *J. Environ. Manag.* **2021**, *286*, 112226. [CrossRef] [PubMed]
57. Rodríguez, E.M.; Fernández, G.; Klamerth, N.; Maldonado, M.I.; Álvarez, P.M.; Malato, S. Efficiency of different solar advanced oxidation processes on the oxidation of bisphenol A in water. *Appl. Catal. B Environ.* **2010**, *95*, 228–237. [CrossRef]
58. Fenoll, J.; Hellín, P.; Martínez, C.M.; Flores, P.; Navarro, S. Semiconductor oxides-sensitized photodegradation of fenamiphos in leaching water under natural sunlight. *Appl. Catal. B Environ.* **2012**, *115–116*, 31–37. [CrossRef]
59. Barwal, A.; Chaudhary, R. Feasibility study for the treatment of municipal wastewater by using a hybrid bio-solar process. *J. Environ. Manag.* **2016**, *177*, 271–277. [CrossRef]
60. Aguas, Y.; Hincapie, M.; Fernández-Ibáñez, P.; Polo-López, M.I. Solar photocatalytic disinfection of agricultural pathogenic fungi (Curvularia sp.) in real urban wastewater. *Sci. Total. Environ.* **2017**, *607–608*, 1213–1224. [CrossRef]
61. Rodríguez-Chueca, J.; Polo-López, M.I.; Mosteo, R.; Ormad, M.P.; Fernández-Ibáñez, P. Disinfection of real and simulated urban wastewater effluents using a mild solar photo-Fenton. *Appl. Catal. B Environ.* **2014**, *150–151*, 619–629. [CrossRef]
62. Onotri, L.; Race, M.; Clarizia, L.; Guida, M.; Alfè, M.; Andreozzi, R.; Marotta, R. Solar photocatalytic processes for treatment of soil washing wastewater. *Chem. Eng. J.* **2017**, *318*, 10–18. [CrossRef]
63. Vela, N.; Calín, M.; Yáñez-Gascón, M.J.; Garrido, I.; Pérez-Lucas, G.; Fenoll, J.; Navarro, S. Solar reclamation of wastewater effluent polluted with bisphenols, phthalates and parabens by photocatalytic treatment with $TiO_2/Na_2S_2O_8$ at pilot plant scale. *Chemosphere* **2018**, *212*, 95. [CrossRef] [PubMed]

64. Durán, A.; Monteagudo, J.M.; San Martín, I. Photocatalytic treatment of an industrial effluent using artificial and solar UV radiation: An operational cost study on a pilot plant scale. *J. Environ. Manag.* **2012**, *98*, 1–4. [CrossRef] [PubMed]
65. Gimenez, J.; Bayarri, B.; Gonzalez, O.; Malato, S.; Peral, J.; Esplugas, S. Advanced Oxidation Processes at Laboratory Scale: Environmental and Economic Impacts. *ACS Sustain. Chem. Eng.* **2015**, *3*, 3188–3196. [CrossRef]

energies

MDPI

Review

Methanogenesis—General Principles and Application in Wastewater Remediation

Ana-Katarina Marić, Martina Sudar, Zvjezdana Findrik Blažević * and Marija Vuković Domanovac *

Faculty of Chemical Engineering and Technology, University of Zagreb, Trg Marka Marulića 19, HR-10000 Zagreb, Croatia; amaric@fkit.unizg.hr (A.-K.M.); msudar@fkit.unizg.hr (M.S.)
* Correspondence: zfindrik@fkit.unizg.hr (Z.F.B.); mvukovic@fkit.unizg.hr (M.V.D.)

Abstract: The first discovery of methanogens led to the formation of a new domain of life known as Archaea. The Archaea domain exhibits properties vastly different from previously known Bacteria and Eucarya domains. However, for a certain multi-step process, a syntrophic relationship between organisms from all domains is needed. This process is called methanogenesis and is defined as the biological production of methane. Different methanogenic pathways prevail depending on substrate availability and the employed order of methanogenic Archaea. Most methanogens reduce carbon dioxide to methane with hydrogen through a hydrogenotrophic pathway. For hydrogen activation, a group of enzymes called hydrogenases is required. Regardless of the methanogenic pathway, electrons are carried between microorganisms by hydrogen. Naturally occurring processes, such as methanogenesis, can be engineered for industrial use. With the growth and emergence of new industries, the amount of produced industrial waste is an ever-growing environmental problem. For successful wastewater remediation, a syntrophic correlation between various microorganisms is needed. The composition of microorganisms depends on wastewater type, organic loading rates, anaerobic reactor design, pH, and temperature. The last step of anaerobic wastewater treatment is production of biomethane by methanogenesis, which is thought to be a cost-effective means of energy production for this renewable biogas.

Keywords: methanogenesis; methanogens; hydrogenase; wastewater remediation

1. Introduction

In nature, methane can be produced in anaerobic environments by methanogenic microorganisms. The starting point of methane production is the degradation of biomass [1,2]. Sources of renewable biomass vary from food industry waste to agricultural waste, animal manure, and forest industry waste [3]. Carbon dioxide acquired from biomass degradation is reduced with hydrogen in a process called methanogenesis. As methanogenesis is a multi-step process, a syntrophic association between two or more organisms is needed [4]. Methanogenesis plays a crucial role in the global carbon cycle. Moreover, processes for waste conversion to biomethane by methanogenesis are being implemented in industry, and more than 25% of all bioenergy is predicted to originate from waste-produced biogas [1,3].

2. Hydrogen for Methane Formation

In Earth's lower atmosphere, molecular hydrogen is present in trace concentrations. Nonetheless, H_2 plays a key role in the biogeochemical cycles of numerous elements, thus connecting different parts of ecosystems, with the most prominent function in anoxic environments. In anaerobic environments, hydrogen occurs in very low concentrations and exhibits a fast turnover rate (turnover times are in a matter of minutes) [5,6]. Despite low concentrations, annual formation and consumption of hydrogen from biomass add up to 0.3 Gt. Half of this quantity is used for methanogenesis from carbon dioxide. In environments where the reduction of carbon dioxide to methane is the predominant reaction, H_2 steady-state partial pressure is around 10 Pa. Because of the ability to diffuse through cytoplasmic membranes, H_2 is the best electron carrier between microorganisms [7, 8]. Consumption of hydrogen in methanogenic environments (where no inorganic electron acceptors other than carbon dioxide are available) is possible by methanogenic Archaea. More than 80 species from the Archaea domain are capable of conducting methanogenesis. Almost all of the hydrogen utilized by methanogenic microorganisms is of biological origin. However, a fraction of the hydrogen for sustaining methanogen growth is of geochemical origin [6,8,9].

3. Global Methane Cycle

Methane gas (CH_4) is one of the main greenhouse gases, with a global warming potential greater than that of carbon dioxide. Alongside CO_2 and N_2O, CH_4 plays a key role in degradation of the ozone layer. Human-related CH_4 emissions make up two-thirds of the total global methane emissions. The concentration of methane in the atmosphere has doubled over the past century [1,4].

Biological production of methane is mediated by methanogenic Archaea (methanogens). This process is called methanogenesis or biomethanation, and it plays a key role in the carbon cycle [1]. In total, 1 Gt of methane is produced annually from biomass in anoxic environments from acetate, carbon dioxide, and hydrogen through methanogen activity. Another 1 Gt is released into the environment by the melting of methane hydrates. From the 2 Gt of CH_4 produced annually, 1 Gt is oxidized by anaerobic Archaea and 0.6 Gt by aerobic bacteria, and 0.4 Gt diffuses to the atmosphere, with an additional 0.2 Gt released from other sources [4]. Biomass consists mostly of carbohydrates (~50%), proteins (around 30 to 40%), and lipids (~10%). Because of the complex biomass composition, different metabolic groups are required for degradation [2]. Methanogens function in a syntrophic manner, meaning that biomass degradation is catalyzed by a combination of multiple organisms that, on their own, cannot catabolize biomass. A combination of bacteria, protozoa, and fungi is needed to decompose biomass. The products are then transformed into acetic acid, carbon dioxide, and hydrogen by syntrophic bacteria [4]. In the last step of the process, methanogenic Archaea use hydrogenases for the activation of hydrogen, resulting in the formation of methane [8]. Methanogenesis is beneficial for treatment of organic waste in industry. Methane harnessed from renewable carbon feedstock by anaerobic digestion with methanogens is a useful tool for biogas production. Biogas is comprised of 30% to 90% CH_4 and can be used as a fuel to generate electricity or to power transportation [1,10]. It is estimated that 1 m^3 of CH_4 has a calorific value around 9.17 kWh [11], meaning that biogas containing the aforementioned ranges of CH_4 can have calorific values up to 9 kWh. When produced by anaerobic digestion, biomethane is a non-poisonous, colorless, and odorless gas. In comparison to other liquid biofuels, the separation of biomethane from the liquid phase is relatively easy and therefore has reduced production costs [3].

4. Methanogens

By studying prokaryotic relationships, Woese and Fox [12] discovered that microbes producing methane (methanogens) are Archaea. The discovery of this domain allowed better understanding of different properties methanogens exhibit in comparison to members of the Bacteria and Eukarya domains [13]. Nowadays, in the Archaea domain, there are seven methanogenic orders: *Methanobacteriales*, *Methanocellales*, *Methanococcales*, *Methanomicrobiales*, *Methanoplasmatales*, *Methanopyrales*, and, lastly, *Methanosarcinales* [14]. Methanogens populate anaerobic environments, especially low-sulfate habitats, and can be isolated from a broad range of thermochemical gradients. It is important to note that the presence of oxygen inhibits methanogen growth [1]. Methanogenic microbes can originate from acidophilic to alkaliphilic environments (pH 3.0 to 10.2) and psychrophilic to hyperthermophilic temperatures (−2 °C to 122 °C) [10]. In recent years, many methanogens have been discovered to be hyperthermophiles with optimum growth temperatures such as 98 °C (e.g., *Methanopyrus kandleri*) and 85 °C (e.g., *M. jannaschii*) [4]. Freshwater and marine sediments, marshes, geothermal systems, rice paddies, and human and animal gastrointestinal tracts are some of the typical methanogenic habitats [1,15]. In many environments, methanogens are consumers of hydrogen (hydrogenotrophs). They are the most numerous and the fastest-growing microbes [13]. Generation times can range from minutes (e.g., 25 min for *Methanocaldococcus jannaschii*) to a few hours (e.g., 6 h for *Methanosarcina barkeri*) [10]. The highest growth rates are observed in the *Methanococcales* order [16]. Methanogens grow by reducing coal, acetate, or one-carbon compounds, such as carbon monoxide, carbon dioxide, and methanol to methane. Generally, methanogen growth can be described by Equation (1). By producing CH_4, methanogenic microorganisms can also conserve energy [10].

$$e^{-}\text{ donor} + e^{-}\text{ acceptor} \rightarrow CH_4 + \text{byproduct} \tag{1}$$

Methanogenic Archaea required for biological CH_4 production can be with or without cytochromes. An overview of the differences between these methanogens is presented in Table 1 [4].

Table 1. Differences between methanogens.

Methanogens with Cytochromes	Methanogens Without Cytochromes	Ref.
contain methanophenazine	do not contain methanophenazine	[4]
only some grow on CO_2 and H_2; most grow on methylamines and acetate; cannot grow on formate	grow on CO_2 and H_2; cannot grow on methylamines and acetate; most can grow on formate	
growth yields on CO_2 and $H_2 \leq 7$ g/mol CH_4	growth yields on CO_2 and $H_2 \leq 3$ g/mol CH_4	
H_2 partial pressure threshold is >10 Pa	H_2 partial pressure threshold is <10 Pa	
not hyperthermophilic	hyperthermophilic	

Hydrogenases from Methanogenic Origin and Their Function

Hydrogenases are widely distributed in microbes. These enzymes serve different physiological functions in various metabolic pathways, e.g., in remediation of toxic heavy metals, parasites, and pathogenic bacteria, as well as in methane formation [17]. Organisms from three domains of life (Archaea, Bacteria, and Eucarya) utilize hydrogenase activity to produce or consume hydrogen [16]. For example, hydrogen gas for methanogenesis is formed in anaerobic environments with the help of hydrogenases [2].

Hydrogenases were first discovered by Stephenson and Stickland [18,19] and are classified based on their active site structure as [NiFe]-hydrogenase, [FeFe]-hydrogenase, and [Fe]-hydrogenases. Besides the main classification based on active site cofactors, hydrogenases can be further categorized by location within the harboring cell (cytoplasmic or membrane-bound), ability to conserve energy, and their electron-carrying redox partners [16]. From hydrogenase classes, the most studied are [NiFe]-hydrogenases [17].

As Stephenson and Stickland [20] studied methane-forming microorganisms from river sediments, the correlation between hydrogen-forming bacteria and hydrogen-consuming methanogens became evident. Additionally, the nickel requirement for hydrogenase activity was first discovered in methanogens. In methanogenic Archaea, only [NiFe]-hydrogenases and [Fe]-hydrogenases were found. [FeFe]-hydrogenases, present in Eucarya and Bacteria domains, have not yet been discovered in Archaea. Genetic and biochemical studies have been conducted on a limited number of methanogen species, namely *Methanococcus maripaludis*, *Methanosarcina barkeri*, and *Methanosarcina mazei* [8].

Functional hydrogenases are not expressed in all methanogens [10]. However, in methanogens expressing hydrogenases, mostly [NiFe]-hydrogenases are found [8]. *M. maripaludis* is a typical representative of methanogens without cytochromes [4]. In methanogens without cytochromes, four types of hydrogenases produced are as follows: (a) membrane-bound, ferredoxin-dependent energy-converting hydrogenase (Ech), (b) cytoplasmic, F_{420}-dependent hydrogenase–F_{420}-reducing hydrogenase (Frh), (c) cytoplasmic, electron-bifurcating Mvh hydrogenase, and (d) cytoplasmic, [Fe]-hydrogenase [21].

On the other hand, all *Methanosarcinales* organisms, with representatives such as *M. barkeri* and *M. mazei*, belong to methanogens with cytochromes [4]. Methanogens with cytochromes encode three types of hydrogenases: (a) membrane-bound, ferredoxin-dependent energy-converting hydrogenase (Ech), (b) cytoplasmic, F_{420}-dependent hydrogenase–F_{420}-reducing hydrogenase (Frh), and (c) membrane-bound, methanophenazine-dependent hydrogenase (Vht) [21].

Even though methanogens with and without cytochromes have different mechanisms for energy conservation, both contain [NiFe] ferredoxin-dependent energy-converting hydrogenases and F_{420}-reducing hydrogenases. With the exception of the bimetallic [NiFe] active site and three FeS clusters found in all [NiFe]-hydrogenases, membrane-bound energy-converting hydrogenases (Ech) have few sequence similarities to other hydrogenases of the [NiFe] family [16,22]. Ech contains six subunits, EchA–F. The EchE subunit harbors the bimetallic active center for the formation and oxidation of H_2. The remaining subunits are in charge of transferring electrons between ferredoxin (Fd) and H_2/H^+ [16].

Furthermore, the coenzyme F_{420}-reducing hydrogenase (Frh) is a heterotrimeric enzyme with three subunits. The FrhA subunit harbors the [NiFe] active site responsible for hydrogen oxidation. Flavin adenine dinucleotide (FAD) containing subunit FrhB functions as an active site for oxidation and reduction of F_{420} coenzyme. Electrons travel from the FrhA subunit through the FrhG subunit to the active site in the FrhB subunit. This hydrogenase is a requirement in all methanogens possessing the ability of CO_2 reduction with H_2-derived electrons [16]. The methanophenazine-dependent hydrogenase was initially identified as viologen-reducing hydrogenase (two) and thus has the Vht abbreviation. The VhtA subunit contains a [NiFe] active center for hydrogen oxidation. Electrons from H_2 oxidation reduce methanophenazine (MP), which is later used for the reduction of the terminal electron acceptor CoM-CoB, a disulfide of coenzyme M (CoM) and coenzyme B (CoB), common in all methane-forming pathways. This type of hydrogenase is found only in methanogens from the *Methanosarcinales* order. One cytoplasmic [Fe]-hydrogenase, limited to methanogens without cytochromes, is a substitute for F_{420}-reducing [NiFe]-hydrogenase in nickel-limiting conditions [8,16].

5. Methanogenic Pathways—Dependence on Substrates and Mechanism

Production of methane is a multi-step process classified based on the substrate utilized in the production pathway on the hydrogenotrophic Equation (2) and acetoclastic Equation (3) pathways. The hydrogenotrophic pathway utilizes inorganic carbon dioxide as a substrate, while acetoclastic involves the use of acetic acid [1]. Another less common pathway is the methylotrophic production of methane, where the substrates are methanol or methylamines [1,14]. Equation (4) describes the methylotrophic production pathway [11].

$$4H_2 + CO_2 \rightarrow CH_4 + H_2O \quad (2)$$

$$CH_3COOH \rightarrow CH_4 + CO_2 \quad (3)$$

$$4CH_3OH \rightarrow 3CH_4 + CO_2 + H_2O \quad (4)$$

The determination of the substrate used for successful methanogenesis is mostly guided by pH and temperature [1]. The substrate abundance affects total methanogenesis yield but does not play a role in controlling the rate of reactions [9]. Depending on the methanogenic order, different methanogenesis pathways are predominant. With the exception of *Methanoplasmatales* (also referred to as *Methanomassiliioccales*), which are only methylotrophic methanogens, all orders support the hydrogenotrophic methanogenesis pathway [10]. This is because the hydrogenotrophic pathway is more energetically favorable than acetoclastic and methylotrophic production pathways [14]. Nevertheless, around two-thirds of biologically produced methane comes from acetoclastic methanogenesis [16]. Alongside hydrogenotrophic, *Methanosarcinales* are capable of acetoclastic and methylotrophic methanogenesis [10].

To understand existing and predicting new methanogenic pathways, thermodynamic studies are the key. As shown in Equation (1), methanogen metabolism can be simply described as electron flow from electron donor to electron acceptor. Thermodynamic studies focus on determining whether the reaction is feasible or not, i.e., whether an organism can conserve energy for growth or not. This estimation is done by calculating Gibbs free energy. The value of Gibbs free energy is used to predict the amount of ATP synthesized depending on the amount of substrate consumed and should be negative for a reaction to be favorable [10]. By comparing Gibbs free energy values, general conclusions about microbial growth and feasibility of reactions can be drawn. These studies have introduced a spectrum of new substrates available for methanogenesis beyond the three previously mentioned and commonly described in literature. A case study conducted by Cozannet et al. based on thermodynamic calculations elucidated new substrates for methanogenesis, bringing the number of known substrates from 10 (until 1979) up to 152 (in 2023). Alongside 152 proven substrates, 41 putative substrates were predicted on account of thermodynamic calculations [23]. To tie in the notion of Gibbs free energy with the metabolic productivity of methanogen, it is worth mentioning that methanogenic reactions with these (152 + 41) substrates have value ranges of >0 kJ mol^{-1} CH_4 (for 6.4% of substrates), 0 to −30 kJ mol^{-1} CH_4 (62.8% of substrates), −30 to −100 kJ mol^{-1} CH_4 (5.2% of substrates), −100 to −200 kJ mol^{-1} CH_4 (62.8% of substrates), and <−200 kJ mol^{-1} CH_4 (2.9% of substrates). A few examples of substrates promoting methane production, alongside the methanogenic order of a methanogen responsible for conversion of these substrates to methane, are shown in Table 2 [23]. *Methanosarcinales* have been an outstanding order of organisms for studies on methanogenic pathways [24]. Here, this statement is supported with Table 2, seeing that most of methanogens derive from the *Methanosarcinales* order. Methanogenic pathways utilize several cofactors and coenzymes shown in Table 3 [1].

Table 2. Methanogenesis substrates excluding previously mentioned carbon dioxide, acetic acid, and methanol.

Substrate	Methanogen Order	Ref.
pyruvate	*Methanococcales*, *Methanosarcinales*	[23]
carbon monoxide	*Methanosarcinales*	
primary and secondary alcohols (ethanol, 2-propanol and 2-butanol)	*Methanomicrobiales*	
methylated amine compounds (monomethylamine, dimethylamine, trimethylamine, choline and glycine betaine)	*Methanosarcinales*	
organosulfur compounds (methanethiol and dimethylsulfide)	*Methanosarcinales*	
metoxylated aromatic compounds	*Methanosarcinales*, *Methanomicrobiales*	

Hydrogenotrophic pathway is described as a stepwise reduction of CO_2 to CH_4 with electrons acquired from hydrogen oxidation by hydrogenases. Even though methanogenic pathways have been greatly described, the involvement of hydrogenases in electron flow and different aspects of energy conservation need further characterization [16]. In the case of methanogens with cytochromes, energy conservation and production of methane from CO_2 relies on 13 reactions. The energy conservation relies on, among other proteins, energy-converting [NiFe]-hydrogenase and methanophenazine-reducing [NiFe]-hydrogenase, in a manner briefly explained previously. The methanogenic pathway in methanogens without cytochromes generally proceeds in the same manner as methanogenesis with cytochromes, with most of the enzymes and coenzymes being the same [4]. The final step for methane production in all methanogenic pathways is catalyzed by methyl-coenzyme M reductase [14]. All anaerobic methanogenic pathways produce CoM-CoB disulfide in the final step of methane production. Methanogens with cytochromes utilize multiple mechanisms for electron transfer depending on the electron source (that being H_2, F_{420}, or Fd) and whether the microorganism is capable of utilizing hydrogen as a substrate. In contrast, methanogens without cytochromes utilize only one mechanism called flavin-based electron bifurcation [16]. Due to the dependence on hydrogenase enzymes and thiol cofactors (coenzyme M and coenzyme B), methanogenesis is a seemingly reversible reaction [10].

Table 3. Coenzymes and their role in methanogenesis.

Coenzymes	Functional Role	Ref.
methanofuran (MF) tetrahydromethanopterin (H_4MPT) tetrahydrosarcinapterin (H_4SPT) coenzyme M (HS-CoM)	carriers of carbon moiety for generation of methane	[1]
coenzyme B (HS-CoB) coenzyme F_{420} coenzyme F_{430} methanophenazine	transferring of electrons for carbon reduction	

6. Factors Effecting Methanogenesis

In view of the close interdependence between hydrogen-producing microorganisms and hydrogen-utilizing methanogens, a change in any environmental factor influencing one of the syntrophic microorganism results in a change in the overall extent and rate of methanogenesis [25].

6.1. Hydrogen Concentration

Firstly, methanogenesis is under the control of the environmental hydrogen concentrations. A steady-state H_2 concentration is reached when the production and utilization rates are equal. When the production rate is greater than the rate of utilization, H_2 concentration increases, resulting in higher utilization of H_2. A higher steady-state hydrogen concentration does not persist for long, as H_2 utilizers grow and increase their biomass. From this, a general conclusion that the concentration of hydrogen is controlled by the kinetic characteristics of hydrogen utilizers can be made. Maximum growth rate (μ_{max}), specific maximum hydrogen utilization rate (u_{max}), hydrogen concentration at half-maximum growth rate (K_s), and Michaelis constant (K_m) are specific for each methanogenic microorganism [6]. Hydrogen concentration, i.e., H_2 accumulation, is also one of the factors that can limit hydrogenase activity. H_2 concentration can reverse the metabolism of a microorganism from H_2-oxidising to H_2-reducing [26]. This characteristic explains the 'reversible' metabolism of methanogens [10].

6.2. Oxygen and Sulfate Presence

Methane production from biomass degradation is feasible only in anoxic environments with low sulfate concentrations [4]. Oxygen inhibition is apparent in methanogen growth, so much so that methanogens will not grow or produce CH_4 even if trace levels of O_2 are present. Most of the hydrogenases present in methanogens are rapidly inactivated by oxygen [8,25]. Only membrane-bound [NiFe]-hydrogenases are known to retain catalytic activity in the presence of oxygen. O_2 tolerance may be the key in allowing methanogenic microorganisms to persist in environments with fluctuating anoxic and oxic conditions [25,27].

Furthermore, in the presence of sulfate as an electron acceptor, methanogen activity is out-competed by sulfate reducers. Because of the lower K_s value, sulfate reducers grow faster as opposed to methanogens, resulting in faster utilization of hydrogen. After sulfate reducers outgrow the methanogenic population, a complete reaction inhibition is achieved. For this reason, in environments where sulfate concentrations are high, methanogens are replaced by sulfate reducers. Analogously, the addition of Fe (III) or nitrate inhibits methanogen activity, with iron and nitrate reducers becoming more predominant in the methane-forming pathway [6].

6.3. Environmental Salinity

Higher environmental salinity may have an inhibitory effect on methanogen activity [9]. In hypersaline conditions, characterized by high concentrations of magnesium chloride, sodium chloride, magnesium sulfate, and other salts, halophilic and extremely halophilic methanogens can be found. The prevailing methanogenesis pathway in these environments is methylotrophic [25]. The mechanism of the inhibitory effect salts have on hydrogenase activity is not fully understood [26]. Given that conditions in hypersaline environments are a result of higher temperatures, methanogenic microorganisms inhabiting these environments may face multiple extremes [25].

6.4. Temperature

A key factor in controlling methanogenic reactions, alongside nutrient availability, is the temperature. Methanogens exhibit the highest activity at temperatures between 35 and 45 °C. In contrast to mesophilic methanogens, psychrophiles and (hyper)thermophiles are not as important in the methane formation pathways. Only methanogens without cytochromes exhibit hyperthermophily [4,9]. Interestingly, *M. kandleri* is the only known methanogen with activity above 100 °C. Methanogenesis reaction rates increase more steeply with the increase in temperature than those of other biological processes, such as photosynthesis or heterotrophic respiration [16,25].

6.5. pH Values

Optimal pH values for methanogenic archaea are in a range from 6.8 to 7.5 [11]. Some methanogen activity was recorded in environments with pH values as low as 3. However, for optimal activity, higher values are required. Low temperature ranges frequently accompany acidic habitats. For acidophilic methanogens, pH values can be 4.3, but optimally are above 5. Like acidic environments, alkaline environments are characterized by some extremes for methanogen activity. Alkalinity is commonly associated with higher temperatures and high salt concentrations; therefore, some alkaliphilic methanogen representatives are also halophilic. Alkaliphilic methanogens have an optimum pH of 9.0 to 9.5 [25].

6.6. Pressure

Pressures of 10 to 20 MPa hinder cellular processes, such as nutrient uptake and mobility in mesophilic methanogens. By contrast, elevated pressures enhance solubility of gaseous substrates and hydrogen transport between microorganisms, allowing for greater substrate conversion [25]. Although it is generally considered a denaturant for mesophilic methanogens, high pressure helps with stabilization of proteins at high temperatures [26]. An example of piezophilic methanogen is *M. jannaschii*, which displays optimum activity at a pressure of 75 MPa [25]. Moreover, activity assays of hydrogenases isolated from this piezophilic methanogen at 90 °C show a 4-fold increase in the half-life when the pressure was raised to 50.7 MPa, compared to the half-life at 1 MPa and the same temperature [26].

7. Application of Methanogenesis in Wastewater Remediation

The amount of industrial wastewater and organic waste is an ever-growing problem [3]. Since the end of the 19th century, systems for anaerobic wastewater treatment have been in use. Recently, extensive research has been conducted to gather a deeper understanding of anaerobic processes employed in wastewater treatment. Methanogenic pathways are the primary pathways for most organic matter disposal [28]. Methanogens reduce wastewater oxygen demand and, as a result, produce renewable biogas in the form of biomethane [10]. However, this remediation method can be utilized only when a high level of readily biodegradable soluble organic matter is present. Anaerobic remediation techniques are usually used for high-loaded wastewaters with COD (chemical oxygen demand) greater than 1.5–2 g L^{-1}. Besides substantial results in the treatment of heavily polluted wastewaters with a COD between 3 and 4 g L^{-1}, anaerobic treatment has some success in the remediation of wastewaters with a COD lower than 1.5 g L^{-1} [28]. In wastewater, organic matter is decomposed by a variety of microorganisms. Primary producers decompose raw waste into organic fatty acids or other substrates, which are then consumed by methanogens to produce biomethane. A simplified representation of these reactions is shown in Figure 1. The composition of syntrophic microorganisms utilized in wastewater remediation is guided by wastewater type, pH, temperature, reactor design, and organic loading rates [3,28]. In anaerobic reactors, methanogens take up 5 to 6% of the total microbial population, while the rest is occupied by the bacterial population [28].

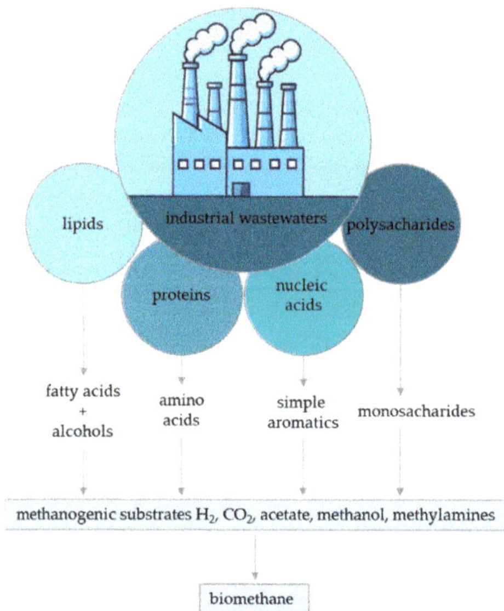

Figure 1. Simplified representation of reactions during wastewater treatment. Modified by authors from Ref. [28].

7.1. Classification of Anaerobic Reactors

Anaerobic wastewater treatments have, in recent years, evolved from lab-scale trials to successful application in various industries. Numerous anaerobic reactors are implemented in wastewater management. The classification of anaerobic reactors is based on the retention characteristics of components involved in biomethane production [3,28]. Type A is the simplest reactor design, with equal retention times of microorganisms (RT_m), solid (RT_s), and liquid (RT_l). This type of reactor allows uniform substrate and temperature distribution. However, it suffers from potential microorganism washout and incomplete homogenization at larger scales. The type B anaerobic reactor has higher RT_m and RT_s than RT_l. Because of this, higher process efficiencies can be achieved. One of the drawbacks is the loss of microorganisms at high loading rates. Lastly, the type C reactor has a microorganism retention time that is higher than the retention times of solid and liquid. The highest biogas performance and overall reaction efficiency are achieved by the type C reactor. Adaptability to a variety of loading rates is a key advantage of type C reactors. Like type B reactors, type C reactors do not need extensive mixing systems. In contrast to the large volume requirements of type A reactors, type C reactors function with smaller tank volumes. Wastewaters with high solid and grease contents are not suitable for treatment in type C reactors, but for diluted wastewaters this reactor is the best choice. A few examples of reactors representing each type are shown in Table 4 and Figure 2 [3].

Table 4. Reactor examples by type. Adapted from Ref. [3].

Reactor Type	Retention Times	Examples	Ref.
Type A	$RT_m = RT_s = RT_l$	closed digester tank, continuously stirred tank reactor	[3]
Type B	RT_m, $RT_s > RT_l$	upflow anaerobic sludge bed reactor, closed digester tank with solid recycle	
Type C	$RT_m > RT_s$, RT_l	membrane bioreactor, upflow anaerobic sludge fixed film reactor	

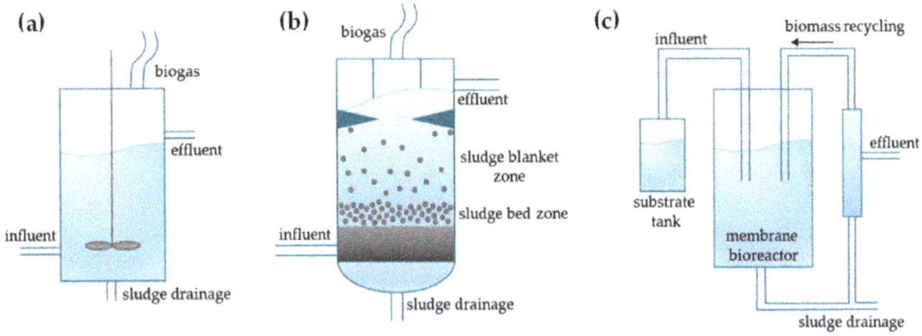

Figure 2. Schematic representations of reactors: (**a**) continuously stirred tank reactor; (**b**) upflow anaerobic sludge bed reactor; (**c**) membrane bioreactor. Modified by authors from Refs. [3,28].

7.2. Microbial Interactions Within Reactors

If the steps of anaerobic wastewater treatment, illustrated in Figure 1, are viewed in more depth, then a question about microbial interactions within the reactor arises. Conversion of organic waste is carried out in four consecutive steps: hydrolysis, acidogenesis, acetogenesis, and methanogenesis. Hydrolysis is a process controlled by the presence of hydrolytic bacteria, such as Bacteroidetes and Firmicutes. The ratio of Firmicutes to Bacteroidetes serves as indication of process performance, with higher values indicating a balanced function. Products from hydrolysis are converted by acidogenesis, followed by the step of acetogenesis. These steps rely on bacterial microorganisms. The last step of the anaerobic process highlights the importance of methanogens belonging to the Archaea domain [29]. One of the most important microbial interactions is granule formation, especially in type B reactors, such as the upflow anaerobic sludge bed reactor. Mechanisms of granule formation differ depending on reactor type. The outer layer of microbial granule is formed by methanogens, such as Methanosatcinales and Methanococcales, while internal cavities contain non-methanogenic microorganisms. Even though the mechanism of microbial granule formation is not fully elucidated, their role in stabilizing anaerobic systems and improving wastewater treatment efficiency has been observed. Acetate-forming bacteria, such as Acetobacterium and Clostridium, form a considerable syntrophic relationship with methanogens by providing acetate and H_2 for methane formation [30]. Depending on the substrates present, the co-existence of fungi is needed. For example, if waste contains lignocellulosic materials, a syntrophic relationship between anaerobic fungi and methanogens is needed for efficient methane production [31]. Additionally, to prevent accumulation of hydrogen above a certain threshold that would disrupt the system balance, removal of hydrogen is conducted by methanogens, sulfate-reducing bacteria, and homoacetogens. This hydrogen interspecies transfer is considered a key factor in the syntrophic relationship between microorganisms in anaerobic reactors. Moreover, interspecies formate transfer has been observed between acetogens and methanogens [30].

7.3. Wastewater Types and the Presence of Heavy Metals

Wastewater compositions vary depending on the industry branch and greatly affect microbiome diversity in the anaerobic reactor [28]. A few examples of wastewater types, their main characteristics, the dominant methanogens used for remediation, and the most used reactor types can be found in Table 5 [3,11,28,32]. With the emergence of industries such as mining, ceramic and glass, metal plating, an increase in heavy metal concentrations in wastewaters has been observed. Heavy metals are not biodegradable and tend to accumulate in high concentrations, causing toxicity and carcinogenicity in living organisms. Because of this toxicity, the removal of heavy metals from industrial wastewaters has gained major attention. Even though heavy metals have an inhibitory effect, the interactions between syntrophic microorganisms utilized in wastewater treatments may offer a protective effect. Because of this, heavy metal removal from wastewater is possible [33]. Methanogenic archaea have the ability to methylate almost all heavy metals from Groups IV, V, and VI in the periodic system of elements [13].

Table 5. Main characteristics, methanogens, and reactors used for wastewater remediation.

Industry Branch	Amount of Waste Produced	Characteristics of Wastewater	Methanogens	Reactor Type	Ref.
Beer industry	3–10 L of wastewater per 1 L of beer	high level of oxygen demand pH from 4.5 to 12 temperatures from 18–40 °C	*Methanosaeta concilii* *Methanosarcina mazei*	B	[3,32]
Oil extraction	0.5–0.75 t of palm oil mill effluent per 1 t of oil palm branch	high level of oxygen demand pH from 4 to 5 temperatures from 80 to 90 °C	*Methanosaeta concilii*	A, B, C	[3]
	1200–1800 L of olive mill wastewater per 1 t of olives	high level of oxygen demand pH around 5 high temperatures	*Methanosaeta concilii* *Methanobacterium formicicum*	B	[3]
Paper industry	at least 30,000 L of wastewater per 1 t of paper pulp	high level of oxygen demand fluctuating pH temperatures from 50 to 60 °C	*Methanosarcina barkeri*	B	[3]
Dairy industry	500–2000 L of wastewater per 1 L of milk	high level of oxygen demand pH from 5.7 to 7.8 temperatures from 30 to 40 °C	*Methanosaeta* spp.	B	[3,11,28]
Fruit and vegetable processing industry	n.a.	low level of oxygen demand low pH temperature range—n.a.	*Methanothrix* spp.	B	[28]
Slaughterhouse wastewater	90–140 L of wastewater per 1 slaughtered pig; although variable with type of animal	high level of oxygen demand fluctuating pH temperature range—n.a.	*Methanobacterium* *Methanosarcinales*	n.a.	[28]
Chemical industry	variable with the type of chemical produced	variable with the type of chemical produced	*Methanothrix* spp.	B, C	[28]

7.4. Effect of Methanogenesis on Wastewater Treatments

Methanogenesis implemented in waste management stops methane leakages to the atmosphere. Waste conversion to biomethane is thought to be a cost-effective means of energy production [3]. Anaerobic wastewater treatment is considerably cheaper than an aerobic process. The lower economic expenses stem from the fact that the wastewater does not need additional aeration [11]. Moreover, excess sludge formed by anaerobic treatment does not need further stabilization, unlike in aerobic treatments where this step is a necessity. It is estimated that up to 90% savings in operational costs can be achieved, as well as 40 to 60% savings in investment costs, by applying anaerobic wastewater treatments [28].

Methanogenic microorganisms are a key group in anaerobic wastewater treatments. If methanogen activity is inhibited, the process stops at acidogenesis, ultimately leading to incomplete organic waste degradation [33]. By controlling methanogenic activity during remedial action and maintaining lower levels of methanogenesis, some benefits were observed. One of these is the maintenance of anoxic conditions in treatment zones, and another is the regulation of co-metabolic activity between all microorganisms involved in wastewater treatment. On the contrary, excessive methanogenesis during wastewater remediation results in efficiency and cost issues, regulatory issues, and, most importantly, potential health and safety issues [13].

Anaerobic microorganisms, especially methanogens, have high sensitivity to environmental conditions and are inhibited by toxic compounds present in wastewater. Some microorganisms have long generation times, which result in long start-up times of the remediation process. These properties hinder the optimization of methanogenesis and its application [28,33].

To conclude, methane production is an indisputable indication that hydrogen generated in previous steps is used by methanogens instead of other microorganisms present in the reactor [13]. The assessment of the methane formation potential from industrial wastewaters relies on concentrations of organic substrates in any given type of wastewater. High CH_4 production potential is identified in numerous industrial wastewaters [28].

8. Methanogenesis for Renewable Energy Production

8.1. Methane for Hydrogen Storage

Specific activities of methanogenic Archaea are high enough for methanogens to be viewed as useful catalysts in industrial energy transformation. An attractive means of storing hydrogen, via the formation of methane from hydrogen and carbon dioxide catalyzed by methanogenic archaea (Equation (2)), is being considered. Methanogens catalyze this reduction process at hydrogen pressures well under 1 bar and at room temperatures. Methane is easily stored/transported, and almost all of the hydrogen combustion energy is conserved in methane. However, the feasibility of this hydrogen storage application is hindered by the difficulty of employing methanogenic hydrogenases in large-scale processes, the economic costs, and hydrogenase deactivation in oxic conditions [8].

8.2. Power-to-Gas

Organic waste in the form of biomass contains stored energy that can be converted into more useful forms, which could potentially replace fossil fuels. This energy is recovered by anaerobic digestion. Microbial populations in anaerobic reactors support high conversion efficiencies for methane production [34]. Hydrogenotrophic methanogens have gained attention for their ability to produce CH_4 from hydrogen and carbon dioxide, in a process called power-to-gas, by mediating the biomethanation step [35]. The foundation of the power-to-gas concept is the transformation of electricity into different products, with methane and hydrogen being the most common products [36]. If methane is the desired product, electric power is utilized for hydrogen production by water electrolysis. This conversion is followed by a biomethanation step, where H_2 and CO_2 are converted to methane by methanogenic Archaea. An example of the successful implementation of the power-to-gas process is a pilot plant in Denmark, where methane gas is produced with a purity grade of more than 98 vol%. Biomethane production from the power-to-gas process is in development phases and further research for commercial-size usage is needed [35].

8.3. Bioelectrochemical Methane Production

As methanogens are electroactive microorganisms, they can be utilized in bio-electrochemical methane production. Bio-electrochemical methanogenic systems rely on exploiting the capability of methanogens to interact with electrodes in order to utilize or produce electric power [35]. Electrochemistry can be implemented in the methanogenic process by integrating electrodes in reactors for wastewater treatments or treatments of other biodegradable wastes, such as solid wastes found in the agricultural sector. Furthermore, the methanogenic bio-electrochemical process can be a separate process integrated in biogas production plants. The conversion mediated by methanogenic Archaea takes place at the cathode. The second option of separating processes reduces inhibitory effects from anodic reactions, such as oxygen generation. Electromethanogenesis allows for increased methane production and biogas purity. When implemented into wastewater treatment, the CH_4 content of produced biogas reaches up to 98.1% [37].

8.4. Micro Biogas Plants

Micro biogas plants are an interesting methanogen application that has recently gained attention in developing countries. These plants convert household organic wastes to biogas, that can later be used as a direct source of energy needed for heating or electrical power. These systems have only a small positive impact on solving the energy problem. However, in some countries, micro biogas production is welcomed and supported by the government. This results in further optimization and reactor engineering of the micro plants in the hopes of producing higher electricity volumes [37].

8.5. Methanogens in the Agricultural Sector

The agricultural sector is the most abundant source of biodegradable organic waste for biogas production. Methanogenic treatments of agricultural wastes, such as production slurry and animal manure, are advantageous not only for biogas production. These treatments help in reducing pathogens and odors present in fertilizers, thus increasing the quality of the fertilizer. Biogas production by methanogenic Archaea does not have to be linked to waste treatments. The source of biodegradable material can also be maize or sugar beet silage, as well as grass and microalgae. Biogas produced by methanogens in the agricultural sector has a methane content ranging from 50 to 70%, with food waste being the most abundant biogas source [37].

9. Conclusions

Methanogenic habitats are anaerobic and range from acidophilic to alkaliphilic environments with psychrophilic to hyperthermophilic temperatures. Even though methanogens are a diverse group of microorganisms, they are capable of utilizing only inorganic carbon dioxide, acetic acid, methanol, or methylamines as a substrate for methane production. Recently, new substrates were shown to benefit methanogenic activity, with the number of available substrates increasing up to 193. As an electron source for reactions, methanogenic Archaea use hydrogen, which is activated by hydrogenase enzymes.

More anaerobic processes are being employed for wastewater remediation. These processes are more economically and environmentally friendly, due to reduction in operational costs and high conversion of waste materials to biogas. Anaerobic wastewater treatments consist of four steps, three of which mediated by bacterial populations. For the last step, that is, biogas formation, anaerobic wastewater treatments use methanogens in a process called methanogenesis. The dominant methanogen used in anaerobic treatment depends on the wastewater type. Methanogenesis, as a complex multi-step process, greatly depends on environmental factors such as pH, temperature, composition of wastewater, and the syntrophic interactions between Bacteria and Archaea populations present. Depending on the wastewater type, different reactor configurations are needed. Each type has benefits and disadvantages, although generally the biggest problem is potential microorganism washout, which would prevent biomass conversion to the final product, i.e., methane. Furthermore, methanogenic Archaea require anaerobic conditions. This is because hydrogenases found in methanogens are rapidly inactivated by oxygen, thus stopping the wastewater treatment process. Additionally, remediation by methanogenesis can only be used in the purification of wastewaters with high levels of soluble organic content. Despite these challenges, methanogenesis is of significant use in wastewater remediation. While most efforts of implementing methanogens in industrial processes are applied in wastewater remediation, methanogenesis can be implemented in other biogas production processes. Examples of such processes are power-to-gas or bio-electrochemical methane production.

Author Contributions: A.-K.M.: data curation, investigation, and writing—original draft; M.S.: validation, conceptualization, and supervision; Z.F.B.: validation, conceptualization, supervision, and funding acquisition; M.V.D.: validation, conceptualization, and supervision. All authors have read and agreed to the published version of the manuscript.

Funding: This research has received funding from the European Union's Horizon Europe programme under the grant agreement No 101135542 (CirculH2).

Data Availability Statement: No new data were created or analyzed in this study. Data sharing is not applicable to this article.

Conflicts of Interest: The authors declare no conflicts of interest.

References

1. Getabalew, M.; Alemneh, T.; Bzuneh, E. Review on Methanogenesis and its Role. *World J. Agric. Soil Sci.* **2020**, *6*, 1–7.
2. Schink, B.; Montag, D.; Keller, A.; Müller, N. Hydrogen or formate: Alternative key players in methanogenic degradation. *Environ. Microbiol. Rep.* **2017**, *9*, 189–202. [CrossRef] [PubMed]
3. Tabatabaei, M.; Rahim, R.A.; Abdullah, N.; Wright, A.D.G.; Shirai, Y.; Sakai, K.; Sulaiman, A.; Hassan, M.A. Importance of the methanogenic archaea populations in anaerobic wastewater treatments. *Process Biochem.* **2010**, *45*, 1214–1225. [CrossRef]
4. Thauer, R.K.; Kaster, A.K.; Seedorf, H.; Buckel, W.; Hedderich, R. Methanogenic archaea: Ecologically relevant differences in energy conservation. *Nat. Rev. Microbiol.* **2008**, *6*, 579–591. [CrossRef]
5. Schuchmann, K.; Chowdhury, N.P.; Müller, V. Complex Multimeric [FeFe] Hydrogenases: Biochemistry, Physiology and New Opportunities for the Hydrogen Economy. *Front. Microbiol.* **2018**, *9*, 2911. [CrossRef]
6. Conrad, R. Contribution of hydrogen to methane production and control of hydrogen concentrations in methanogenic soils and sediments. *FEMS Microbiol. Ecol.* **1999**, *28*, 193–202. [CrossRef]
7. Thauer, R.K. Hydrogenases and the global H_2 cycle. *Eur. J. Inorg. Chem.* **2011**, *2011*, 919–921. [CrossRef]
8. Thauer, R.K.; Kaster, A.K.; Goenrich, M.; Schick, M.; Hiromoto, T.; Shima, S. Hydrogenases from methanogenic archaea, nickel, a novel cofactor, and H_2 storage. *Annu. Rev. Biochem.* **2010**, *79*, 507–536. [CrossRef]
9. Katz, B.J. Microbial Processes and Natural Gas Accumulations. *Open Geol. J.* **2011**, *5*, 75–83. [CrossRef]
10. Buan, N.R. Methanogens: Pushing the boundaries of biology. *Emerg. Top. Life Sci.* **2018**, *2*, 629–646.
11. Zieliński, M.; Kazimierowicz, J.; Dębowski, M. Advantages and Limitations of Anaerobic Wastewater Treatment—Technological Basics, Development Directions, and Technological Innovations. *Energies* **2023**, *16*, 83. [CrossRef]
12. Woese, C.R.; Fox, G.E. Phylogenetic structure of the prokaryotic domain: The primary kingdoms. *Proc. Nat. Acad. Sci. USA* **1977**, *74*, 5088–5090. [CrossRef] [PubMed]
13. Mueller, J.; Booth, J.G. Managing Excessive Methanogenesis During ERD/ISCR Remedial. *Remediat. J.* **2016**, *26*, 53–71. [CrossRef]
14. Kietäväinen, R.; Purkamo, L. The origin, source, and cycling of methane in deep crystalline rock biosphere. *Front. Microbiol.* **2015**, *6*, 725. [CrossRef]
15. Liu, Y.; Whitman, W.B. Metabolic, Phylogenetic, and Ecological Diversity of the Methanogenic Archaea. *Ann. N. Y. Acad. Sci.* **2008**, *1125*, 171–189. [CrossRef]
16. Mand, T.D.; Metcalf, W.W. Energy Conservation and Hydrogenase Function in Methanogenic Archaea, in Particular the Genus Methanosarcina. *Microbiol. Mol. Biol. Rev.* **2019**, *83*, e00020-19. [CrossRef]
17. Xuan, J.; He, L.; Wen, W.; Feng, Y. Hydrogenase and Nitrogenase: Key Catalysts in Biohydrogen Production. *Molecules* **2023**, *28*, 1392. [CrossRef]
18. Stephenson, M.; Stickland, L.H. Hydrogenase: A bacterial enzyme activating molecular hydrogen: The properties of the enzyme. *Biochem. J.* **1931**, *25*, 205–214. [CrossRef]
19. Stephenson, M.; Stickland, L.H. Hydrogenase: The reduction of sulphate to sulphide by molecular hydrogen. *Biochem. J.* **1931**, *25*, 215–220. [CrossRef]
20. Stephenson, M.; Stickland, L.H. Hydrogenase: The bacterial formation of methane by the reduction of one-carbon compounds by molecular hydrogen. *Biochem. J.* **1933**, *27*, 1517–1527. [CrossRef]
21. Mand, T.D.; Kulkarni, G.; Metcalf, W.W. Genetic, Biochemical, and Molecular Characterization of *Methanosarcina barkeri* Mutants Lacking Three Distinct Classes of Hydrogenase. *J. Bacteriol.* **2018**, *200*, e00342-18. [CrossRef] [PubMed]
22. Qiu, S.; Li, Q.; Xu, Y.; Shen, S.; Sun, C. Learning from nature: Understanding hydrogenase enzyme using computational approach. *WIREs Comp. Mol. Sci.* **2019**, *10*, e1422. [CrossRef]

23. Cozannet, M.; Le Guellec, S.; Alain, K. A variety of substrates for methanogenesis. *Case Stud. Chem. Environ. Eng.* **2023**, *8*, 100533. [CrossRef]
24. Kulkarni, G.; Kridelbaugh, D.M.; Guss, A.M.; Metcalf, W.W. Hydrogen is a preferred intermediate in the energy-conserving electron transport chain of Methanosarcina barkeri. *Proc. Nat. Acad. Sci. USA* **2009**, *106*, 15915–15920. [CrossRef] [PubMed]
25. Hoehler, T.; Losey, N.A.; Gunsalus, R.P.; McInerney, M.J. Environmental Constraints that Limit Methanogenesis. In *Biogenesis of Hydrocarbons*; Stams, A., Sousa, D., Eds.; Handbook of Hydrocarbon and Lipid Microbiology; Springer: Cham, Switzerland, 2018; pp. 1–26.
26. Eberly, J.O.; Ely, E.L. Thermotolerant Hydrogenases: Biological Diversity, Properties, and Biotechnological Applications. *Crit. Rev. Microbiol.* **2008**, *34*, 117–130. [CrossRef]
27. Ogata, H.; Lubitz, W.; Higuchi, Q. Structure and function of [NiFe] hydrogenases. *J. Biochem.* **2016**, *106*, 251–258. [CrossRef]
28. Vítězová, M.; Kohoutová, A.; Vítěz, T.; Hanišáková, N.; Kushkevych, I. Methanogenic Microorganisms in Industrial Wastewater Anaerobic Treatment. *Processes* **2020**, *8*, 1546. [CrossRef]
29. Chen, S.; Cheng, H.; Wyckoff, K.N.; He, Q. Linkages of Firmicutes and Bacteroidetes popu-lations to methanogenic process performance. *J. Ind. Microbiol. Biotechnol.* **2016**, *43*, 771–781. [CrossRef]
30. Harirchi, S.; Wainaina, S.; Sar, T.; Nojoumi, S.A.; Parchami, M.; Parchami, M.; Varjani, S.; Khanal, S.K.; Wong, J.; Awasthi, M.K.; et al. Microbiological insights into an-aerobic digestion for biogas, hydrogen or volatile fatty acids (VFAs): A review. *Bioengineered* **2022**, *13*, 6521–6557. [CrossRef]
31. Li, Y.; Meng, Z.; Xu, Y.; Shi, Q.; Ma, Y.; Aung, M.; Cheng, Y.; Zhu, W. Interactions between Anaerobic Fungi and Methanogens in the Rumen and Their Biotechnological Potential in Biogas Production from Lignocellulosic Materials. *Microorganisms* **2021**, *9*, 190. [CrossRef]
32. Rao, A.G.; Reddy, T.S.K.; Parakash, S.S.; Vanajakshi, J.; Joseph, J.; Sarma, P.N. pH regulation of alkaline wastewater with carbon dioxide: A case study of treatment of brewery wastewater in UASB reactor coupled with absorber. *Bioresour. Technol.* **2007**, *98*, 2131–2136. [CrossRef] [PubMed]
33. Paulo, L.M.; Stams, A.J.M.; Sousa, D.Z. Methanogens, sulphate and heavy metals: A complex system. *Rev. Environ. Sci. Biotechnol.* **2015**, *14*, 537–553. [CrossRef]
34. Martin, M.R.; Fornero, J.J.; Stark, R.; Mets, L.; Angenent, L.T. A Single-Culture Bioprocess of *Methanothermobacter thermautotrophicus* to Upgrade Digester Biogas by CO_2-to-CH_4 Conversion with H_2. *Archaea* **2013**, *2013*, 157529. [CrossRef] [PubMed]
35. Contreras, G.; Thomsen, J.; Pfitzer, M.; Hafenbradl, D.; Kostner, D.; Holtmann, D.; Scmitz, R.A.; Rother, M.; Molitor, B. New perspectives for biotechnological applications of methanogens. *Curr. Res. Biotechnol.* **2022**, *4*, 468–474. [CrossRef]
36. Wulf, C.; Zapp, P.; Schreiber, A. Review of Power-to-X Demonstration Projects in Europe. *Front. Energy Res.* **2020**, *8*, 191. [CrossRef]
37. Enzmann, F.; Mayer, F.; Rother, M.; Holtmann, D. Methanogens: Biochemical background and biotechnological applications. *ABM Express* **2018**, *8*, 1. [CrossRef]

energies

MDPI

Article

Evaluating Energy Efficiency Parameters of Municipal Wastewater Treatment Plants in Terms of Management Strategies and Carbon Footprint Reduction: Insights from Three Polish Facilities

Iwona Kłosok-Bazan [1], Adam Rak [1], Joanna Boguniewicz-Zabłocka [1,*], Anna Kuczuk [1] and Andrea G. Capodaglio [2]

1 Department of Thermal Engineering and Industrial Facilities, Faculty of Mechanical Engineering, Opole University of Technology, 45-758 Opole, Poland; i.klosok-bazan@po.edu.pl (I.K.-B.); a.rak@po.edu.pl (A.R.); a.kuczuk@po.edu.pl (A.K.)

2 Department of Civil Engineering & Architecture, University of Pavia, 27100 Pavia, Italy; a.capodaglio@unipv.it

* Correspondence: j.boguniewicz-zablocka@po.edu.pl

Abstract: Water management in cities is a critical factor for sustainable growth and development. Satisfying the current needs with respect for the future is not possible without properly managed water and wastewater systems. An essential element of wastewater systems is the wastewater treatment plant (WWTP). The nexus between wastewater treatments and energy demand is a well-known problem. In times of energy crisis, effective energy management in this critical infrastructure is a key task. The purpose of this article is to analyze WWTPs' energy consumption with regard to proposed management strategies for managers, designers and decision makers. A detailed analysis of WWTP operational parameters and a proposal of improvement actions will be useful for applicability and benchmarking studies. Estimating the carbon footprint (CF) of selected WWTPs considering the indirect emissions due to energy consumption is an important step for developing energy neutrality of WWTPs. Due to the desire to deepen research in the area of a complex phenomenon, which is the energy management system in WWTPs, the research undertaken herein is based on the case study method of three water and sewage companies operating southwestern Poland. Each urban area has different specificities, natural conditions and needs. The presented results of the analyses may be the basis for developing directions for changes in national policy, other benchmarking studies, and improving the energy management system in WWTPs.

Keywords: municipal wastewater treatment plant; energy consumption; efficiency indicators

1. Introduction

The water supply is one of the most important services that ensures the efficient functioning of a community. Water is a non-substitutable foodstuff and guarantees minimum standards of hygiene; as such, it is an indispensable resource for meeting basic human needs. Water supply systems are an infrastructure designed to collect, treat and supply water for people and industry. These systems have their own specifics, because they are complicated technical structures in which individual objects play different roles. What is more, water supply systems belong to the so-called "critical infrastructure", which is so essential that their continued operation is required to ensure the security of a given nation, its economy and the public's health and/or safety [1]. Cities devote significant resources and efforts to the implementation of such essential services that are also major energy consumers, as they are responsible for nearly 35% of the energy consumption of all municipal public facilities [2,3], and are important sources of global greenhouse gas emissions; on average, CO_2eq emissions for water services are estimated to be 10.6 kg/m^3

of the water consumed [4]. Due to current economic, social, and regulatory pressures, it is essential to assess and improve the energy efficiency of existing and new service facilities. Concerning only wastewater treatment, the specific electric energy consumption of a wastewater treatment plant (WWTP) depends on many factors, including inflow quality, the WWTP's capacity, the technology used, and the climate. WWTP energy consumption generally ranges from 0.4 to 0.9 kWh/m^3 treated wastewater [2,5,6]; their energy efficiency has become increasingly critical, both due to energy costs and GHG emission reduction factors. For continued sustainability of the water sector, the most energy-intensive processes need to be identified, on a case-by-case basis, and either optimized or substituted with more efficient technologies [7]. Wastewater treatment plants (WWTPs) could potentially be transformed into self-sufficient and resource recovery facilities [8]. Several significant challenges have recently accumulated along the line for municipal water services. EU member states (along with other nations) have abruptly discovered their degree of dependence on fossil fuel imports from other nations [9]; therefore, energy efficiency, savings and recovery are also becoming primary goals in the water sector [10]. Geopolitical situations in Europe have considerably impacted the global economy, not just in the energy sector, but also in the availability of essential commodities. For example, as indicated by an EU Parliament resolution, recovery of ammonia, an essential constituent of fertilizer, from wastewater could partly substitute imported industrial sources while avoiding considerable GHG emissions [11–13]. The optimization of WWTP energy balance and of the related carbon footprint is a key contemporary task for facility operators that can contribute to increasing the reliability of operation of such critical infrastructures. This paper deals with energy management issues, and related GHG emissions, in WWTPs, based on a case study analysis of three facilities operating in southwestern Poland. The results of this analysis could indicate solutions for better energy management systems in these and other similar WWTPs and constitute the basis for developing guidelines for national policy improvement. The analysis could also lead to the introduction of measures to minimize individual WWTPs' impact on GHG emission reduction.

Energy is an important component of WWTPs' overall operational cost; it is generally the second-largest factor after personnel costs. WWTPs' energy requirements are influenced by a multitude of factors such as demand increases with flowrate and pollutant loads; however, in general, larger facilities (PE > 100,000) have lower specific (per m^3 wastewater treated) consumption rates, although there is no clear linear relationship between capacity and energy cost, as this depends on implemented operational optimization strategies [13].

Technology is also important. Traditional facilities designed as conventional biological activated sludge processes may require over 50–60% of the overall energy requirements, for the mere operation of compressors for air supply; this fraction further increases with the introduction of more performant membrane bioreactor (MBR) processes, which need additional aeration for media scouring, and increased pumping requirements. Nitrogen removal (nitrification/denitrification) implies higher oxygen consumption (e.g., for complete nitrification) [2]. A sizable (up to 15%) contribution to energy consumption can also be due to internal sludge handling (mixers and recirculation pumps). Effluent disinfection might also be quite energy-intensive; UV lamps may require up to 0.11 kWh/m^3 treated [14].

Energy efficiency is also strongly related to a facility's correct design, i.e., the match between planned and actual operational conditions; a plant working according to its design capacity works more efficiently (lower specific energy consumption) than one (with similar technology) with over-estimated design capacity [15]. Regulations also affect a facility's energy demand; increasingly stringent effluent quality limits generally imply increased energy input either by treatment extension or more energy-demanding processes (e.g., MBR, advanced oxidation processes—AOPs), although the latter may show higher performance in terms of specific energy per mass unit of removed pollutants [16].

Studies have shown that WWTP energy consumption could be reduced up to 80% by implementing systematic programs to monitor and manage its use [17], by optimizing process operation [18–20], or by adopting alternative process technologies, e.g., by substituting energy-intensive aerobic processes with anaerobic technologies which simultaneously eliminate energy requirements and allow its recovery in the form of methane-rich biogas [21]. Many studies emphasize that it is essential to assess and improve the energy efficiency (EE) of existing facilities due to social and administrative pressures in order to ensure the long-term sustainability of the sector [22].

As every WWTP is faced with the need to assess critical areas where the most energy is consumed, energy audits should be used to assess the energy intensity of unit processes [23], which will help to identify areas where energy savings are possible; energy management systems, i.e., tools to support energy efficiency improvement in industry, could then be applied. These consist of a set of procedures and practices that ensure systematic planning, analysis, control, monitoring and improvement of energy use [24]. Since 2011, ISO 50001 "Energy management systems—Requirements with guidance for use" has been the global standard for energy management systems and is a key aspect of energy efficiency policy in industry [25]. Its implementation is intended to help organizations and industries to improve energy consumption, protect the environment and save production costs, without restricting the continuation of their activities, while fulfilling their corporate social responsibility [26]. The methodology has been applied to water and wastewater systems, as well as to many industrial sectors [27,28].

The implementation of energy benchmarking at wastewater treatment plants in Australia, as a subset of ISO 50001, and the subsequent optimization of their energy efficiency has brought tangible benefits to public water companies [29]. Results of ISO 50001 implementation in industrial settings also confirm that this approach enables a reduction in electricity consumption in cement production by 25%, which translated into CO_2 emission reduction. In a comprehensive study, McKane et al. [30] state that under ISO 50001 management, cumulative energy savings in the industrial and service sectors of around 105 EJ could be achieved by 2030, with economic savings of nearly USD 700 billion and 6500 million Mt of avoided CO_2 emissions.

The use of renewable energy sources (RES) based on distributed power generation systems could help individual WWTPs or their units to become energy-self-sufficient; such sources are being developed all around the world, and wastewater has been indicated as one such resource [31]. The most commonly exploited source of self-produced energy at WWTPs is biogas, produced by fermentation of excess biological solids. When used in CHP generators, it is seen as a sustainable way to recover energy from wastewater treatment plants [32]. In a case study of a WWTP in Poland, the energy recovered from wastewater and sewage sludge covered about 83% of the plant's demand [33]. In the US, the City of Gresham's WWTP was reported as the first facility nationally to become net energy-positive through biogas generation and recovery, saving about USD 500,000/y in energy bills [34]. WWTP-produced biogas can also contribute to the biofuels for transport EU strategy [35]. Other solutions using alternative sources of WWTP-derived clean energy include heat pumps for thermal energy recovery from wastewater heat [36–38]. Wastewater-recovered thermal energy has also been used as energy exported to district heating systems [39,40]. In Germany, analysis of WWTPs' potential to provide ancillary services to power grids has suggested promising developments [41–43].

The carbon footprint of a WWTP is a measure of the total amount of greenhouse gases (GHGs) emitted as a result of its operations and production processes. It is calculated using data on energy and fuel consumption, emissions from production processes and other factors that affect GHG generation by the WWTP. Onsite direct CO_2 emissions from organic matter oxidation are usually considered climate neutral, due to wastewater organics' biogenic origin; N_2O and CH_4 are also direct GHG emissions from treatment processes. The former is generated during nitrification/denitrification processes, and has a much higher global warming potential (GWP) than CO_2 ($\approx$300 CO_2 equivalents). Several studies on N_2O WWTP emissions found that these accounted for up to 55.6% of the total [44]. Methane, produced by anaerobic digestion, is also a possible GHG emission from WWTPs, with CO_2 equivalent of 25. Approximately 75% of direct onsite CH_4 emissions are assumed to occur from the sludge treatment line, from fugitive losses from digester effluent, sludge storage, and CHP units. Facilities with poor sludge management and/or older infrastructure have higher emissions.

Indirect GHG emissions from externally acquired electricity or thermal power consumed by the WWTP for pumping and treatment operations dominate WWTPs' contributions, accounting for 65–75% of the total [45]. Other "process chain" emissions are caused by chemical use (including chemicals' "embedded" emissions, i.e., emission caused by treatment process chemicals' production and transport), transportation and off-site disposal of residuals and sludge. All these should be factored in according to the "GHG Protocol Corporate Accounting and Reporting Standard" (GHG-PCARS) [46].

The purpose of this article is to analyze WWTPs' energy consumption with regard to proposed management strategies for managers, designers and decision makers. A detailed analysis of WWTP operational parameters and proposal of improvement actions will be useful for applicability for a reduction in carbon footprint. Indicators such as kWh per different unit per year (kWh unit^{-1} year^{-1}) are useful for benchmarking. Comparing energy use against industry standards or similar facilities can highlight areas for improvement.

2. Materials and Methods

This study analyzes the electric energy consumption of three purposely selected municipal WWTPs within a wide range of capacity and loading rates, located in southwestern Poland, in the rural and peri-urban areas of the Opolskie Voivodeship. To understand the drivers of increasing energy use in wastewater treatment plants, this paper focuses on total energy use rather than energy demand and energy costs.

2.1. Description of the Study Areas

The study area and plant locations are represented in Figure 1. The selected plants' capacity ranges from 6000 to 130,000 P.E. and nominal flow capacity ranges between 600 and 18,000 m^3/d. As such, the selected facilities represent the current Polish situation of small rural and urban facilities, adopting different treatment technologies and energy recovery options. The characteristics of the selected WWTPs are summarized in Table 1.

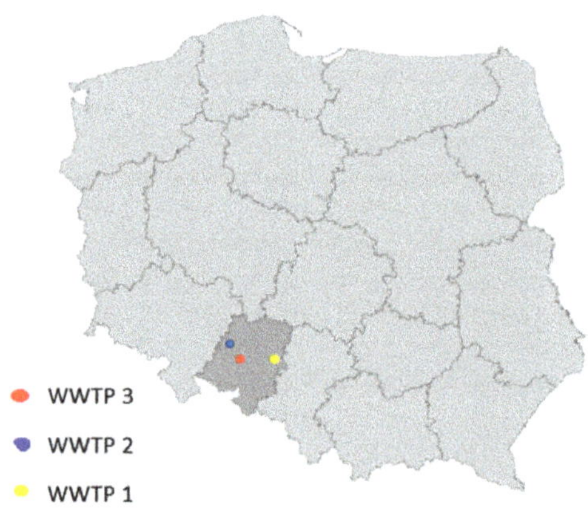

Figure 1. Plant locations.

This wide range of facility capacity (representing a common situation in Poland) was deliberately selected to allow specific analyses and identification of optimal solutions within each plant category, established in the strategic national document "National Urban Wastewater Treatment Program" [47]. WWTP1 is a wastewater treatment facility located in an urban area, utilizing a three-phase activated sludge system that includes nitrification and pre-denitrification processes. This setup allows for effective biological organic matter and nitrogen removal, treating wastewater efficiently in line with modern standards. The plant's sludge treatment line incorporates anaerobic digestion (AD). The AD process generates biogas, which is captured and used for energy recovery to power some of the plant's operations, enhancing its sustainability. WWTP2 uses Sequencing Batch Reactor (SBR) technology, which treats wastewater in batches rather than a continuous flow, which allows for greater control over the treatment stages. SBR systems operate in cycles mainly for domestic wastewater. WWTP3 handles both domestic and industrial wastewater, and is equipped with two Complete Mix Module (CMM) 300 reactors. WWTP3 begins with a pretreatment phase that includes sand removal and a coagulation chamber to remove large particles and non-biodegradable materials, helping to protect the downstream treatment processes from clogging or damage. These three WWTPs demonstrate a range of technologies used to treat urban wastewater, each tailored to meet the unique demands of its setting.

Table 1. Characteristics of analyzed WWTPs.

Facility ID	Design Flow [m^3/d]	Population Equivalent [P.E.]	BOD_5 Loading Rates [kg/d]	COD Loading Rates [kg/d]	Effluent Recipient
WTTP1	18,000	134,400	4029	8044	Oder River
WTTP2	3000	9000	429	1129	Gosciejowice Canal, Ścinawa River
WTTP3	600	6000	230	686	Mała Panew River

Detailed operational data of the facilities can be found in Tables S1–S3 in Supplementary Files.

2.2. Determination of Basic Energy Consumption Indicator (I)

On the basis of operational data provided by the facilities' operators, an assessment of their actual electricity consumption was carried out for three consecutive years (Y1, Y2, Y3). Electricity consumption was determined based on the readings of the main electric meters at the treatment plants (for indicator calculation total for the plant, in case of aeration only for aeration unit). A basic energy consumption indicator (I) is the ratio between the energy consumption and one relevant parameter in the plant. In the study, the basic energy consumption indicators (indices of unit electricity consumption) were determined in relation to the volume of treated wastewater (I_V), the plant size expressed in PE ($I_{P.E.}$) and the removed pollutant load (I_{BOD5}, I_{COD}, I_S), as presented in Table 2. The energy intensity of the treatment plant was analyzed on the basis of quarterly data.

Table 2. Specific energy consumption indicators.

Parameter	Unit	Calculation	Note
I_V	kWh/m^3	Energy consumption kWh: Treated wastewater m^3	ratio between the daily energy consumption and the daily volume treated (annual average)
I_{PE}	kWh/PE $^{-1}$ year $^{-1}$	Energy consumption kWh: Population Equivalent (PE)	ratio between the annual energy consumption and the PE served in the plant
I_{BOD}	kWh/kg BOD removed	Energy consumption: BOD removed	ratio between the annual energy consumption and the BOD removed, expressed in kg
I_{COD}	kWh/kg COD removed	Energy consumption: COD removed	ratio between the annual energy consumption and the BOD removed, expressed in kg
I_S	kWh/kg SS removed	Energy consumption: SS	ratio between the annual energy consumption and the BOD removed, expressed in kg

The calculation of the correlation between pollutant load and electricity consumption was also evaluated. The pollutant load, calculated as the difference between the concentrations of raw and treated wastewater multiplied by the volume of wastewater treated in the period analyzed, was correlated with the average electricity consumption (monthly energy consumption) for the period under review.

2.3. *Carbon Footprint Estimation*

A WWTP's carbon footprint is the sum of GHG emissions from energy use and the technological processes, primarily carbon dioxide (CO_2), emitted directly or indirectly. The studied WWTPs represent facilities with different process configurations, levels of energy neutrality and options for sludge disposal. In this study, CF analysis is based on the analysis of Scope 2 emissions from historical operating data. This means that indirect emissions resulting from the electricity purchased by the WWTP, which is essential to power the equipment and processes in the facility, are analyzed. This electricity is generated off-site and, in Poland, predominantly comes from fossil fuels.

The carbon footprint was then calculated, taking into account that, in Poland, electricity comes entirely from coal combustion; therefore, according to year 2021 conversion factors, 1 kWh of energy implies emissions of 657.1 g CO_2.

3. Results and Discussion

The electricity consumption of the considered WWTPs ranged from 0.6 to 1.5 kWh/m^3, with an average value of electricity consumption rate of 1.26 kWh/m^3; considering that the actual number of PE served was lower than the design assumptions, the effective electricity consumption varied from 22.10 to 112.3 kWh/PE-year. Considering that the optimal target for overall electricity consumption at WWTPs is indicated as 20 kWh/PE-year, with a guide value at 26 kWh/PE-year [2], the observed range ranges from within the norm to very high. Smaller WWTPs generally have higher electricity consumption per population equivalent PE due to reduced economies of scale and less efficient energy use relative to their capacity. Conversely, larger plants tend to achieve better energy efficiency, benefiting from more advanced technological solutions and optimized operational processes. However, even large plants can have higher energy consumption if operating conditions are not optimized, for example, due to inefficient aeration systems, obsolete equipment or increased energy requirements associated with certain sludge treatment technologies. Therefore, factors such as plant size, technological innovation and operational efficiency are critical in determining whether electricity consumption remains within recommended limits.

3.1. *Description of WWTP Performance*

The averages of daily operating parameter measurements for each individual quarter (Q1–Q4) in the considered period of three years are presented in Tables S1–S3. Throughout the first year (Y1) of analyzed data, WWTP 1 operated hydraulically underloaded, as shown by a comparison of the average volumes of wastewater incoming flow (10,400–13,800 m^3/d) to design flow (18,000 m^3/d); for most of the year, the facility operated at just 60% of the project's design. Specifically, for about 1/4 of the year, inflow did not exceed 11,000 m^3/d. An analysis of the hydraulic load highlights that the maximum flows occur in Q1 and Q4; this is due to incoming meltwater and rainwater in spring and winter. The parameters BOD_5 and COD have been chosen as representative indicators of biodegradable and non-biodegradable matter, and their reduction is closely linked to energy consumption. Concentrations of biodegradable compounds (BOD_5) varied in the range of 298–383 mg/L with an average of 356 mg/L. COD was in the range from 610 mg/L to 769 mg/L. Despite a very wide range of observed concentration values of the parameters, the average ratio of COD/BOD_5 was consistently below 2 (e.g., in Y3, 1.78). This qualifies the influent wastewater as easily or moderately easily decomposed. A summary of the effluent monitoring of the treated wastewater indicates that parameters do not exceed permitted values.

A similar situation was observed in the case of WWTP2, with the facility operating at an actual load of about 55% of the design for most of the year. BOD_5 in influent ranged from 258 mg/L to 782 mg/L, with an average of 530 mg/L. Similarly large fluctuations refer to COD, where values in raw wastewater ranged from 290 mg/L to 1472 mg/L, with an average of 324 mg/L. These large fluctuations of pollutants in raw sewage may be related to components other than domestic sewage; discharges from several industrial and service plants are directed into this locality's sewage system, with much higher pollutant loads than domestic users. On the other hand, dilution is affected by rainwater collected by the combined sewage system; therefore, the average effluent BOD_5 discharged into the river lies below 3 mg/L, as much as 80.0% lower than the applicable limit value (15.0 mg/L). The same ration applies to COD discharge, at 30 mg/L, much lower than the limit value (125.0 mg/L). Treated WWTP2 effluents indicate that monitored parameter values are always lower than those allowed in the water discharge permit.

A different situation was observed in WWTP3, which operated at close (90%) to the hydraulic design load, and in several periods, it was actually overloaded, i.e., incoming wastewater flow was greater than the design by around 10%. The observed BOD_5 in the influent is usually around 400 mg/L. Fluctuations observed in autumn and winter measurements are higher than in the case of the mean value due to loads from storm water and snowmelt. Mean effluent concentrations did not exceed the limit value.

3.2. Determination of Basic Energy Consumption Coefficients and Energy Efficiency

There are various indicators to quantify specific energy use by WWTPs. As indicated before, one bulk indicator is kWh/PE served. Decreasing values were observed for specific energy consumption with the increasing PE served, due to economies of scale and more frequent process automation in large plants. Per capita energy consumption is more suitable for general technology benchmarking assessments than specific performance evaluations. A common practice is to express energy use in volumetric terms (kWh/m^3 wastewater treated); however, this can also be misleading, since it considers only water pumped through a facility, and not pollutant mass removed. Therefore, plants subject to significant stormwater influx or parasite clean infiltration/inflow (e.g., from phreatic superficial aquifers) could erroneously appear more performant and energy-efficient. Figure 2 shows calculated volumetric indicators for the three considered facilities. For smaller volumes of wastewater treated, the energy efficiency indicator has the highest value due to the relatively fixed amount of energy associated with running the plant. Smaller plants often have to operate with the same basic infrastructure as larger plants, but with lower wastewater flows. This results in higher energy consumption per unit of wastewater treated, as the basic energy requirements of the plant remain constant regardless of the volume treated. In addition, smaller plants may not have the advanced technologies and optimization strategies available to larger plants, further contributing to lower overall energy efficiency. A more meaningful approach consists of comparing energy consumption in terms of mass of pollutants removed, for example, kWh/kg BOD, COD or TTS.

Calculated mass-based energy efficiency indicators for the considered facilities are summarized in Table 3. The table shows calculated energy efficiency indicators for the three plants, allowing a comparative analysis of their performance. Differences in energy efficiency can be attributed to various factors such as plant size, process configurations (BOD and COD removed) and the volume of wastewater treated. These results are consistent with values reported in the literature, which range from 1.1 to 6.09 kWh/kg BOD_5 [46].

The calculated energy efficiency indicators (I) provide insights into the energy consumption characteristics of the three wastewater treatment plants (WWTPs) across different years. The Iv indicator accounts for the general energy demands of the facility, including pumping, aeration, and other processes that vary with the volume of wastewater. WWTP1 and WWTP2 have lower Iv values, ranging from 0.61 to 0.69 kWh/m^3, suggesting efficient energy use per unit volume of wastewater. In contrast, WWTP3 shows significantly higher Iv values (1.07 to 1.15 kWh/m^3), which may indicate either a higher energy demand due to older equipment or higher treatment standards for the influent wastewater. Also, the climate condition could impact those indicators; as mentioned in other studies, the electrical consumption increased from 0.36 kw/m^3 to 0.51 kw/m^3 when the rainfall intensity increased from 0.8 mm/min to 2.9 mm/min [48].

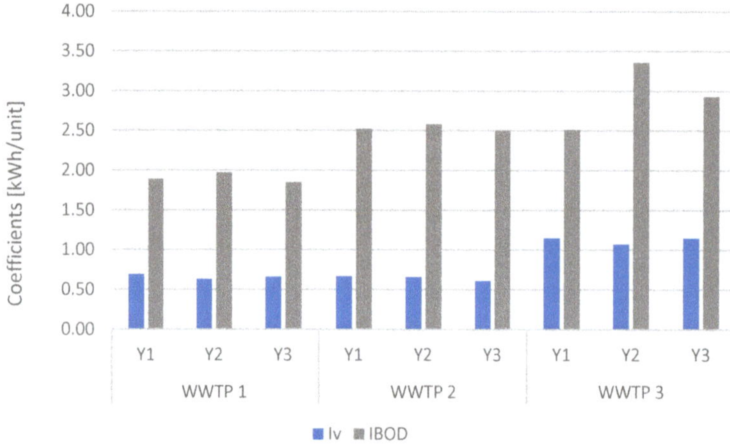

Figure 2. Selected energy performance indicators for the 3 considered facilities.

Table 3. Calculated energy efficiency indicators (I) for the three facilities.

I	WWTP1			WWTP2		WWTP3		
	Y1	**Y2**	**Y3**	**Y2**	**Y3**	**Y1**	**Y2**	**Y3**
I_V	0.69	0.63	0.66	0.66	0.61	1.15	1.07	1.15
I_{PE}	0.11	0.11	0.11	0.15	0.14	0.15	0.21	0.17
I_{BOD}	1.89	1.91	1.85	2.58	2.50	2.51	3.56	2.93
I_{COD}	0.95	0.97	0.95	1.11	0.89	0.93	1.13	0.88
I_S	4.63	4.60	4.78	3.14	3.15	1.55	2.38	1.78

The I_{PE} indicator allows for comparison between facilities of different sizes and helps assess energy use in relation to the plant's capacity. All three plants maintain relatively low I_{PE} values, with WWTP1 at a steady 0.11, WWTP2 between 0.14 and 0.15, and WWTP3 with a slight fluctuation between 0.15 and 0.21. The higher values in WWTP3 may indicate greater energy needs due to increased load or differences in treatment intensity. I_{BOD} focuses specifically on the energy-intensive biological treatment processes. WWTP1 has the lowest values (1.85–1.91 kWh/kg BOD), followed by WWTP2 (2.50–2.58 kWh/kg BOD), with WWTP3 showing the highest values (2.93–3.56 kWh/kg BOD). This indicates that WWTP3 has less efficient biological treatment, which requires more energy per unit of BOD removed. I_{COD} covers both biological and some chemical oxidation processes. WWTP1 and WWTP2 maintain values under 1.0 kWh/kg COD, indicating a balanced energy use for COD reduction. WWTP3 has slightly higher values, ranging from 0.88 to 1.13 kWh/kg COD, which again point to less efficient treatment and higher initial COD loads in the influent. I_S represents the energy associated with mechanical treatment processes. WWTP1 has the highest I_S values (4.60–4.78 kWh/kg SS), suggesting high energy needs for solid removal. WWTP2 follows, with values around 3.14–3.15 kWh/kg SS. WWTP3, on the other hand, has significantly lower values, ranging from 1.55 to 2.38 kWh/kg SS, which could be related to having no need for pumping a large volume of wastewater to a sedimentation tank for solid removal.

Energy consumption in WWTPs is largely determined by location, quality of influent wastewater, process configurations, treatment capacity, required effluent quality, etc. As treatment standards become more stringent, the energy required to remove pollutants increases. In practice, energy demand decreases as plant treatment capacity increases, with a threshold of approximately 37,850 m^3/d, beyond which further capacity increases have minimal impact on in-plant energy consumption [49]. The correlation between Biochemical Oxygen Demand (BOD) load and energy consumption is significant, especially in wastewater treatment processes. Higher BOD loads typically require more energy for treatment because more oxygen is needed to break down the organic matter. This increased energy consumption can be attributed to the need for more intensive aeration and other treatment processes to ensure that the effluent meets environmental standards [50]. Of the WWTPs analyzed, only WWTP 1 showed correlations between electricity consumption for the treatment process and the amount of load removed, expressed as BOD5. The results of the analyses are shown in Figure 3.

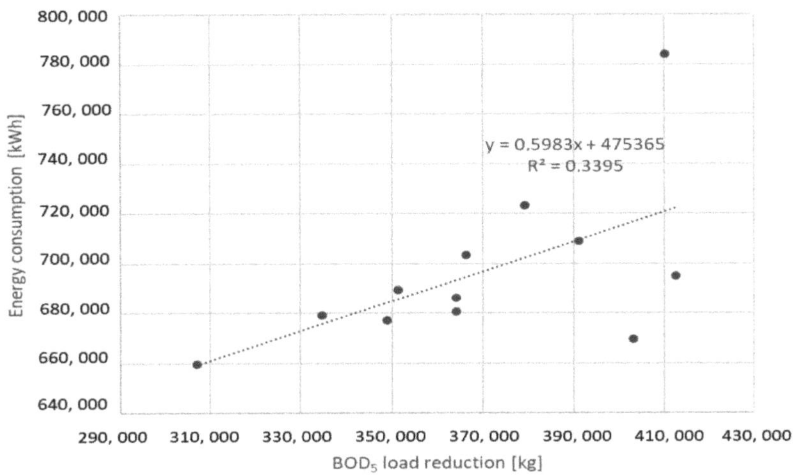

Figure 3. Correlation between energy consumption and BOD_5 load reduction in WWTP 1.

The trendline suggests a positive correlation between BOD_5 load reduction and energy consumption. The positive correlation suggests that as WWTP 1 handles higher BOD_5 loads, its energy requirements generally increase. This is consistent with expectations, as treating larger loads typically demands more energy. However, the relatively low R^2 value indicates that BOD_5 load reduction alone does not fully account for variations in energy consumption, implying that operational efficiency and equipment performance might also be significant.

For small wastewater treatment plants, the correlation between Biochemical Oxygen Demand (BOD) load and energy consumption can still be significant, but there are some unique factors to consider. Smaller plants often have higher specific energy consumption per unit of wastewater treated than larger plants. This is due to less efficient economies of scale and the need for more frequent maintenance and operational adjustments [51].

For example, smaller plants may experience greater variability in wastewater temperature, quality and dissolved oxygen levels, which can affect energy consumption. In addition, the type of treatment technology used can also influence energy consumption [52]. So, while the basic relationship between BOD load and energy consumption holds true, the specifics can be quite different for smaller plants. This can be seen in the other WWTPs analyzed in this study. Although the graphs shown (Figures 4 and 5) indicate a lack of correlation, they may even indicate a high degree of flexibility in the operation of the wastewater treatment process. In Figure 5, most data points are clustered around the middle, where BOD_5 load reduction is between 10,000 and 20,000 kg, and energy consumption ranges from about 55,000 to 65,000 kWh. This suggests that within this range, energy consumption and load reduction might be relatively stable or typical for this WWTP. With a very different composition of the wastewater entering WWTP 2 and WWTP 3, the effect of pollutant removal was satisfactory, so that the load removed was high. This is particularly true for the two points marked in red on the graph.

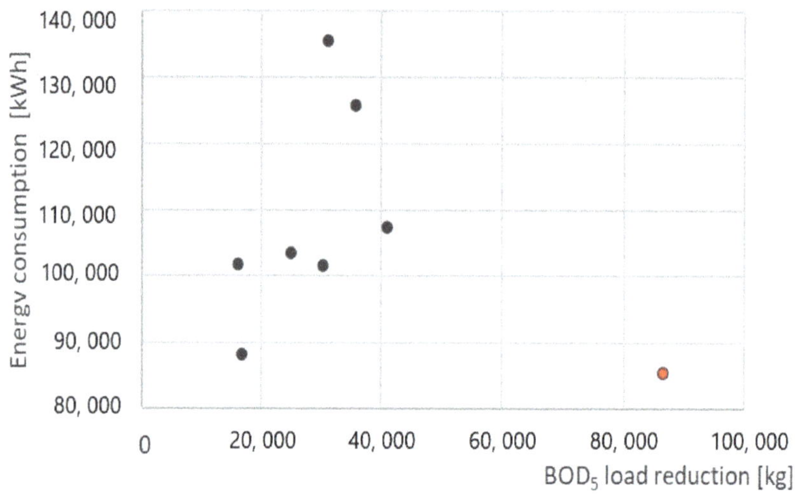

Figure 4. Correlation between energy consumption and BOD_5 load reduction in WWTP 2.

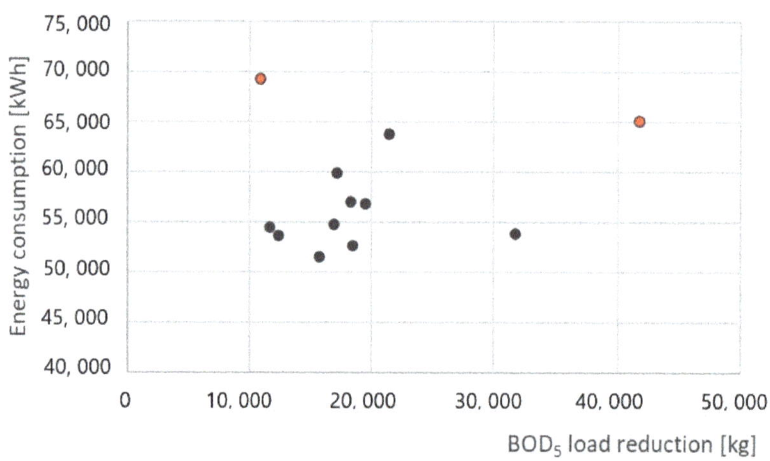

Figure 5. Correlation between energy consumption and BOD_5 load reduction in WWTP3.

3.3. WWTP Footprint Calculations

The carbon footprint was calculated based on energy consumption data and emission factors, as presented in Table 4. The results for Scope 2 indicate total greenhouse gas emissions ranging from 145 to 1909 tons per year of CO_2e, which reflects the emissions associated with purchased electricity during the analyzed period. Scope 2 emissions highlight the company's dependence on energy suppliers and its energy efficiency. Reducing emissions in this area can be achieved by improving energy efficiency or purchasing energy from renewable sources.

Table 4. Carbon footprint calculations.

	CO_2 Emission [kg]		
	Y1	Y2	Y3
WTTP1	1,811,062	1,770,667	1,909,318
WTTP2	-	248,996	228,932
WTTP3	145,688	145,774	163,502
	CO_2 Emission per m^3 Wastewater [kg]		
	Y1	Y2	Y3
WTTP1	0.4521	0.4111	0.4345
WTTP2	-	0.4311	0.3980
WTTP3	0.7556	0.7023	0.7561
	CO_2 Emission per P.E.		
	Y1	Y2	Y3
WTTP1	0.0739	0.0743	0.0721
WTTP2	-	0.0982	0.0931
WTTP3	0.0963	0.1370	0.1125

Specific carbon footprints (scope 2) for WWTP1 and WWTP2 are similar, in the range of 0.4–0.45 kg CO_2/m^3y. WWTP3 had the largest carbon footprint, due to relatively high volume of WW treated (working also overloaded). The two plants with lower carbon footprints had in common that they were both partially self-sufficient in terms of heat and electricity. In general, energy consumption has the greatest impact on the carbon footprint.

Total CO_2 emissions show a fluctuating trend, with emissions ranging from around 1.77 million kg in Y2 to almost 1.91 million kg in Y3 for WWTP1. WWTP3 consistently has the lowest total emissions among the plants. CO_2 emissions per cubic meter of treated wastewater provide an indication of emission efficiency. The emissions are moderate for WWTP1 and WWTP2, ranging from 0.03980 kg/m^3 to 0.4521 kg/m^3. WWTP3 shows the highest emissions per m^3 in all years, with values around 0.70–0.76 kg/m^3, suggesting room for improvement. CO_2 emissions per population equivalent (PE) provide a benchmark in relation to the population served by each plant. In the case of WWTP1, stable CO_2 emissions around 0.073–0.074 kg indicate a consistent performance over the years. WWTP2 shows values around 0.093–0.098 kg per PE, slightly higher than WWTP1. WWTP3 has the highest CO_2 emissions per PE, especially in Y2, with a value of 0.1370 kg, although it decreased to 0.1125 kg in Y3.

Recommendations to reduce the carbon footprint of each WWTP will be different. For plants with anaerobic digestion (such as WWTP1), increasing biogas production from sludge and using combined heat and power (CHP) systems can reduce reliance on grid electricity and thus reduce the carbon footprint. Integrating renewable energy (e.g., solar or wind) into the operation of WWTP2 and WWTP3 can directly reduce the CO_2 emissions associated with fossil fuel-based electricity. As aeration is highly energy-intensive and a significant contributor to CO_2 emissions, optimizing aeration processes can help reduce energy consumption and emissions at each WWTP. Older or poorly maintained equipment consumes more energy. Routine maintenance and replacing outdated components such as pumps and motors with energy-efficient models can result in significant energy savings [52]. Implementing these strategies can help reduce the carbon footprint of each plant by improving energy efficiency, increasing the use of renewable energy, and optimizing treatment processes.

4. Conclusions

Municipal wastewater treatment plants (WWTPs) require significant amounts of energy to operate, with each plant having unique energy consumption characteristics that are influenced by technology, operational practices, and operating conditions such as the type of wastewater being treated, flow modifications, and so on. Based on data from three WWTPs in southwestern Poland, the research highlights the different energy needs and opportunities for optimization in different urban areas. Specific energy performance indicators and their correlation with greenhouse gas (GHG) reductions were identified, providing a basis for benchmarking energy efficiency across the sector. Of the indicators analyzed, the population equivalent (I_{PE}) coefficient offers the greatest comparability, while the I_{BOD} coefficient is the most reliable in reflecting the high energy demands of biological processes. Key factors influencing the energy performance of pretreatment include kWh per kilogram of TSS removed (I_{TSS}). Optimization of the pre-treatment stage can also significantly improve biogas production and energy recovery. For example, the use of chemical pre-treatment (CEPT) or advanced filtration systems (fine screens or drum filters) instead of conventional sedimentation can increase organic matter capture for anaerobic digestion. Plants with higher energy efficiency tend to be those with advanced technologies, optimized processes and energy recovery systems, while those with lower efficiencies may suffer from inefficient aeration systems or outdated equipment. These findings highlight the importance of considering different factors to improve overall energy performance. Knowing the energy consumption indicators for wastewater treatment systems allows the selection of the most efficient technological variant when modernizing the system. A combined analysis of the energy requirements of the WWTP and the carbon footprint is essential, as energy consumption has the largest share in the total CF if the WWTP meets all of its energy requirements from non-renewable sources. Measuring the carbon footprint of a wastewater treatment plant is the first step in reducing it. For the time being, Poland is in a peculiar position; the vast majority of the country's energy is produced on the basis of fossil fuels, that is, on the basis of raw materials whose energy use is particularly harmful to the environment. In response to strong consumption in energy, WWTPs, public agencies, local government and the wastewater industry need to explore and implement measures to ensure the achievement of the targets set out in the 2020 Climate and Energy Package. The best solution will be, on the one hand, to reduce energy consumption by modernizing the infrastructure, and on the other hand, to invest in renewable energy sources that can change this unfavorable trend.

Supplementary Materials: The following supporting information can be downloaded at: https://www.mdpi.com/article/10.3390/en17225745/s1, Table S1: Operational parameters of WWTP1. Table S2: Operational parameters of WWTP2. Table S3: Operational parameters of WWTP3.

Author Contributions: Conceptualization, I.K.-B., A.K. and A.R.; methodology, I.K.-B., J.B.-Z. and A.G.C.; writing—original draft preparation I.K.-B., J.B.-Z. and A.G.C.; investigation, J.B.-Z., A.R. and I.K.-B.; resources writing—original draft preparation, A.K. and J.B.-Z.; writing—review and editing, A.G.C. and J.B.-Z. All authors have read and agreed to the published version of the manuscript.

Funding: This research received no external funding.

Data Availability Statement: The authors declare that the data supporting the findings of this study are available within the paper and its Supplementary Files.

Conflicts of Interest: The authors declare no conflicts of interest.

References

1. ASCE Critical Infrastructure Guidance Task Committee. *Guiding Principles for the Nation's Critical Infrastructure*; American Society of Civil Engineers: Reston, VA, USA, 2009.
2. Capodaglio, A.; Olsson, G. Energy Issues in Sustainable Urban Wastewater Management: Use, Demand Re-duction and Recovery in the Urban Water Cycle. *Sustainability* **2019**, *12*, 266. [CrossRef]
3. Masłoń, A. Energy Consumption of Selected Wastewater Treatment Plants Located in South-Eastern Poland. *Eng. Prot. Environ.* **2017**, *20*, 331–342. [CrossRef]
4. Water Intelligence. The Carbon Footprint of Water. 2022. Available online: https://wint.ai/solutions/carbon-emissions/ (accessed on 19 March 2024).
5. Siatou, A.; Manali, A.; Gikas, P. Energy Consumption and Internal Distribution in Activated Sludge Wastewater Treatment Plants of Greece. *Water* **2020**, *12*, 1204. [CrossRef]

6. Zaborowska, E.; Czerwionka, K.; Makinia, J. Strategies for achieving energy neutrality in biological nutrient removal systems—A case study of the Slupsk WWTP (northern Poland). *Water Sci. Technol.* **2017**, *75*, 727–740. [CrossRef]
7. Maziotis, A.; Sala-Garrido, R.; Molinos-Senante, M. Assessing Energy Efficiency and Its Dynamic Changes in The Water Sector Integrating Heterogeneity and Carbon Emissions. *J. Clean. Prod.* **2024**, *480*, 144137. [CrossRef]
8. Hamawand, I. Energy Consumption in Water/Wastewater Treatment Industry—Optimisation Potentials. *Energies* **2023**, *16*, 2433. [CrossRef]
9. Costantini, V.; Morando, V.; Olk, C.; Tausch, L. Fuelling the Fire: Rethinking European Policy in Times of Energy and Climate Crises. *Energies* **2022**, *15*, 7781. [CrossRef]
10. Capodaglio, A.G.; Callegari, A. Energy and resources recovery from excess sewage sludge: A holistic analysis of opportunities and strategies. *Resour. Conserv. Recycl. Adv.* **2023**, *19*, 200184. [CrossRef]
11. Zakeri, B.; Paulavets, K.; Barreto-Gomez, L.; Echeverri, L.G.; Pachauri, S.; Boza-Kiss, B.; Zimm, C.; Rogelj, J.; Creutzig, F.; Ürge-Vorsatz, D.; et al. Pandemic, War, and Global Energy Transitions. *Energies* **2022**, *15*, 6114. [CrossRef]
12. Capodaglio, A.G. Urban Wastewater Mining for Circular Resource Recovery: Approaches and Technology Analysis. *Water* **2023**, *15*, 3967. [CrossRef]
13. European Parliament. *European Parliament Resolution, 24th March 2022 "Need for an Urgent EU Action Plan to Ensure Food Security Inside and Outside the EU in Light of the Russian Invasion of Ukraine"*; P9_TA(2022)0099; European Parliament: Strasbourg, France, 2022.
14. Cardoso, B.J.; Rodrigues, E.; Gaspar, A.R.; Gomes, Á. Energy performance factors in wastewater treatment plants: A review. *J. Clean. Prod.* **2021**, *322*, 129107. [CrossRef]
15. Torregrossa, D.; Castellet-Viciano, L.; Hernández-Sancho, F. A data analysis approach to evaluate the impact of the capacity utilization on the energy consumption of wastewater treatment plants. Sustain. *Cities Soc.* **2019**, *45*, 307–313. [CrossRef]
16. Callegari, A.; Boguniewicz-Zablocka, J.; Capodaglio, A.G. Energy recovery and efficiency improvement for an activated sludge, agro-food WWTP upgrade. *Water Pract. Technol.* **2018**, *13*, 909–921. [CrossRef]
17. Longo, S.; d'Antoni, B.M.; Bongards, M.; Chaparro, A.; Cronrath, A.; Fatone, F.; Lema, J.M.; Mauricio-Iglesias, M.; Soares, A.; Hospido, A. Monitoring and diagnosis of energy consumption in wastewater treatment plants. A state of the art and proposals for improvement. *Appl. Energy* **2016**, *179*, 1251–1268. [CrossRef]
18. Lozano Avilés, A.B.; del Cerro Velázquez, F.; Llorens Pascual del Riquelme, M. Methodology for Energy Optimization in Wastewater Treatment Plants. Phase I: Control of the Best Operating Conditions. *Sustainability* **2019**, *11*, 3919. [CrossRef]
19. Dairi, A.; Cheng, T.; Harrou, F.; Sun, Y.; Leiknes, T. Deep learning approach for sustainable WWTP operation: A case study on data-driven influent conditions monitoring. *Sustain. Cities Soc.* **2019**, *50*, 101670. [CrossRef]
20. Ráduly, B.; Gernaey, K.V.; Capodaglio, A.G.; Mikkelsen, P.S.; Henze, M. Artificial neural networks for rapid WWTP performance evaluation: Methodology and case study. *Environ. Model. Softw.* **2007**, *22*, 1208–1216. [CrossRef]
21. Cecconet, D.; Mainardis, M.; Callegari, A.; Capodaglio, A.G. Psychrophilic treatment of municipal wastewater with a combined UASB/ASD system, and perspectives for improving urban WWTP sustainability. *Chemosphere* **2022**, *297*, 134228. [CrossRef]
22. Castellet, L.; Molinos-Senante, M. Efficiency assessment of wastewater treatment plants: A data envelopment analysis approach integrating technical, economic, and environmental issues. *J. Environ. Manag.* **2016**, *167*, 160–166. [CrossRef]
23. Zhang, S.; Tang, Y. A POET Based Energy Audit Methodology for WWTPs. In Proceedings of the 2018 37th Chinese Control Con-ference (CCC), Wuhan, China, 25–27 July 2018; pp. 7399–7404.
24. Ritchie, J. Energy Management Systems and Digital Technologies for Industrial Energy Efficiency and Productivity. Report from IEA Workshop; 2017. Available online: www.iea.org (accessed on 2 November 2022).
25. Marimon, F.; Casadesús, M. Reasons to Adopt ISO 50001 Energy Management System. *Sustainability* **2017**, *9*, 1740. [CrossRef]
26. Pawlowska, E.; Machnik-Slomka, J.; Klosok-Bazan, I.; Gono, M.; Gono, R. Corporate Social Responsibility of Water and Sanitation Company in the Czech Republic—Case Study. *Energies* **2021**, *14*, 3981. [CrossRef]
27. Walsh, B.P.; Murray, S.N.; O'Sullivan, D.T.J. The water energy nexus, an ISO50001 water case study and the need for a water value system. *Water Resour. Ind.* **2015**, *10*, 15–28. [CrossRef]
28. Longo, S.; Hospido, A.; Mauricio-Iglesias, M. Energy efficiency in wastewater treatment plants: A framework for benchmarking method selection and application. *J. Environ. Manag.* **2023**, *344*, 118624. [CrossRef] [PubMed]
29. Clos, I.; Alvarez-Gaitan, J.P.; Saint, C.P.; Short, M.D. Energy Benchmarking for Efficient, Lower Carbon Wastewater Treatment Operations in Australia. In *Decarbonising the Built Environment*; Springer: Singapore, 2019; pp. 305–320.
30. McKane, A.; Therkelsen, P.; Scodel, A.; Rao, P.; Aghajanzadeh, A.; Hirzel, S.; Zhang, R.; Prem, R.; Fossa, A.; Lazarevska, A.M.; et al. Predicting the quantifiable impacts of ISO 50001 on climate change mitigation. *Energy Policy* **2017**, *107*, 278–288. [CrossRef]
31. Shizas, I.; Bagley, D.M. Experimental Determination of Energy Content of Unknown Organics in Municipal Wastewater Streams. *J. Energy Eng.* **2004**, *130*, 45–53. [CrossRef]
32. Mo, W.; Zhang, Q. Can municipal wastewater treatment systems be carbon neutral? *J. Environ. Manag.* **2012**, *112*, 360–367. [CrossRef]
33. Zaborowska, E. Wykorzystanie ścieków i osadów ściekowych jako źródła energii w oczyszczalni ścieków. In Proceedings of the AQUA 2006: Problemy Inżynierii Środowiska: XXVI Międzynarodowe Sympozjum im. Bolesława Krzysztofika, Płock, Poland, 8–9 June 2006; Volume 8, pp. 104–107.

34. USEPA. Examples of Innovation in the Water Sector. US Environmental Protection Agency. 2023. Available online: https://19january2017snapshot.epa.gov/innovation/examples-innovation-water-sector_.html (accessed on 19 March 2024).
35. Raboni, M.; Viotti, P.; Capodaglio, A.G. A comprehensive analysis of the current and future role of biofuels for transport in the European union (EU). *Rev. Ambiente E Agua* **2015**, *10*, 9–21. [CrossRef]
36. Alekseiko, L.N.; Slesarenko, V.V.; Yudakov, A.A. Combination of wastewater treatment plants and heat pumps. *Pac. Sci. Rev.* **2014**, *16*, 36–39. [CrossRef]
37. Kowalik, R.; Bąk-Patyna, P. Analysis of heat recovery from wastewater using a heat pump on the example of a wastewater treatment plant in the świętokrzyskie voivodeship in Polish. *Struct. Environ.* **2021**, *13*, 90–95. [CrossRef]
38. Tinivella, R.; Bargiggia, R.; Zanoni, G.; Callegari, A.; Capodaglio, A.G. High-Strength, Chemical Industry Wastewater Treatment Feasibility Study for Energy Recovery. *Sustainability* **2023**, *15*, 16285. [CrossRef]
39. Schäfer, M.; Gretzschel, O.; Steinmetz, H. The Possible Roles of Wastewater Treatment Plants in Sector Coupling. *Energies* **2020**, *13*, 2088. [CrossRef]
40. Cecconet, D.; Raček, J.; Callegari, A.; Hlavínek, P. Energy recovery from wastewater: A study on heating and cooling of a multipurpose building with sewage-reclaimed heat energy. *Sustainability* **2020**, *12*, 116. [CrossRef]
41. Kollmann, R.; Neugebauer, G.; Kretschmer, F.; Truger, B.; Kindermann, H.; Stoeglehner, G.; Ertl, T.; Narodoslawsky, M. Renewable energy from wastewater—Practical aspects of integrating a wastewater treatment plant into local energy supply concepts. *J. Clean. Prod.* **2017**, *155*, 119–129. [CrossRef]
42. Schäfer, M.; Gretzschel, O.; Schmitt, T.G.; Knerr, H. Wastewater Treatment Plants as System Service Provider for Renewable Energy Storage and Control Energy in Virtual Power Plants—A Potential Analysis. *Energy Procedia* **2015**, *73*, 87–93. [CrossRef]
43. Vrzala, M.; Goňo, M.; Goňo, R.; Kotulla, M.; Wzorek, M.; Leonowicz, Z. Distributed Generation Power Sys-tems in Wastewater Management. *Energies* **2022**, *15*, 6283. [CrossRef]
44. Bao, Z.; Sun, S.; Sun, D. Assessment of greenhouse gas emission from A/O and SBR wastewater treatment plants in Beijing, China. *Int. Biodeterior. Biodegrad.* **2016**, *108*, 108–114. [CrossRef]
45. Liao, X.; Tian, J.; Gan, Y.; Ji, J. Quantifying urban wastewater treatment sector's greenhouse gas emissions us-ing a hybrid life cycle analysis method—An application on Shenzhen city in China. *Sci. Total Environ.* **2020**, *745*, 141176. [CrossRef]
46. WRI & WBCSD. *The Greenhouse Gas Protocol: A Corporate Accounting and Reporting Standard*; World Resources Institute and World Business Council for Sustainable Development: Washington, DC, USA, 2004.
47. The National Urban Wastewater Treatment Program, Regulation of the Ministry of Environment, Republic of Poland, 2016. IV AKPOŚK (AKPOŚK2015)—Dokument zatwierdzony przez Radę Ministrów 21.04.2016r. Available online: https://www.gov.pl/web/wody-polskie/krajowy-program-oczyszczania-sciekow-komunalnych (accessed on 19 March 2024).
48. Ranieri, E.; D'Onghia, G.; Lopopolo, L.; Gikas, P.; Ranieri, F.; Gika, E.; Spagnolo, V.; Herrera, J.A.; Ranieri, A.C. Influence of climate change on wastewater treatment plants performances and energy costs in Apulia, south Italy. *Chemosphere* **2024**, *350*, 141087. [CrossRef]
49. Liu, Y.; Gu, J.; Zhang, M. *A-B Processes: Towards Energy Self-Sufficient Municipal Wastewater Treatment*; IWA Publishing: London, UK, 2019; pp. 1–28. [CrossRef]
50. Cardenes, I.; Hall, J.W.; Eyre, N.; Majid, A.; Jarvis, S. Quantifying the energy consumption and greenhouse gas emissions of changing wastewater quality standards. *Water Sci. Technol.* **2020**, *81*, 1283–1295. [CrossRef]
51. Vaccari, M.; Foladori, P.; Nembrini, S.; Vitali, F. Benchmarking of energy consumption in municipal wastewater treatment plants—A survey of over 200 plants in Italy. *Water Sci. Technol.* **2018**, *77*, 2242–2252. [CrossRef]
52. Tsalas, N.; Golfinopoulos, S.K.; Samios, S.; Katsouras, G.; Peroulis, K. Optimization of Energy Consumption in a Wastewater Treatment Plant: An Overview. *Energies* **2024**, *17*, 2808. [CrossRef]

energies

MDPI

Review

Recent Advances in Wastewater Electrocoagulation Technologies: Beyond Chemical Coagulation

Sangyeol Jo, Rahul Kadam, Heewon Jang, Dongyun Seo and Jungyu Park *

Department of Advanced Energy Engineering, Chosun University, Gwangju 61452, Republic of Korea; jsywin71@chosun.ac.kr (S.J.); rahulkadam@chosun.kr (R.K.); jheewon95726@chosun.kr (H.J.); sdy4035@chosun.ac.kr (D.S.)

* Correspondence: jp@chosun.ac.kr; Tel.: +82-62-230-7119; Fax: +82-62-230-7110

Abstract: This review provides a comprehensive analysis of the recent research trends and application cases of chemical coagulation (CC) and electrocoagulation (EC), which play a crucial role in wastewater treatment. In particular, the principles and process performances of the EC technologies are comparably reviewed with traditional CC technologies. EC offers the advantage of reducing the use of chemical agents and minimizing sludge generation compared to CC. Moreover, recent research cases have demonstrated its effectiveness in removing pollutants from wastewater. With increasing water consumption due to industrial development, the application of coagulation processes in wastewater and sludge treatment is expected to expand to minimize environmental impact. This review provides insights into the current status and future development direction of CC and EC technologies and can serve as foundational information for more efficient and environmentally friendly coagulation systems.

Keywords: wastewater treatment; electrocoagulation; chemical coagulation; sustainability

1. Introduction

In the 21st century, water and energy have emerged as two of the most pressing challenges in the world. The rapid growth of the global population, coupled with industrialization aimed at meeting these demands, has significantly compromised both the quality and availability of water resources. Around 40% of the population is affected by water contamination, and over 20% faces a shortage of freshwater [1]. Additionally, industries, particularly those known for high water consumption such as textiles, consume around 300 L of freshwater for each kilogram of product. This results in the discharge of substantial quantities of heavily pigmented wastewater into the environment [2]. This leads to pollution of both surface- and groundwater due to contaminants and other harmful organic compounds. Many of these compounds can be carcinogenic, mutagenic, and sometimes teratogenic, posing a risk to living organisms [3]. Moreover, industrial wastewater typically contains very fine suspended solids (SS), dissolved solids, and inorganic/organic particles. Due to their small size and surface charge, aggregating these particles into a heavier mass for effective removal becomes a considerable challenge. A range of both traditional and advanced technologies has been employed to eliminate colloidal particles from wastewater. These methods include ion exchange, membrane filtration, precipitation, flotation, solvent extraction, adsorption, and coagulation, as well as biological and electrolytic approaches [4,5].

Coagulation is one of the most widely used methods in water and wastewater treatment processes, involving various chemicals [6,7]. This approach aligns with the growing trend toward wastewater-to-energy technology, which emphasizes the direct recovery of organic matter from wastewater. To minimize energy input and maximize energy production, these technologies aim to transform waste into resources, such as treated water for reuse and recoverable materials from residuals. This technique comprises three distinct stages:

(1) the rapid mixing of the dispersed coagulant into water or wastewater through vigorous agitation, (2) the gentle stirring of the mixtures to encourage the formation of larger flocs by aggregating tiny particles, and (3) subsequent settling of these flocs to the bottom [8]. Coagulation is primarily induced by inorganic metal salts, including aluminum (Al) and iron (Fe). The most frequently used Al-based coagulants are aluminum sulfate, aluminum chloride, and sodium aluminate. For Fe-based coagulants, the common types include ferric sulfate, ferrous sulfate, ferric chloride, and ferric chloride sulfate [9]. However, the use of inorganic coagulants has decreased due to environmental concerns. The presence of chemical compounds can result in toxic residues in sludge, which may pose health risks after extended exposure. Additionally, chemical processes often require significant chemical inputs, which not only elevate the overall cost but also complicate downstream operations. As a result, both synthetic polymeric and natural flocculants have gained high attention in water and wastewater treatment processes. This is attributed to their inherent stability against pH fluctuations, high effectiveness at low dosages, and ease of use [10]. Recently, synthetic polymers like polyaluminum chloride (PAC), polyferric sulfate (PFS), and polyacrylamide (PAM) have become the most commonly utilized coagulants in wastewater treatment because they can produce large shear-stable flocs [11]. Similarly, natural coagulants present a viable alternative to reduce environmental pollution and health risks linked to chemical coagulants. Numerous studies have highlighted the use of natural coagulants derived from various plant species, including Moringa oleifera, Jatropha curcas, banana peels, and bagasse [12]. Both polymeric and natural coagulants demonstrate effective treatment capabilities, positioning them as promising replacements for traditional chemical coagulants. Despite these benefits, the acceptance of synthetic and natural coagulants in water and wastewater treatment facilities has been limited. This is largely due to a lack of industrial confidence in their use, driven by concerns regarding their effectiveness and consistency when utilized on a large scale [13].

In recent years, electrochemical technologies have gained considerable interest for their capacity to utilize various parameters to enhance performance efficiency. One notable method, electrocoagulation (EC) has become increasingly popular as an effective and economical solution for treating wastewater, all while reducing sludge generation. It provides versatility, user-friendliness, and the capacity to manage a range of contaminants [14]. One of the key advantages of EC is its ease of control over the dosing of necessary chemicals. Unlike traditional methods that require the addition of stabilizing agents like chlorides and sulfates, EC utilizes fixed chemical electrodes such as Al and Fe, which simplifies the process and minimizes the need for additional chemical inputs. This reduction in chemical concentration not only lowers the generation of inorganic sludge but also mitigates the risk of producing potentially toxic byproducts. Furthermore, the electrochemical reactions involved in EC allow for the use of various materials, including waste. This approach enhances the stability of coagulants and counteracts the slow corrosion often seen with synthesized coagulants [15]. The application of voltage during the EC process can effectively oxidize soluble organic matter, reducing the energy requirements for subsequent biological treatments. Additionally, the inevitable byproduct, hydrogen ions can be converted to hydrogen gas, which aids in maintaining pH levels and contributes to energy production [16]. Due to these advantages, EC has been extensively used in the treatment of various industrial wastewater, including those from the textile, pharmaceutical, municipal, dairy, petroleum, and paper industries as well as in drinking water treatment [17]. Ongoing advancements in EC research have consistently highlighted the importance of EC and its potential benefits in wastewater treatment.

Therefore, this review paper seeks to engage in a critical analysis of the coagulation processes involved in water and wastewater treatment, with a particular emphasis on chemical coagulation (CC) and EC. Notably, there is a lack of published literature that specifically addresses these two methods in tandem. CC has been a cornerstone of wastewater treatment for a while, evolving through various process modifications and the introduction of innovative natural coagulants. Meanwhile, EC has emerged as a noteworthy alternative to conventional CC techniques. This review aims to deliver a thorough overview of the EC process, including its reaction mechanisms, challenges, and potential avenues for future research. This review will provide a deeper understanding of both CC and EC processes, highlighting their limitations and suggesting possible solutions.

2. Chemical Coagulation

2.1. Fundamentals of Chemical Coagulation Process

Coagulation is a three-phase process designed to eliminate stable particles by creating larger aggregates. As illustrated in Figure 1, the key stages of coagulation involve destabilizing colloidal particles through the addition of coagulants, facilitating the aggregation of these particles into larger flocs, and, finally, allowing these flocs to settle at the bottom. In aqueous environments, the majority of solids or suspended particles are usually relatively small and often have a negative charge. To enhance the sedimentation process, particles need to aggregate into larger flocs. However, this aggregation is hindered by electrostatic repulsive forces that keep negatively charged particles from coming together. As a result, settling takes longer. This issue can be addressed by using a coagulant to destabilize the particles [18]. Destabilization can occur through one or more of the following mechanisms following the addition of a coagulant:

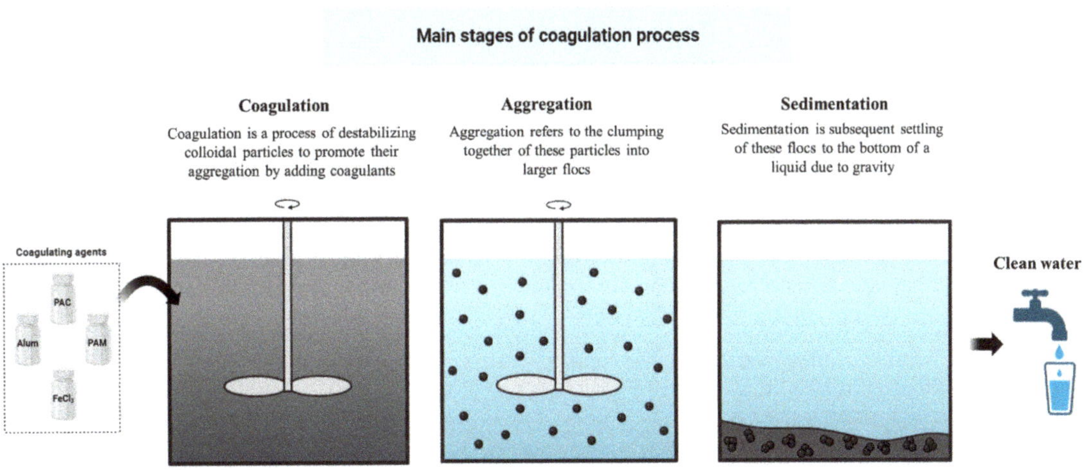

Figure 1. Graphical illustration depicting the main stages of chemical coagulation process.

(a) Double-layer compression: It is a mechanism that involves countercharged ions penetrating the double layer around colloids. These counter ions modify the double layer, making it thinner and less voluminous. Continuous double-layer compression by electrolytes diminishes electrostatic repulsion and enhances the van der Waals force, promoting the aggregation of destabilized colloids. The continuously increased aggregation rate makes the tiny flocs formed at the double-layer compression step larger and denser [12,19]. However, the stability of the flocs is influenced by the ionic charge of the coagulant used. Monovalent ions, which carry a weak charge, tend to create large but loosely bound flocs that take longer to settle. In the presence of weakly charged ions, the double layer remains significantly charged, resulting in a strong repulsive force that hinders the likelihood of agglomeration [20].

(b) Sweep flocculation: It is a process that eliminates colloids by trapping them within a net-like framework. This framework is formed by the precipitation of amorphous metal hydroxides during hydrolysis. Theoretically, a higher fractal dimension leads to stronger flocs that are more resistant to breakage. However, flocs formed through sweep flocculation tend to be larger and have a faster formation rate, making them more susceptible to breakage [12,19,21]. The presence of repulsive forces between flocs contributes to this phenomenon. Sweep flocculation primarily entraps colloids within a net-like structure but does not neutralize the repulsive forces among them, leading to the formation of weak flocs [22].

(c) Charge neutralization: It takes place when oppositely charged coagulants adhere to the surfaces of colloids through adsorption. Charge neutralization happens on the colloid surfaces in a patch-wise fashion, referred to as the electrostatic patch mechanism. Different cations attached to the colloid surface create regions with both positive and negative charges. The mixed charge distribution diminishes repulsive forces and enhances the van der Waals interaction between particles. Flocs generated through the charge neutralization mechanism are stronger than those formed by sweep flocculation, yet weaker than those created by interparticle bridging [12,23,24]. Flocs formed through charge neutralization are indeed strong, but their strength is limited because they depend on physical bonds, which are weaker than chemical bonds [22].

(d) Interparticle bridging: It involves polymer chains that are long and highly reactive, extending into wastewater. One end of the polymer chain binds to colloids, while the free ends connect to other colloidal particles. The resultant structure is known as colloid–polymer–colloid, where polymer acts as a bridge. Multiple colloid–polymer–colloids can intertwine, resulting in easily settleable flocs. Theoretically, a low fractal dimension typically leads to weak flocs that are susceptible to breakage. However, these flocs can be quite robust and resistant to fragmentation into smaller clusters. This is due to the presence of various polymers that act as strong bridges formed by numerous chemical bonds among flocs [12,22,25]. A recent study indicated that employing a natural coagulant with an interparticle bridging mechanism can increase floc growth by at least three times compared to using a chemical coagulant. This enhancement is attributed to the capacity of polymeric chains to extend and bind to multiple colloids effectively [26]. Overall, coagulation relies on these mechanisms to destabilize particles and facilitate aggregation, leading to larger flocs that settle more easily. Each mechanism contributes uniquely to the formation of flocs with varying strengths and resistance to breakage.

2.2. Importance in Wastewater Treatment Plants

Wastewater treatment (WWTP) is a critical element of modern urban infrastructure, playing an important role in public health and environmental protection. The most widely used sewage treatment process is the activated sludge, accounting for over 90% of all sewage treatment processes worldwide [27]. This method is based on primary treatment through gravity sedimentation of raw sewage and secondary treatment through the microorganism's metabolism with primarily treated sewage. The widespread use of the activated sludge process is due to its process stability, ease of maintenance, and effective performance in removing various pollutants [28].

However, the activated sludge has several major drawbacks. First, a significant amount of energy is consumed in the aeration process for aerobic microorganism activity during biological treatment. It has been shown that approximately 50–75% of the total energy consumption in WWTP is used in the aeration process [29]. Second, large amounts of activated sludge are generated during the organic matter treatment process, which incurs additional costs and environmental burdens for treatment and disposal [30].

Recently, the field of wastewater treatment has been facing new challenges. Effluent regulation standards are becoming more stringent, and there is an increasing demand for carbon neutrality in wastewater treatment facilities. In these changing environmental conditions, it has become necessary to explore new wastewater treatment methods that can complement or replace the existing activated sludge method [31,32].

In this context, the application of the Chemical Enhanced Primary Treatment (CEPT) process is gaining attention. CEPT recognizes wastewater not as mere waste but as an energy source, aiming to recover it effectively and reduce the load on biological treatment processes. The CEPT process significantly improves the removal efficiency of organic matter and SS in the primary sedimentation stage by using chemical coagulants. The application of CEPT to existing activated sludge processes can be implemented without separate civil engineering work, and its main advantages are as follows [33,34]:

1. Improved Energy Efficiency: By removing more organic matter in the primary treatment stage through CEPT, the load on subsequent biological treatment stages is reduced. This has the positive effect of reducing the energy consumption required for aeration in the bio-processes linked with the CEPT. Compared to the conventional activated sludge process, previous studies reported that the application of CEPT could reduce energy consumption for aeration by up to 50% [35,36]. Moreover, when comparing the entire treatment process, it has been reported that energy savings of up to 71% can be achieved [37].
2. Energy Source Recovery: The primary sludge generated through CEPT has a higher organic content than the primary sludge formed by conventional gravity sedimentation. As a result, CEPT sludge can be used as a more effective raw material for energy recovery processes, such as anaerobic digestion, and it has been reported that methane yield can be increased by 30–40% [38]. Additionally, when CEPT is applied, the amount of low-biodegradability-activated sludge formed in the subsequent biological treatment process decreases, which can reduce sludge treatment costs. This ultimately has the advantage of increasing the value of sewage sludge as an energy source.
3. Greenhouse Gas Reduction: The application of the CEPT process significantly reduces greenhouse gas emissions from wastewater treatment plants. According to recent studies, when CEPT is applied, greenhouse gas emissions from wastewater treatment plants can be reduced by up to 70% compared to the conventional activated sludge method [37]. This is attributed to two factors: reduced aeration due to efficient organic matter removal from the influent, and increased biogas production through anaerobic digestion of recovered sludge.

However, there are still several challenges remaining for the widespread application of the CEPT process. Due to the continuous increase in wastewater generation caused by population growth and industrial development, large quantities of chemicals are required, which can lead to increased operational costs and environmental impacts. Excessive use of chemicals can adversely affect subsequent biological treatment processes, requiring a cautious approach. Additionally, further research is needed on the impact of changes in the physicochemical characteristics of recovered sludge on subsequent treatment processes.

Therefore, it is essential to optimize factors such as the type and amount of coagulant, and operating conditions for the successful application of CEPT (pH, mixing intensity, retention time, etc.). Furthermore, along with optimal conditions, it is necessary to quantitatively verify the impact on subsequent processes.

In summary, while CEPT offers promising advantages in terms of energy efficiency, resource recovery, and greenhouse gas reduction, its effective application requires careful optimization of coagulant type, dosage, and operating conditions. A quantitative assessment of CEPT's impact on subsequent treatment stages is essential to maximize its benefits and minimize potential drawbacks. The following section will explore these aspects in greater detail.

2.3. *Types of Chemical Coagulants*

WWTPs primarily utilize inorganic coagulants, which can be broadly classified into Al-based and Fe-based coagulants. When these inorganic coagulants are introduced into water or wastewater, they undergo three main reaction stages: dissociation, hydrolysis, and polymerization [39].

1. Dissociation: This initial stage involves the breakdown of the metal salts into their constituent ions. For example, when Al or Fe salts are added to the water, they dissociate into ions such as Al^{3+}, SO_4^{2-}, Fe^{3+}, and Cl^-.
2. Hydrolysis: In this stage, the dissociated ions react with water, leading to the formation of hydroxyl complexes like $Al(OH)_3$ and $Fe(OH)_3$. These hydroxides play a crucial role in the coagulation process by helping to aggregate suspended particles.
3. Polymerization: The final stage involves the reaction of hydrolyzed Al and Fe ions to form larger, more complex structures known as polymeric aluminum and polymeric ferric species. These polymers possess a higher charge density, which effectively bridges and agglomerates suspended particles, thereby enhancing the coagulation process.

Inorganic coagulants offer several advantages over other types of coagulants. They demonstrate high efficiency in removing various pollutants, including heavy metals, turbidity, Chemical Oxygen Demand (COD), and Biological Oxygen Demand (BOD). Indeed, these metal-based coagulants can achieve impressive removal efficiencies of up to 99.3% for phosphate, 99.5% for SS, and 95.6% for COD [40]. Moreover, these coagulants are commercially available and possess the ability to effectively inactivate bacteria.

Al-based coagulants include $Al_2(SO_4)_3$, PAC, and composite PAC. $Al_2(SO_4)_3$ is the most widely used coagulant, preferred for its cost-effectiveness and efficiency. While it is readily available and inexpensive, it is only effective within a limited pH range (5.5–7.0). PAC is effective over a broader pH range (4.0–8.0) than alum and performs well at low temperatures. It also offers advantages such as rapid reaction and large floc formation but may be insufficient for removing high molecular weight particles and hydrophobic particles [9,41,42].

Fe-based coagulants include $Fe_2(SO_4)_3$, $FeCl_3$, and $FeSO_4$. These are effective over an even wider pH range (4–11) compared to Al-based coagulants and exhibit excellent performance in phosphorus removal. However, they may cause color issues in treated water due to residual iron. $Fe_2(SO_4)_3$ and $FeCl_3$ show high coagulation efficiency but can be corrosive. $FeSO_4$ is relatively inexpensive and effective in phosphorus removal but may require an additional oxidation step [40,42].

As outlined above, each coagulant has unique characteristics along with advantages/disadvantages depending on its type, necessitating appropriate selection based on the application scenario. The main advantages and disadvantages of each coagulant type are summarized in Table 1.

Table 1. Overview of common metal coagulants: merits and demerits as reported in previous studies.

Coagulant	Merits	Demerits	Reference
Aluminum sulfate	Easily accessible, low cost, and low sludge volume index.	Effective in limited pH range, and residual Al ions in the sludge.	[43]
PAC	Fast reaction and bigger flocs.	Insufficient in removing higher molar mass particles and hydrophobic particles.	[44]
Sodium aluminate	Small dosage and efficient in highly contaminated wastewater.	Costly and limited application due to inefficiency in treating low-polluted wastewater.	[45]
Ferrous sulfate	Easily available and less costly.	Alkalinity addition is needed and corrosive.	[46]
Ferric sulfate	Less sensitivity to overdosage.	Alkalinity addition is needed and leaves visible rust-colored strains.	[47]
Ferric chloride	Effective in a broad pH range, easily available, and less costly.	Corrosive and hazardous.	[48]

While these inorganic coagulants with diverse characteristics demonstrate high efficiency, their performance is significantly influenced by various factors. Since optimal operation conditions depend on the coagulant type and dose, coagulation efficiency can be greatly reduced if these conditions are not accurately controlled. Therefore, it is crucial to understand the characteristics of each coagulant and the interactions of various variables that need to be considered in the optimization process to derive optimal operating conditions. This approach can maximize the effectiveness of inorganic coagulants in the wastewater treatment process and enhance treatment efficiency.

2.4. Key Factors Influencing Chemical Coagulation

The efficiency and effectiveness of CC processes in wastewater treatment are significantly influenced by various factors. Understanding and appropriately controlling these factors are essential for achieving optimal wastewater treatment results. The main influencing factors include pH, coagulant dosage, mixing intensity and time, temperature, and wastewater characteristics.

pH is one of the most critical factors in the coagulation process, directly affecting coagulant hydrolysis, charge characteristics, and the surface charge of pollutants. Each coagulant has the recommended optimal pH range, outside of which coagulation efficiency can significantly decrease. For instance, Al-based coagulants are generally most effective in the pH range of 6.5–8.0, while Fe-based coagulants perform optimally in the pH range of 4.5–6.0, where charge neutralization and adsorption bridging mechanisms operate simultaneously [6]. Recent research by Zhou et al. (2023) used molecular dynamics simulations to analyze the impact of pH on coagulation mechanisms, revealing how pH changes alter the structure and charge distribution of coagulant molecules, directly influencing their interaction with pollutants [49].

Coagulant dosage directly impacts both treatment efficiency and economic viability. Li et al. (2022) elucidated the non-linear relationship between coagulant dosage and removal efficiency, demonstrating that increasing dosage beyond a certain threshold does not significantly improve removal efficiency and may lead to problems associated with overdosing [50]. Kurniawan et al. (2020) found that underdosing results in the formation of unstable small flocs, leading to poor sedimentation efficiency and reduced treatment effectiveness [51]. Conversely, recent studies have detailed the problems arising from coagulant overdosing, including increased sludge production, elevated residual metal concentrations, pH changes, and increased electrical conductivity of treated water [52].

Mixing intensity and duration also significantly influences coagulation efficiency. The process typically involves two stages: rapid mixing and slow mixing. Suzuki et al. (2023) reported that the optimal rapid mixing intensity and duration could be varied by the type of coagulant and pollutant characteristics. For instance, polymer coagulants require longer rapid mixing times (60–90 s) compared to metal salt coagulants (30–60 s) [53]. Liu et al. (2024) emphasized the importance of the slow mixing stage, showing that appropriate slow mixing conditions (30 rpm, 20 min) optimize floc growth and structure, enhancing final sedimentation efficiency [54]. Sun et al. (2019) analyzed the impact of mixing intensity on floc morphology and strength, reporting that excessive mixing intensity can lead to floc breakage and reduced treatment efficiency, while insufficient mixing results in inadequate coagulation [55]. However, contrasting these findings, Abbas et al. (2021) used design expert software to analyze optimal conditions for $FeCl_3$ coagulant and found that while pH and dosage significantly affected COD, turbidity, and total suspended solids (TSS) removal, mixing time had no significant impact, and mixing speed only slightly influenced COD removal [56].

Temperature is another crucial factor affecting the coagulation process. Generally, increased temperature enhances molecular motion and chemical reaction rates, potentially improving coagulation efficiency [39]. However, Zhang et al. (2023) observed that coagulation efficiency decreased when temperatures exceeded 30 °C. This phenomenon is attributed to accelerated hydrolysis reactions at high temperatures, resulting in looser floc formation and consequently reduced pollutant removal rates. Conversely, at low temperatures, hydrolysis of coagulant and aggregation of destabilized colloid particles could be inefficient due to weakened Brownian motion. Therefore, maintaining an appropriate temperature range is crucial for achieving optimal coagulation efficiency [57].

Wastewater characteristics, including SS concentration, particle size distribution, and organic matter content, significantly influence the coagulation process. The alkalinity of wastewater affects pH buffering capacity, playing a crucial role in coagulation efficiency. Various ions in wastewater (e.g., Ca^{2+}, Mg^{2+}, SO_4^{2-}) can affect the coagulation efficiency by directly reacting with coagulants or indirectly interfering with floc formation [58]. Previous studies have investigated the effect of molecular weight distribution of organic matter in wastewater on coagulation efficiency. Results showed that higher molecular weight organic matter tends to improve coagulation efficiency, as it more readily combines with coagulants to form larger flocs [59,60].

Recent research has employed various statistical techniques and artificial intelligence-based methodologies to optimize the coagulation process. In particular, response surface methodology (RSM), artificial neural network (ANN), and genetic algorithm (GA) are widely used.

RSM remains an important optimization tool. Abbas et al. (2021) used design expert software to analyze optimal conditions for $FeCl_3$ coagulant. They employed Central Composite Design to design experiments and developed a second-order polynomial model to determine the effects and optimal conditions of independent variables such as pH, dosage, mixing time, and mixing speed [56].

Furthermore, hybrid models combining ANN and GA are being increasingly used for coagulation process optimization. Ejimofor et al. (2021) used an ANN-GA model to optimize the removal of colloidal particles from paint wastewater using clay [61]. Kusuma et al. (2021) employed ANN-GA to optimize turbid water treatment using Ipomoea batatas leaf extract as a green coagulant. These studies demonstrate that ANN-GA hybrid models can provide more accurate predictions and optimal solutions for various coagulants and wastewater types compared to RSM or other optimization techniques [62]. These models can predict optimal operating conditions by considering various process parameters such as pH, coagulant dosage, mixing time, and temperature. They have high potential for application in real-time control systems as they can predict optimal coagulant dosage and operating conditions based on various water quality parameters [39].

While these optimization techniques can significantly enhance the efficiency of the coagulation process, they still have some limitations. Firstly, these models have primarily been validated at laboratory or pilot scales, and further research is needed regarding their applicability in large-scale wastewater treatment plants. Secondly, these optimization techniques mainly focus on pollutant removal efficiency, potentially overlooking other important factors in terms of overall water treatment system efficiency and sustainability.

Therefore, future research should focus on overcoming these limitations and developing more comprehensive and sustainable coagulation process optimization methods. Additionally, the development of adaptive coagulation systems that can rapidly respond to changing wastewater characteristics by combining real-time monitoring and control technologies is necessary. Alongside this, continuous research into new technologies and methods that can reduce or replace the use of CC should be pursued.

2.5. Limitation of Chemical Coagulant

Chemical coagulants, particularly inorganic metal salts, are widely used in wastewater treatment. However, their use presents several significant limitations that manifest in environmental, operational, and health aspects. The use of chemical coagulants significantly affects the pH and alkalinity of water. Al- and Fe-based coagulants, in particular, release hydrogen ions, lowering pH and reducing alkalinity. This can impact other stages of the treatment process and may necessitate additional chemical use for pH adjustment.

Since the CC process generates substantial amounts of sludge containing hazardous and toxic substances, process costs would be raised for efficient treatment and disposal. The use of inorganic salt coagulants can increase sludge volume by 37–97%, negatively affecting subsequent processes such as anaerobic sludge digestion [40,63]. For instance, Al can reduce the specific methanogenic activity of methanogenic and acetogenic bacteria by 50–72% [64].

Excessive use of coagulants can leave residual metals in treated water, potentially harming aquatic life. These metals can disrupt physiological processes, leading to impaired growth, reproduction, and survival rates in fish and other aquatic animals. They may also bioaccumulate in the tissues of organisms, affecting the entire food chain [42]. Al-based coagulants, in particular, may leave Al salts in residuals, and long-term exposure to water containing these residuals has been associated with an increased risk of Alzheimer's disease [41], but whether the link is causal is still open to debate. However, previously published epidemiological studies of Al and Alzheimer's disease have shown statistically significant positive relations [65]. In response to these concerns, the US Environmental Protection Agency (EPA) has established a secondary standard of 0.05 to 0.2 mg/L for Al in drinking water [66]. Fe-based coagulants, while effective over a wider pH range and excellent for phosphorus removal, can leave residual Fe causing color issues in treated water, necessitating additional treatment [67]. This residual Fe can impart a yellowish or reddish tint to the water, which may affect consumer perception and acceptance. Although it does not pose health risks, the EPA has set a threshold of 0.3 mg/L for Fe in domestic water use, including drinking water, due to concerns about aesthetic quality [68].

Coagulated sludge can cause problems in both subsequent aerobic biological treatment and sludge treatment. The aggregated and dense structure of coagulated sludge can limit the access of bacteria and enzymes to organic matter within flocs, while residual coagulant concentrations can alter microbial activity in activated sludge processes, acting as inhibitors and affecting the settleability of activated sludge [34,69].

Al-based coagulants, in particular, can have severe impacts on aquatic environments depending on pH levels. Al has the lowest solubility between pH 5.7 and 6.2 but tends to exist in solution outside this range. Studies have shown that the combination of pH below 5.5 and dissolved Al concentrations above 0.5 mg/L can pose a serious threat to aquatic ecosystems [69].

In the case of Fe-based coagulants, while they primarily precipitate as $Fe(OH)_3$, resulting in relatively low concentrations of dissolved Fe^{3+}, the residual Fe^{3+} can damage important biomolecules such as DNA, proteins, and lipids through the following reactions:

$$Fe^{2+} + H_2O_2 \rightarrow Fe^{3+} + OH\bullet + OH^- \quad (1)$$

$$Fe^{3+} + H_2O_2 \rightarrow Fe^{2+} + HOO\bullet + H^+ \quad (2)$$

Fe^{3+} ions do not directly form oxygen-free radicals but play a crucial role in promoting and amplifying their formation. This underscores the importance of managing Fe^{3+} concentrations in water treatment systems, as excessive Fe^{3+} can negatively impact microbial activity and overall treatment efficiency [70].

Chemical coagulants are highly corrosive and can shorten the lifespan of treatment facilities, leading to increased long-term maintenance costs. Additionally, the effectiveness of chemical coagulants is sensitive to changes in water quality, requiring continuous adjustment of dosage in response to changing wastewater characteristics, which increases operational complexity [71].

Lastly, the environmental impacts of producing and transporting chemical coagulants should also be considered. This can increase the overall environmental footprint of the wastewater treatment process [26].

In conclusion, while chemical coagulants are effective in wastewater treatment, their use presents several significant limitations. Considering the environmental impacts, operational issues, and potential risks to health and ecosystems. The use of these coagulants should be approached with caution, and where possible, more sustainable alternatives should be explored.

3. Electrocoagulation

EC is a method for treating wastewater that utilizes electrical currents to eliminate contaminants. This process neutralizes negatively charged particles by forming hydroxide complexes in the water. These complexes aid in aggregating SS, which strengthens the floc that eventually settles under the force of gravity. Additionally, it generates coagulants onsite through the electrical dissolution of specific metal electrodes such as Al, Fe, copper (Cu), or stainless steel (STS). At the anode, metal ions are released, while hydrogen gas is generated at the cathode [72]. The core principle of EC is based on "electrolysis", which involves using electricity to decompose compounds. This concept was introduced by Michael Faraday in 1820. The process occurs in an electrolyte solution, facilitating the transfer of ions between the electrodes. In the EC cell, positive ions move toward the cathode, where they undergo reduction, while negative ions migrate toward the anode and experience oxidation [73,74]. EC has a rich history as a water treatment technology. It was first proposed in London in 1889 and implemented in sewage treatment facilities for a decade [75]. In 1909, J.T. Harries received a patent in the United States for electrolysis technology that employed Al and Fe electrodes for wastewater treatment [76]. At that time, EC was not commonly used for water treatment, primarily due to the high costs associated with power and investment. However, it began to gain recognition for its effectiveness in removing both organic and inorganic pollutants from groundwater and surface water, especially when compared to traditional CC methods. It was not until 1984 that EC was implemented to treat significant volumes of drinking water in the United States [77]. Following extensive research in the late 20th century, the adoption of EC has grown due to advancements in technology that reduce electrical power consumption while enhancing effluent throughput rates. Today, the EC is recognized as a cost-effective solution for treating surface and wastewater. It boasts numerous benefits, such as environmental stability and versatility. EC is also energy-efficient and safe, offering selectivity and ease of automation, all of which enhance its overall cost-effectiveness.

3.1. Fundamentals of Electrocoagulation

The EC unit comprises an electrolytic cell containing a cathode and an anode, which are connected to an external power source and submerged in an electrolytic solution. In this setup, the anode acts as the coagulant, releasing metal cations when DC is applied to the cell. When the cell is connected to an external power supply, the anode material undergoes electrochemical corrosion due to oxidation. Consumable metal plates are commonly utilized as sacrificial electrodes to continuously generate ions in the water. These ions neutralize the charges of various particles, triggering the coagulation process. The released ions eliminate unwanted contaminants through either chemical reactions and precipitation or by facilitating the aggregation of colloidal particles, which can then be removed by flotation. As water containing colloidal particles, oils, or other impurities passes through the applied electric field, processes such as ionization, electrolysis, hydrolysis, and free-radical generation may occur. The occurrence and efficiency of these processes depend on specific factors, including the type of electrode material, current density, pH levels, and the presence of contaminants. These reactions can modify the physical and chemical characteristics of both water and contaminants. Consequently, this reactive and energized state allows for the release and destruction or reduction in solubility of contaminants in the water [73,78]. The mechanism of EC is quite intricate, involving multiple processes that work together to eliminate pollutants from water. The literature presents a diverse range of perspectives regarding the primary mechanism and reactor configurations involved. However, as per the literature, the EC process typically includes the following mechanisms. Figure 2 represents the EC process, illustrating the key reactions, including the generation of metallic cations, formation of metal hydroxides, oxidation of pollutants, and aggregation and removal of contaminants through sweep flotation.

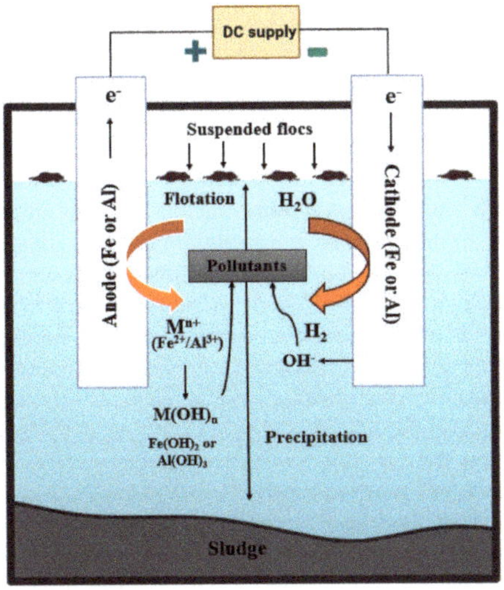

Figure 2. Schematic representation of reactions occurring in the electrocoagulation process.

(1) Sacrification of anode material: When the anode material is exposed to electrical current, it undergoes sacrification, leading to the generation of metallic cations.
(2) Hydrolysis at the cathode: Hydrolysis occurs at the cathode, generating hydroxyl ions (OH^-).
(3) Formation of metal hydroxides: The metallic cations interact with hydroxyl ions to form metal hydroxides, which have strong adsorption capabilities to bind pollutants.
(4) Oxidation of pollutants: Pollutants are oxidized to form fewer toxic species.
(5) Reaction with metal oxides: Metal oxides react with pollutants, facilitating the formation of neutralized matter.
(6) Aggregation and coagulation: The neutralized matter aggregates and adsorbs onto the metal hydroxides, followed by sweep coagulation and removal from the water.
(7) Sweep flotation: Some of the neutralized matter interacts with gasses generated in the system, leading to the entrapment of gas within the flocculated structures, which follows the process of sweep flotation [79]. Overall, the EC process generally includes chemical reactions that take place when an electric current flows through electrodes. The primary electrode reactions are as follows [14].

Anode:

$$M_{(s)} \rightarrow M^{n+}{}_{(aq)} + ne^- \quad (3)$$

$$2H_2O \rightarrow O_2 + 4H^+ + 4e^- \quad (4)$$

Cathode:

$$2H_2O + 2e^- \rightarrow 2OH^-{}_{(aq)} + H_2\,(g) \quad (5)$$

$$M^{n+} + nOH^- \leftrightarrow M\,(OH)_n \quad (6)$$

3.2. *Application of Electrocoagulation in Wastewater Treatment and Limitations*

EC equipment is engineered for ease of use, affordability, and versatility in treating various types of effluents. This section provides a broad overview of how EC is utilized in the treatment of different types of wastewater. To date, extensive research has been conducted on its effectiveness in treating wastewater from diverse sources, including municipal facilities, textile industries, oil-related operations, urban environments, industrial sectors, and palm oil effluents. The successful removal of pollutants from these diverse wastewater sources highlights its reliability as an alternative technology for wastewater treatment [80].

For biologically resistant wastewater, such as hospital effluent, traditional biological treatments may not be effective. In these cases, EC has emerged as a promising solution, which can remove up to 88.75% of the antibacterial drug Ciprofloxacin [81]. Additionally, EC has proven effective in treating wastewater containing pollutants, such as TiO_2 nanoparticles, commonly used in pharmaceutical and culinary industries. A recent study reported that EC can remove 95% of these nanoparticles [77]. Microplastics, tiny plastic particles resulting from consumer products and industrial waste, are another significant emerging pollutant. Researchers have investigated the use of EC to treat wastewater containing microplastics, finding removal rates of 93.2% for polyethylene, 91.7% for polymethyl methacrylate, 98.2% for cellulose acetate, and 98.4% for polypropylene [82]. Table 2 summarizes the applications of EC in treating various organic and inorganic pollutants across different sectors. Recent studies indicate that EC is a highly effective method for treating diverse wastewaters, including those with elevated COD, demonstrating significant removal efficiencies while generating minimal solid waste. Consequently, implementing EC treatment systems offers numerous advantages, such as faster organic matter separation, stronger pH buffering capability, reduced secondary pollution, and easier automation [83]. Furthermore, it can operate effectively across a broad spectrum of conditions, including high salinity and varying pH levels, throughout electrochemically supported potential redox reactions. It is clear that EC technology is rapidly advancing and holds significant potential to supplant traditional wastewater treatment approaches.

Table 2. Application of electrocoagulation for treating various industry effluents.

Type of Wastewater	Pollutants	Electrode Configuration (Anode–Cathode)	Operational Conditions	Removal Efficiency	Reference
Tannery wastewater	COD, BOD, TDS, and Cr	Fe-Al	Continuous (flow rate = 6 lpm; current density = 20 mA/cm^2)	46% Cr, 42% COD, 42% TDS, and 35% BOD	[84]
Textile wastewater	COD and TOC	Al-Al	pH = 5.6; current density = −52.5 mA/m^2; time = 33.9 min	68% COD and 69% TOC	[85]
Refinery wastewater	Selenium	Fe-Fe	pH = 5 to 8; current density = 15.3 mA/cm^2; voltage = 3.20 V and time = 360 min	90% selenium	[86]
Oily wastewater	COD and Turbidity	Al-STS	pH = 7; electrode spacing = 1 cm; temperature = 20–22 °C	90% COD and Turbidity = 99%	[87]
Palm oil mill effluent	COD	Al-Al	pH = 4.5; current density = 56 mA/cm^2; time = 65 min	75.4% COD	[88]
Food industry wastewater	COD and TSS	Fe-Fe	pH = 10; current density = 86.4 mA/cm^2; time = 20 min	100% TSS and 98.94% COD	[89]
Pharmaceutical wastewater	Color and COD	Al-STS	pH = 4.5–10.5; current density = 3.47–12.15 mA/cm^2; voltage = 1–15 V; time = 30–150 min	58.35% COD and 97.83% color	[90]
Municipal wastewater	TSS and BOD	STS-STS	pH = 7.0; current = 0.80 A; time = 5 min	95.4% TSS and 99% BOD	[91]

Although EC has proven effective for treating wastewater pollutants at the laboratory scale, scaling this technology for industrial applications presents challenges. One significant challenge is the high capital investment required, primarily due to the cost associated with construction and electrode materials. Commonly used electrode materials include Fe, Al, and STS. These electrodes are sacrificial and hence necessitate regular replacement. Moreover, electrode passivation can diminish overall process efficiency and increase electricity consumption. Therefore, further investigation is needed into aspects such as the proton–electron transfer mechanism, consistent performance improvements, and the development of more cost-effective anode and cathode materials [92]. The other critical challenges in scaling up EC technology are reactor operation mode and configuration. The batch and continuous operation modes are the most representative in various laboratory-scale studies. Since real wastewater continuously flows into the system, continuous mode should be introduced for successful practical application. The design of reactor configurations mainly depends on electrode arrangement. Various reactor configurations can be classified into monopolar or dipolar setups, either in series or parallel connections [93]. Among them, the most available reactor configuration for practical application is a monopolar parallel connection due to its high performance, easy electrode maintainability, and low energy consumption [77]. A future study has to be directed to demonstrate how to select and integrate appropriate reactor configuration and operation mode. Optimum standardization of comprehensive reactor designs will help the EC accelerate achieving successful practical applications and unravel reaction complexities in real wastewater [94]. Long-term operational stability is also a significant concern in EC treatment systems. Performance failures have been observed in pilot-scale studies due to various factors, including electrode degradation, floc clogging, insufficient biofilm development, and reduced activity. Addressing these issues is essential for the successful large-scale implementation of EC technology. Although both small- and large-scale EC technologies can achieve simultaneous hydrogen production, the rates are generally lower compared to laboratory studies, highlighting the need for further optimization. High electricity consumption is another major challenge that directly impacts operating costs. The success of EC technology will rely heavily on minimizing both operational and management expenses [77]. Finally, the characteristics of flocs, including their strength, size, weight, and compaction, significantly influence the EC process. Denser flocs may settle over time, while floc size tends to increase with current intensity and duration due to enhanced coagulation. However, prolonged exposure can lead to floc breakage due to shear forces [95]. In summary, while EC offers promising results for various wastewater types, further advancements in reactor design, operational stability, energy efficiency, and floc management are needed to support its widespread industrial application. Additionally, integrating the EC process with other treatment methods is advisable to address these limitations and improve the overall efficiency of the treatment system.

3.3. *Synergistic Approaches: Integration of Electrocoagulation with Conventional Techniques*

Over the years, extensive research has focused on addressing the challenges associated with EC and enhancing its performance for large-scale applications. One effective strategy has been integrating the EC with traditional methods such as CC, adsorption, and biological treatments. Numerous studies have explored the feasibility of this integration, aiming to achieve faster reaction rates, higher removal efficiency, and reduced operational costs. Figure 3 provides a graphical illustration of the integration of EC with traditional methods.

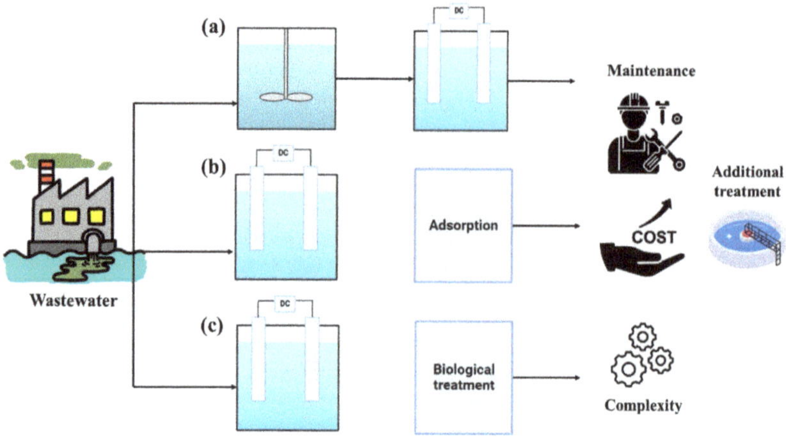

Figure 3. Integration of electrocoagulation with (**a**) chemical coagulation, (**b**) adsorption, and (**c**) biological pretreatment.

3.3.1. Electrocoagulation and Chemical Coagulation-Based Process

CC remains a prevalent method for treating wastewater from various industries but often leaves behind dissolved and SS, which need further treatment. The combination of CC with EC could enhance process efficiency and lower overall costs. Numerous studies have explored the use of CC with appropriate coagulants prior to EC. This combination aims to enhance pollutant removal while also minimizing both the cost and duration of the EC treatment phase. For example, a hybrid CC-EC method was employed to treat wastewater from the slaughterhouse industry. The method utilized PACl as a chemical coagulant up to 100 mg/L in the first stage (CC), followed by the application of Al electrodes at 40 V in the EC stage. The findings indicated that this hybrid approach effectively treated slaughterhouse wastewater, achieving over 99% removal of COD and BOD [96]. In another study, a combined EC-CC process was used to treat brewery wastewater. Operating the EC-CC system at 5 W for 20 min with Al electrodes and dissolved aluminum sulfate yielded consistent removal rates of 26% for COD, 74% for reactive phosphorus, 76% for TP, and 85% for TSS. Despite significant energy consumption, the EC-CC treatment proved economically viable when considering potential savings from reduced discharge fees. Notably, at the lower power setting of 5 W, recovery costs were 23% higher than at a 10 W supply, indicating effective processing at reduced power input [97]. A recent study investigated a combined approach of CC and EC at the pilot scale to eliminate fluoride from tungsten-melting wastewater. The pilot tests effectively reduced fluoride concentrations, with operational costs ranging from USD 0.99 to 1.51/m^3 of wastewater. The cost analysis revealed that the three most significant expenses in the CC process were liquid caustic soda, aluminum sulfate, and solid waste management. In contrast, the primary costs associated with the EC advanced treatment included the Al electrodes fee, electricity charges, and solid waste disposal [98]. These studies indicate that the combined application of CC and EC greatly improves pollutant removal from various types of wastewater. However, it is crucial to highlight that in the combined CC-EC process, CC is implemented first to manage the pollutant entering the subsequent EC stage. While the high effectiveness of the combined CC-EC method has been extensively examined in lab- scale studies, it has been less frequently evaluated at the pilot scale. This is primarily due to the significant drawback of elevated operational costs associated with CC, electrode replacement, and electricity usage.

3.3.2. Electrocoagulation and Adsorption-Based Process

Adsorption is a promising alternative technique for achieving substantial reductions in pollutants from various industrial effluents. However, the operating cost tends to be high due to the need for adsorbent regeneration and disposal. Incorporating EC prior to the adsorption process can enhance pollutant removal efficiency and lower overall operating costs. An investigation was carried out using EC as the sole treatment method, allowing for simultaneous EC and adsorption with Al electrodes. The study utilized a central composite design to optimize pH, current density, and reaction time to assess their effects on the removal of COD, total organic carbon (TOC), and TSS from landfill leachate. Under optimal conditions of pH 7.35, a current density of 15.29 mA/cm^2, and a reaction time of 57 min, the removal efficiencies were 83.56% for COD, 73.12% for TSS, and 85.58% for TOC [99]. Similarly, another study explored the effectiveness of combining EC with adsorption using natural zeolite for treating industrial wastewater. The study utilized response surface methodology to optimize various operational parameters, including pH, current density, reaction time, and zeolite dosage. Optimal conditions of pH 7, a current density of 38 mA/m^2, an electrolysis duration of 20 min, and a zeolite dosage of 0.183 g/mol resulted in 92% COD removal and 97% turbidity reduction [100]. Additionally, a study assessed a solar photovoltaic power EC-assisted adsorption system for treating pharmaceutical wastewater. A COD removal rate of 95.5% was achieved at optimum conditions of 20 min reaction time, 6.656 mA/cm^2 current density, and a temperature of 45 °C with an operating cost of USD 0.273/m^3 of wastewater [101]. It is crucial to note that all these studies employed EC as a preliminary treatment before adsorption, ultimately enhancing overall process efficacy. However, the integration of EC and adsorption presents significant challenges, including the need for electrode replacement, adsorbent regeneration, and process optimization. As a result, this combined approach is infrequently examined in pilot-scale studies.

3.3.3. Electrocoagulation and Bioprocess-Based Treatment

EC combined with biological treatments has emerged as an effective technology for addressing heavily contaminated industrial effluents. While biological treatments have demonstrated significant potential in removing organic pollutants from wastewater, they often struggle with non-biodegradable and toxic substances. In this context, EC presents a promising solution that can be integrated with biological processes to improve contaminant removal. A study was conducted to assess the effectiveness of integrating EC with biological fungal treatment to treat tannery wastewater. RSM was employed to examine how different conditions influenced treatment efficiency. The findings revealed that using Al-Fe electrodes under optimal conditions resulted in removal efficiencies of 96% for COD and 97% for Cr^{6+} [102]. In another study, EC combined with biological treatment effectively addressed wastewater from the oil industry. The process of EC followed by aerobic biofiltration significantly increased the biodegradability of total petroleum hydrocarbons, achieving reductions of 95% in COD and 98% in total petroleum hydrocarbons [103]. A more recent study explored a modified biological integrated EC method for treating municipal wastewater, aiming to reuse the treated water for irrigation. This combined approach achieved notable removal efficiencies: 78.8% for turbidity, 56.8% for hardness, 28.4% for conductivity, 37.4% for total dissolved solids (TDS), 98.3% for TSS, 27.6% for chloride, 26.7% for NH_3-N, 78% for BOD, 81% for COD, and an impressive 99.9% for total coliforms. The process consumed 9.9 Wh/L of energy, with an operational cost of USD 0.76/m^3 of municipal wastewater [104]. However, this approach faces challenges, including the complexity of the integrated system, the need for additional equipment, higher maintenance costs, and the optimization of operational parameters, which limits its broader application.

3.4. Next-Generation Approaches: Electrocoagulation Integration with Emerging Technologies

The integration of EC with traditional approaches such as biological treatment, CC, and adsorption often falls short of effectively removing contaminants like dyes, salts, and surfactants. Furthermore, these conventional methods are complex and require significant investment. Considering these challenges, researchers have focused their efforts on developing innovative hybrid technologies that can integrate with EC. This advancement aims to simplify processes, lower operational costs, and enhance the removal of contaminants from wastewater.

3.4.1. Electrocoagulation-Membrane-Based Treatment

Membrane separation processes offer great simplicity and energy efficiency compared to other separation techniques, making them ideal for eliminating contaminants. Several membrane-based methods, including ultrafiltration (UF), reverse osmosis (RO), forward osmosis (FO), membrane distillation (MD), nanofiltration (NF), electrodialysis (ED), and membrane bioreactors (MBR) have demonstrated superior effectiveness in removing pollutants from different industrial wastewater. However, these technologies face significant challenges, mainly fouling, which can lead to increased operational costs despite their ability to eliminate nearly all contaminants. To mitigate this issue, pretreatment methods are often employed before the membrane processes to improve efficiency and reduce costs. In this context, integrating the EC as a pretreatment step before membrane technology presents a versatile and promising solution for water treatment characterized by reduced space requirements, minimal chemical usage, and enhanced efficiency. For example, a hybrid EC-membrane technology for industrial wastewater treatment is illustrated in Figure 4. Numerous studies have investigated integrating various membrane technologies with EC to improve the removal of contaminants from industrial effluents. One study, for instance, combined EC with UF and RO to remove pollutants from hospital wastewater. UF uses a membrane with pore sizes of 1 to 100 nanometers to retain larger molecules, while RO employs a membrane with pore sizes of approximately 0.0001 microns to remove dissolved salts and small contaminants. The EC-UF configuration resulted in TSS, TDS, BOD, and COD removal efficiency of 95.12%, 97.53%, 95.18%, and 97.88%, respectively, with an operating cost of USD 3.92/m^3 of wastewater. In contrast, the EC-RO configuration demonstrated even higher efficiencies with TSS, TDS, BOD, and COD removal rates of 97.64%, 99.85%, 97.88%, and 98.38%, respectively, at an operating cost of USD 4.02/m^3 of wastewater. When the treatment scale was increased to 50 m^3/day, the operating costs dropped to USD 0.89/m^3 for EC-UF configuration and USD 0.93/m^3 for EC-RO configuration [105]. In another study, EC was integrated with UF membrane to treat palm oil mill effluent, revealing that using a bipolar electrode significantly enhanced coagulation efficiency. The removal efficiencies for TDS, TSS, COD, and BOD were 59.1%, 99.9%, 96.8%, and 96%, respectively, with an estimated operating cost of USD 2.71/m^3 and an energy requirement of 6.20 kWh/m^3 [106]. FO is recognized as an innovative membrane technology for freshwater production, operating on the principle of natural osmotic pressure gradients. A lab-scale hybrid EC-FO setup was tested for treating produced water, achieving a maximum flux of 1.2 LPM under optimal conditions with a current density of 10 mA/cm^2 and a residence time of 10 min for the EC process. This configuration effectively removed 99% of TSS, 98% of turbidity, and 16% of conductivity [107]. MD, another separation technique, involves vapor molecules passing through a porous hydrophobic membrane. Utilizing EC as a pretreatment for MD offers several advantages, particularly in preventing membrane wetting, which occurs when water permeates through the membrane pores. A study on a hybrid EC-MD process for treating produced water from hydraulic fracturing found that EC effectively reduces TOC, which is crucial for minimizing membrane fouling during the MD process [108]. ED is an electrically powered membrane process that drives ions through a membrane using an electric current. It boasts high selectivity, excellent recovery rates, and the ability to remove most contaminants from raw water. However, additional treatment steps are often necessary for effective wastewater treatment. One study demonstrated the success of a hybrid EC-ED process in treating tannery wastewater, achieving a COD removal efficiency of 92% when Al electrodes were used while eliminating NH_3-N, chromium (Cr), and color. When Fe electrodes were employed, the COD removal efficiency was 87%, with complete removal of NH_3-N, Cr, and color [109]. Additionally, a submerged membrane reactor was integrated with EC for greywater treatment. With a voltage gradient of 1.26 V/cm, the removal efficiencies for turbidity, color, COD, NH_3-N, and total phosphorous (TP) were found to be 100%, 99.7%, 92%, 94.1%, and 96.5%, respectively [110]. While utilizing EC as a pretreatment method marks a significant advancement in water treatment technology, membrane fouling remains a critical challenge for long-term operation. Researchers have extensively studied the various types of fouling, their causes, and potential mitigation strategies. By examining particle size distribution, differing influx ratios, and the condition of fouled membranes, it has been found that approximately 75–85% of the initial flux can be recovered when filtering EC-treated wastewater [111]. Furthermore, pore-blocking models have been developed to analyze the nature of fouling on membranes exposed to various effluents. Hermia's models are notably recognized for their effectiveness in evaluating types of fouling in membrane filtration processes [112]. Ultimately, a fully optimized and controlled EC process can significantly enhance membrane filtration performance by minimizing fouling, presenting substantial potential for efficiently treating industrial effluents while reducing energy consumption and costs.

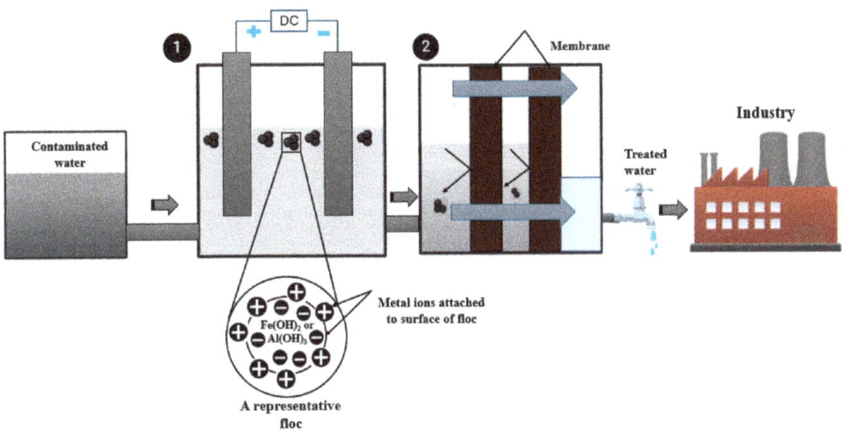

Figure 4. Schematic depiction of hybrid electrocoagulation-membrane technology for treating industrial effluents.

3.4.2. Electrocoagulation–Electrochemical Processes

Hybrid EC-membrane-based processes have shown great potential for wastewater treatment. Several electrochemical-based processes such as electrooxidation, ozonation, Fenton, and their integration with EC have also shown promise. However, there is a growing need to explore and adapt these technologies to improve treatment efficiency and effectively address the full spectrum of contaminants found in wastewater. EC has already proven effective in removing numerous pollutants and microbial pathogens. When combined with advanced electrochemical methods, it not only improves treatment efficiency but also supports environmental sustainability. This combination is user-friendly and characterized by low energy consumption and ease of operation [113]. Advanced oxidation processes, in particular, are highly effective as they generate a variety of reactive species capable of breaking down organic pollutants. Different methods for generating radicals include electroreduction, electro-Fenton, photo-Fenton, and ozonation [114]. Overall, integrating EC with various electrochemical processes offers a holistic approach to wastewater treatment, leveraging the strengths of both technologies.

Among the various electrochemical treatment systems, the combination of EC and electrooxidation (EO) is frequently utilized for wastewater treatment. EO is a sophisticated oxidation method that relies on electrolytic reactions occurring at the surface of the electrodes. In this process, pollutants present in the wastewater are drawn to the electrode surface, where they undergo oxidation after adsorption. Using EC as a preliminary treatment, colloidal and suspended particles can be rapidly coagulated, enabling the EO system to eliminate any remaining pollutants. A combined EC and EO process has been employed to treat industrial wastewater. Using the EC system alone, the removal efficiencies for COD, BOD, total coliforms, color, and turbidity were recorded at 85.6%, 46.4%, 99%, 52%, and 83.8%, respectively, at a current density of 80 mA/cm^2. In contrast, the hybrid EC-EO system rapidly coagulated and eliminated colloidal particles, while the EO component effectively addressed the remaining particles, achieving over 99% removal of all contaminants [115]. In another investigation, the combined EC-EO method was utilized for water treatment. Although EO effectively removed certain trace organic compounds, its performance was limited by the presence of dissolved organic carbon. However, when paired with EC, the removal of these trace contaminants improved, and the interference from dissolved organic carbon was mitigated. This synergistic approach resulted in higher pollutant removal rates than when each method was applied individually [116]. Additionally, research on the EC-EO integration for treating industrial container wash wastewater, characterized by high levels of COD and phosphorus, demonstrated significant removal rates of 97% for phosphorus and 95% for COD, successfully meeting discharge regulations [117]. These findings highlight the potential of the combined EC-EO process to enhance wastewater treatment by effectively addressing a variety of pollutants. Nonetheless, it is essential to note that this hybrid system may not be universally effective for all types of wastewater, particularly those containing surfactants and cleaning agents. Therefore, further research is essential to optimize operational parameters, assess the applicability of these processes across different wastewater types, and evaluate the scalability and cost-effectiveness of this combined approach.

Ozonation is a highly effective technology that utilizes ozone as a powerful oxidant to break down various pollutants, including dyes. This process generates hydroxyl radicals, solid and non-selective oxidants, allowing for the indirect oxidation of a wide range of contaminants. Although ozonation has been extensively researched on a pilot scale due to its high efficiency, it has limitations when dealing with high-molecular-weight pollutants. To address this, a synergistic method that combines EC with ozonation has emerged as a promising solution. One study evaluated the effectiveness of the EC-ozonation process in treating greywater. Under optimal conditions—using Fe electrodes, a neutral pH, an ozone concentration of 47.4 mg/L, and a current density of 15 mA/cm^2—significant reductions were achieved: 85% for COD and 70% for TOC [118]. Another investigation proposed a combined ozone-assisted EC approach for degrading color and COD in distillery spent wash. This continuous process attained impressive removal rates of 97.3% for COD and 98.7% for color [119]. Additionally, a study introduced a method that integrates sedimentation, EC, and ozonation for wastewater treatment, achieving remarkable removal efficiencies of 90.9% for COD, 100% for color, 73.7% for total solids (TS), and 99.7% for SS [120]. These findings collectively illustrate the potential of the combined EC-ozonation approach for enhancing wastewater treatment by leveraging the strengths of both methods. Nonetheless, further research is essential to optimize the process and assess its applicability to various wastewaters, aiming for the cost-effective scalability of this combined treatment strategy.

Electro-Fenton (EF) is an advanced oxidation technique designed to eliminate persistent organic pollutants by generating hydroxyl radicals through specific mechanisms. This process is categorized into homogeneous and heterogeneous EF based on the mechanism of pollutant degradation. In homogeneous EF, the catalyst dissolves, leading to the generation of hydroxyl radicals, which subsequently oxidize organic compounds. Conversely, heterogeneous EF occurs when Fe^{2+} is replaced with Fe^{3+} or other transition metal ions. While EF processes are highly effective for treating various pollutants, they require pH adjustments, which can increase operational costs. To address these limitations, combining EC with EF has been proposed to enhance efficacy. One study introduced a dual anode system that integrates EC and EF processes to treat leachate concentrate, achieving a removal rate of 57% for organic materials and 60% for NH_3 under specific conditions. The radicals produced by this dual-electrode system effectively oxidized both organic substances and other persistent pollutants [121]. Another research effort implemented a combined EC and EF approach to treat tannery wastewater, resulting in a COD removal rate of 88.1 ± 4.8% and complete elimination of Cr. This combined method demonstrated a lower global environmental impact compared to using each process individually [122]. Additionally, a recent study explored a novel integration of heterogeneous EF and EC for the efficient removal of Cu-ciprofloxacin complexes. Under optimal conditions, removal efficiencies of 99.6% for Cu, 96.4% for ciprofloxacin, and 83.6% for TOC were achieved. The primary mechanism involved the degradation of Cu-ciprofloxacin complexes into smaller molecules, releasing Cu^{2+} ions [123]. These findings collectively highlight the potential of combined EC-EF processes for effectively removing contaminants from wastewater. However, further research is necessary to optimize the parameters, evaluate scalability, and assess the cost-effectiveness of this integrated approach.

In summary, integrating EC with emerging technologies such as membrane processes and advanced electrochemical techniques presents a promising approach for addressing highly contaminated wastewater. This combination reduces the formation of passive films on electrode surfaces, boosts radical generation, and improves the degradation of pollutants. As outlined in Table 3, comprehensive research highlights the effectiveness of EC combined with membrane and electrochemical technologies in eliminating a range of contaminants, including phenols, dyes, oils, heavy metals, COD, BOD, and TDS, achieving high removal rates. These integrated processes demonstrate enhanced treatment efficiency compared to traditional methods, making them well suited for managing residual effluent and promoting sustainable practices in water and wastewater treatment. However, it is crucial to address the long-term sustainability and cost-effectiveness of these integrated systems, particularly concerning electrode replacement and power consumption. Therefore, long-term pilot-scale studies should be conducted to better understand the economic viability and operational implications of these advanced technologies in real-world applications.

Table 3. Summary of electrocoagulation integrated with membrane and electrochemical-based processes.

New Integration Technology	Hybrid Process	Wastewater	Operating Conditions	Results	Reference
Membrane-based integration	EC-UF and EC-RO	Hospital wastewater	Electrode configuration: 2A-2C-2B) and (4A-2C-2B); current density: 88.5 A/m^2.	EC-UF: 95.12% TSS, 97.53% TDS; 95.18% BOD and 97.88% COD removal. EC-RO: 97.64% TSS, 99.85% TDS; 97.88% BOD, and 98.38% COD removal.	[105]
	EC-UF	Palm oil mill effluent	Electrode configuration: 2A-2C-2B.	59.1% TDS, 99.9% TSS, 96.8% COD, and 96% BOD removal.	[106]
	EC-FO	Produced water	EC: 10 mA/cm^2, 10 min residence time. FO: 1.2 LPM FS-DS flowrate in PRO mode.	99% TSS, 98% turbidity, 16% conductivity.	[107]
	EC-MD	Hydraulic-fracturing-produced water	Al/Fe electrodes	Effective TOC removal, insoluble species were effectively coagulated rather than dissolved.	[108]
	EC-ED	Tannery wastewater	Al: 0.371 mS/cm at 45 min. Fe: 1.5 mS/cm at 75 min.	92% COD and 100% NH_3-N, Cr, color for Al electrodes. 87% COD and 100% NH_3-N, Cr, color for Fe electrodes.	[109]
	EC-Submerged MBR	Gray water	EC: AL electrodes at voltage gradient of 1.26 V/cm.	100% turbidity, 99.7% color, 92% COD, 94.1% NH_3-N and 96.5% TP.	[110]

Table 3. *Cont.*

New Integration Technology	Hybrid Process	Wastewater	Operating Conditions	Results	Reference
Electrochemical integration	EC-EO	Industrial wastewater	EC: current density of 80 mA/cm^2. EO: boron-doped diamond anode and Fe cathode.	EC: 85.6% COD, 46.4% BOD, 99% total coliforms, 52% color, 83.8% turbidity. Hybrid EC-EO: 99% contaminant removal.	[115]
	EC-EO	Drinking water	EO: boron-doped diamond electrodes, current density of 14.8 mA/cm^2. EC: Fe electrodes, current density of 1.85–11.1 mA/cm^2.	EO is affected by the presence of dissolved organic carbon. EC removed 74 ± 7% dissolved organic carbon from water. Hybrid EC-EO effectively removed all contaminants.	[116]
	EC-EO	Industrial container wash water	EC followed by EO with boron-doped electrodes and current density of 0.12 A/cm^2.	The EC-EO hybrid process removed 97% phosphorus and 95% COD. EC-EO reduced treatment cost around 3.3 times more than individual.	[117]
	EC-ozonation	Gray water	60 min electrolysis time, pH 7.0, 47.4 mg/L ozone, 15 mA/cm^2 current density.	85% COD, 70% COD were removed. EC with Fe electrodes exhibited high catalytic activity.	[118]
	EC-ozonation	Distillery spent wash	Al electrodes	Combined ozone-assisted EC approach achieved 97.3% COD and 98.7% color removal.	[119]
	EC-ozonation	Fiberboard industry wastewater	EC: Al anode and Fe cathode with current density of 214.3 A/m^2. Ozonation: flow of 2.5 g/h at initial pH of 7.4.	Hybrid process achieved 90.9% COD, 100% color, 73.7 TS, and 99.7% SS.	[120]
	EC-EF	Leachate concentrate	Dual anode system, electrical charge of 7 Ah/L, potential of 7 V, pH 7.	57% organics, 60% NH_3 were removed. Most of the fulvic and humic substances were removed.	[121]
	EC-EF	Tannery wastewater	EC: mild steel electrodes. EF: air diffusion cathode and boron-doped diamond anode.	The hybrid 2h EF-5 h EC process achieved 88.1 ± 4.8 COD and total elimination of Cr. The electrical energy consumption was reduced to 1.7 times.	[122]
	EC-EF	Heavy metal wastewater	N-Co/Fe-PC cathode and graphite anode. Current: 2–150 mA, pH: 3–7.	99.69% Cu, 96.40% Ciprofloxacin, and 83.62% TOC removed.	[123]

4. Conclusions and Future Perspective

Coagulation is a method used to eliminate contaminants from wastewater effectively. Both synthetic and natural coagulants have been recognized as effective materials in this process. Natural coagulants have been utilized for decades, even before the advent of chemical coagulants. However, the latter have gained significant popularity recently despite their drawbacks, such as high costs and the production of toxic byproducts. This has prompted a shift towards natural coagulants, which are more cost-effective and environmentally friendly. Nonetheless, concerns regarding their effectiveness on a larger scale often limit their use.

EC has emerged as a highly efficient and eco-friendly technology for contaminant removal in water and wastewater treatment. Like other treatment methods, EC faces challenges, including electrode passivation and substantial energy consumption, mainly due to the lower conductivity of specific wastewater. Additionally, EC is less effective at removing stable, persistent organic compounds to levels that meet discharge regulations. Our findings suggest that integrating EC with established treatment methods, such as CC, adsorption, and biological treatment, presents a promising avenue for enhancing wastewater treatment efficiency. However, this integrated approach faces several challenges related to operational economics, including high operational costs, the need for adsorbent regeneration, process optimization, and increased maintenance expenses. Furthermore, these methods have yet to be studied at the pilot scale. In this context, emerging technologies like membrane processes and advanced electrochemical techniques, particularly when integrated with EC, offer significant potential. Utilizing EC as a pretreatment method before these advanced techniques provides several advantages, including increased permeate water volume, improved water quality, reduced membrane fouling, and lower operational costs for hybrid systems. However, for the commercialization of hybrid EC processes, several challenges must be addressed. These include minimizing the consumption of sacrificial electrodes, which necessitates further research into cost-effective alternatives. Additionally, more pilot-scale studies are needed to evaluate the effectiveness of hybrid EC processes on various industrial wastewater. Long-term assessments should also be conducted to determine the efficacy of different hybrid EC systems. Addressing these research gaps and challenges will significantly enhance the field of hybrid EC processes, providing sustainable and cost-effective solutions for water and wastewater treatment.

Author Contributions: S.J., writing—original draft preparation, writing—review, and editing; R.K., writing—original draft preparation and writing—review and editing; H.J. and D.S., writing—review and editing; J.P., conceptualization, writing—review and editing, supervision, and funding acquisition. All authors have read and agreed to the published version of the manuscript.

Funding: This research was funded by Chosun University 2023 [grant number K208702003] and "Regional Innovation Strategy (RIS)" through the National Research Foundation of Korea (NRF), funded by the Ministry of Education (MOE) [2021RIS-002].

Conflicts of Interest: The authors declare no conflicts of interest.

Abbreviations

The following abbreviations are used in this manuscript.

CC: chemical coagulation; EC: electrocoagulation; SS: suspended solids; Al: aluminum; Fe: iron; PAC: polyaluminum chloride; WWTP: wastewater treatment plant; CEPT: chemical enhanced primary treatment; COD: chemical oxygen demand; BOD: biological oxygen demand; TSS: total suspended solids; RSM: response surface methodology; ANN: artificial neural network; GA: genetic algorithm; Cu: copper; STS: stainless steel; TOC: total organic carbon; TDS: total dissolved solids; UF: ultrafiltration; RO: reverse osmosis; FO: forward osmosis; MD: membrane distillation; NF: nanofiltration; ED: electrodialysis; MBR: membrane bioreactor; TP: total phosphorous; EO: electrooxidation; TS: total solids; Cr: chromium; EF: electro-Fenton; EPA: US Environmental Protection Agency.

References

1. Kishor, R.; Purchase, D.; Saratale, G.D.; Saratale, R.G.; Ferreira, L.F.R.; Bilal, M.; Chandra, R.; Bharagava, R.N. Ecotoxicological and health concerns of persistent coloring pollutants of textile industry wastewater and treatment approaches for environmental safety. *J. Environ. Chem. Eng.* **2021**, *9*, 105012. [CrossRef]
2. Kishor, R.; Raj, A.; Bharagava, R.N. Synergistic role of bacterial consortium (RKS-AMP) for treatment of recalcitrant coloring pollutants of textile industry wastewater. *J. Water Process Eng.* **2022**, *47*, 102700. [CrossRef]
3. Kodešová, R.; Švecová, H.; Klement, A.; Fér, M.; Nikodem, A.; Fedorova, G.; Rieznyk, O.; Kočárek, M.; Sadchenko, A.; Chroňáková, A. Contamination of water, soil, and plants by micropollutants from reclaimed wastewater and sludge from a wastewater treatment plant. *Sci. Total Environ.* **2024**, *907*, 167965. [CrossRef] [PubMed]
4. Liu, W.; Huang, F.; Liao, Y.; Zhang, J.; Ren, G.; Zhuang, Z.; Zhen, J.; Lin, Z.; Wang, C. Treatment of CrVI-containing Mg (OH) 2 nanowaste. *Angew. Chem.* **2008**, *120*, 5701–5704. [CrossRef]
5. Radoiu, M.T.; Martin, D.I.; Calinescu, I.; Iovu, H. Preparation of polyelectrolytes for wastewater treatment. *J. Hazard. Mater.* **2004**, *106*, 27–37. [CrossRef]
6. Abujazar, M.S.S.; Karaağaç, S.U.; Amr, S.S.A.; Alazaiza, M.Y.; Bashir, M.J. Recent advancement in the application of hybrid coagulants in coagulation-flocculation of wastewater: A review. *J. Clean. Prod.* **2022**, *345*, 131133. [CrossRef]

7. Khan, R.; Inam, M.A.; Park, D.R.; Zam Zam, S.; Shin, S.; Khan, S.; Akram, M.; Yeom, I.T. Influence of organic ligands on the colloidal stability and removal of ZnO nanoparticles from synthetic waters by coagulation. *Processes* **2018**, *6*, 170. [CrossRef]
8. Noor, M.H.M.; Ngadi, N. Global research landscape on coagulation-flocculation for wastewater treatment: A 2000–2023 bibliometric analysis. *J. Water Process Eng.* **2024**, *64*, 105696. [CrossRef]
9. Bratby, J. *Coagulation and Flocculation in Water and Wastewater Treatment*; IWA Publishing: London, UK, 2016.
10. Singh, R.P.; Karmakar, G.; Rath, S.; Karmakar, N.; Pandey, S.; Tripathy, T.; Panda, J.; Kanan, K.; Jain, S.; Lan, N. Biodegradable drag reducing agents and flocculants based on polysaccharides: Materials and applications. *Polym. Eng. Sci.* **2000**, *40*, 46–60. [CrossRef]
11. Zhao, S.; Huang, G.; Fu, H.; Wang, Y. Enhanced coagulation/flocculation by combining diatomite with synthetic polymers for oily wastewater treatment. *Sep. Sci. Technol.* **2014**, *49*, 999–1007. [CrossRef]
12. Bahrodin, M.B.; Zaidi, N.S.; Hussein, N.; Sillanpää, M.; Prasetyo, D.D.; Syafiuddin, A. Recent advances on coagulation-based treatment of wastewater: Transition from chemical to natural coagulant. *Curr. Pollut. Rep.* **2021**, *7*, 379–391. [CrossRef]
13. Diver, D.; Nhapi, I.; Ruziwa, W.R. The potential and constraints of replacing conventional chemical coagulants with natural plant extracts in water and wastewater treatment. *Environ. Adv.* **2023**, *13*, 100421. [CrossRef]
14. Shah, A.A.; Walia, S.; Kazemian, H. Advancements in combined electrocoagulation processes for sustainable wastewater treatment: A comprehensive review of mechanisms, performance, and emerging applications. *Water Res.* **2024**, *252*, 121248. [CrossRef] [PubMed]
15. Vepsäläinen, M.; Sillanpää, M. Electrocoagulation in the treatment of industrial waters and wastewaters. In *Advanced Water Treatment*; Elsevier: Amsterdam, The Netherlands, 2020; pp. 1–78.
16. Nidheesh, P.; Singh, T.A. Arsenic removal by electrocoagulation process: Recent trends and removal mechanism. *Chemosphere* **2017**, *181*, 418–432. [CrossRef]
17. Akter, S.; Suhan, M.B.K.; Islam, M.S. Recent advances and perspective of electrocoagulation in the treatment of wastewater: A review. *Environ. Nanotechnol. Monit. Manag.* **2022**, *17*, 100643. [CrossRef]
18. Agunbiade, M.O.; Pohl, C.H.; Ashafa, A.O. A Review of the Application of Biofloccualnts in Wastewater Treatment. *Pol. J. Environ. Stud.* **2016**, *25*, 1381–1389. [CrossRef]
19. Ghernaout, D.; Ghernaout, B. Sweep flocculation as a second form of charge neutralisation—A review. *Desalin. Water Treat.* **2012**, *44*, 15–28. [CrossRef]
20. Xia, X.; Lan, S.; Li, X.; Xie, Y.; Liang, Y.; Yan, P.; Chen, Z.; Xing, Y. Characterization and coagulation-flocculation performance of a composite flocculant in high-turbidity drinking water treatment. *Chemosphere* **2018**, *206*, 701–708. [CrossRef]
21. Jiao, R.; Fabris, R.; Chow, C.W.; Drikas, M.; van Leeuwen, J.; Wang, D.; Xu, Z. Influence of coagulation mechanisms and floc formation on filterability. *J. Environ. Sci.* **2017**, *57*, 338–345. [CrossRef]
22. Li, T.; Zhu, Z.; Wang, D.; Yao, C.; Tang, H. Characterization of floc size, strength and structure under various coagulation mechanisms. *Powder Technol.* **2006**, *168*, 104–110. [CrossRef]
23. Lek, B.L.C.; Peter, A.P.; Chong, K.H.Q.; Ragu, P.; Sethu, V.; Selvarajoo, A.; Arumugasamy, S.K. Treatment of palm oil mill effluent (POME) using chickpea (*Cicer arietinum*) as a natural coagulant and flocculant: Evaluation, process optimization and characterization of chickpea powder. *J. Environ. Chem. Eng.* **2018**, *6*, 6243–6255.
24. Ma, C.; Hu, W.; Pei, H.; Xu, H.; Pei, R. Enhancing integrated removal of Microcystis aeruginosa and adsorption of microcystins using chitosan-aluminum chloride combined coagulants: Effect of chemical dosing orders and coagulation mechanisms. *Colloids Surf. A Physicochem. Eng. Asp.* **2016**, *490*, 258–267. [CrossRef]
25. Wang, B.; Shui, Y.; He, M.; Liu, P. Comparison of flocs characteristics using before and after composite coagulants under different coagulation mechanisms. *Biochem. Eng. J.* **2017**, *121*, 107–117. [CrossRef]
26. Choy, S.Y.; Prasad, K.N.; Wu, T.Y.; Raghunandan, M.E.; Ramanan, R.N. Performance of conventional starches as natural coagulants for turbidity removal. *Ecol. Eng.* **2016**, *94*, 352–364. [CrossRef]
27. Luo, B.; He, H.; Yan, Y.; Wang, Y.; Yang, X.; Liu, Y.; Xu, J.; Huang, W. Flocculants for the High-Concentration Activated Sludge Method and the Effectiveness of Urban Wastewater Treatment. *Water* **2024**, *16*, 2281. [CrossRef]
28. Khalidi-Idrissi, A.; Madinzi, A.; Anouzla, A.; Pala, A.; Mouhir, L.; Kadmi, Y.; Souabi, S. Recent advances in the biological treatment of wastewater rich in emerging pollutants produced by pharmaceutical industrial discharges. *Int. J. Environ. Sci. Technol.* **2023**, *20*, 11719–11740. [CrossRef]
29. Luo, L.; Dzakpasu, M.; Yang, B.; Zhang, W.; Yang, Y.; Wang, X.C. A novel index of total oxygen demand for the comprehensive evaluation of energy consumption for urban wastewater treatment. *Appl. Energy* **2019**, *236*, 253–261. [CrossRef]
30. Samuel, O.; Othman, M.H.D.; Kamaludin, R.; Sinsamphanh, O.; Abdullah, H.; Puteh, M.H.; Kurniawan, T.A.; Li, T.; Ismail, A.F.; Rahman, M.A. Oilfield-produced water treatment using conventional and membrane-based technologies for beneficial reuse: A critical review. *J. Environ. Manag.* **2022**, *308*, 114556. [CrossRef]
31. Khatri, N.; Khatri, K.K.; Sharma, A. Enhanced energy saving in wastewater treatment plant using dissolved oxygen control and hydrocyclone. *Environ. Technol. Innov.* **2020**, *18*, 100678. [CrossRef]
32. Lichtmannegger, T.; Hell, M.; Wehner, M.; Ebner, C.; Bockreis, A. Seasonal tourism's impact on wastewater composition: Evaluating the potential of alternating activated adsorption in primary treatment. *Sci. Total Environ.* **2024**, *926*, 171869. [CrossRef]

33. Abdelrahman, A.M.; Kosar, S.; Gulhan, H.; Cicekalan, B.; Ucas, G.; Atli, E.; Guven, H.; Ozgun, H.; Ozturk, I.; Koyuncu, I. Impact of primary treatment methods on sludge characteristics and digestibility, and wastewater treatment plant-wide economics. *Water Res.* **2023**, *235*, 119920. [CrossRef] [PubMed]
34. Checa-Fernández, A.; Ruiz, L.; Torre-Marín, J.; Muñoz-Ubina, A.; Pérez, J.; Gómez, M. Direct application of chemically enhanced primary treatment in a municipal wastewater treatment plant: A case study. *Chem. Eng. Res. Des.* **2024**, *204*, 183–192. [CrossRef]
35. Budych-Gorzna, M.; Szatkowska, B.; Jaroszynski, L.; Paulsrud, B.; Jankowska, E.; Jaroszynski, T.; Oleskowicz-Popiel, P. Towards an energy self-sufficient resource recovery facility by improving energy and economic balance of a municipal WWTP with chemically enhanced primary treatment. *Energies* **2021**, *14*, 1445. [CrossRef]
36. Maktabifard, M.; Zaborowska, E.; Makinia, J. Achieving energy neutrality in wastewater treatment plants through energy savings and enhancing renewable energy production. *Rev. Environ. Sci. Bio/Technol.* **2018**, *17*, 655–689. [CrossRef]
37. Zhuang, H.; Guan, J.; Leu, S.-Y.; Wang, Y.; Wang, H. Carbon footprint analysis of chemical enhanced primary treatment and sludge incineration for sewage treatment in Hong Kong. *J. Clean. Prod.* **2020**, *272*, 122630. [CrossRef]
38. Mu, W.; Dagnew, M. Enhancing biomethane production and phosphorus recovery from CEPT sludge through a low temperature thermal alkali and ozonation pretreatment processes. *Case Stud. Chem. Environ. Eng.* **2022**, *5*, 100178. [CrossRef]
39. El-taweel, R.M.; Mohamed, N.; Alrefaey, K.A.; Husien, S.; Abdel-Aziz, A.; Salim, A.I.; Mostafa, N.G.; Said, L.A.; Fahim, I.S.; Radwan, A.G. A review of coagulation explaining its definition, mechanism, coagulant types, and optimization models; RSM, and ANN. *Curr. Res. Green Sustain. Chem.* **2023**, *6*, 100358. [CrossRef]
40. Shewa, W.A.; Dagnew, M. Revisiting chemically enhanced primary treatment of wastewater: A review. *Sustainability* **2020**, *12*, 5928. [CrossRef]
41. Akinnawo, S.O.; Ayadi, P.O.; Oluwalope, M.T. Chemical coagulation and biological techniques for wastewater treatment. *Ovidius Univ. Ann. Chem.* **2023**, *34*, 14–21. [CrossRef]
42. Luo, H.; Sun, Y.; Taylor, M.; Nguyen, C.; Strawn, M.; Broderick, T.; Wang, Z.W. Impacts of aluminum-and iron-based coagulants on municipal sludge anaerobic digestibility, dewaterability, and odor emission. *Water Environ. Res.* **2022**, *94*, e1684. [CrossRef]
43. Daryabeigi Zand, A.; Hoveidi, H. Comparing aluminium sulfate and poly-aluminium chloride (PAC) performance in turbidity removal from synthetic water. *J. Appl. Biotechnol. Rep.* **2015**, *2*, 287–292.
44. Abdo, S.M.; Mahmoud, R.H.; Youssef, M.; El-Naggar, M.E. Cationic starch and polyaluminum chloride as coagulants for River Nile water treatment. *Groundw. Sustain. Dev.* **2020**, *10*, 100331. [CrossRef]
45. Tripathy, A.; Mahalik, S.; Sarangi, C.; Tripathy, B.; Sanjay, K.; Bhattacharya, I. A pyro-hydrometallurgical process for the recovery of alumina from waste aluminium dross. *Miner. Eng.* **2019**, *137*, 181–186. [CrossRef]
46. Mbaeze, M.C.; Agbazue, V.; Orjioke, N. Comparative assessment of performance of aluminium sulphate (alum) and ferrous sulphate as coagulants in water treatment. *Mod Chem. Appl* **2017**, *5*, 1–14.
47. Sadeghi, N.; Moghaddam, J.; Ilkhchi, M.O. Determination of effective parameters in pilot plant scale direct leaching of a zinc sulfide concentrate. *Physicochem. Probl. Miner. Process* **2017**, *53*, 601–616.
48. Bouchareb, R.; Derbal, K.; Özay, Y.; Bilici, Z.; Dizge, N. Combined natural/chemical coagulation and membrane filtration for wood processing wastewater treatment. *J. Water Process Eng.* **2020**, *37*, 101521. [CrossRef]
49. Zhou, Z.; Zhang, C.; Xi, M.; Ma, H.; Jia, H. Multi-scale modeling of natural organic matter–heavy metal cations interactions: Aggregation and stabilization mechanisms. *Water Res.* **2023**, *238*, 120007. [CrossRef]
50. Li, N.; Hu, Y.; Lu, Y.-Z.; Zeng, R.J.; Sheng, G.-P. Multiple response optimization of the coagulation process for upgrading the quality of effluent from municipal wastewater treatment plant. *Sci. Rep.* **2016**, *6*, 26115. [CrossRef]
51. Kurniawan, S.B.; Abdullah, S.R.S.; Imron, M.F.; Said, N.S.M.; Ismail, N.I.; Hasan, H.A.; Othman, A.R.; Purwanti, I.F. Challenges and opportunities of biocoagulant/bioflocculant application for drinking water and wastewater treatment and its potential for sludge recovery. *Int. J. Environ. Res. Public Health* **2020**, *17*, 9312. [CrossRef]
52. Yaser, A.Z.; Haqim, A.I.; Mijong, J.R. *Water Treatment Sludge as Coagulant and Adsorbent: A Recent Review*; International Symposium on Water Pollution and Treatment; Springer: Singapore, 2022; pp. 1–7.
53. Suzuki, Y.; Kaku, R.; Takahashi, K.; Kanai, M.; Tamai, S.; Annaka, Y.; Chuganji, N. Optimum conditions for high-speed solid–liquid separation by ballasted flocculation. *Water Sci. Technol.* **2023**, *88*, 35–46. [CrossRef]
54. El Mouhri, G.; Elmansouri, I.; Amakdouf, H.; Belhassan, H.; Kachkoul, R.; Merzouki, M.; Lahrichi, A. Evaluating the effectiveness of coagulation–flocculation treatment on a wastewater from the moroccan leather tanning industry: An ecological approach. *Heliyon* **2024**, *10*, e27056. [CrossRef] [PubMed]
55. Sun, Y.; Zhou, S.; Chiang, P.-C.; Shah, K.J. Evaluation and optimization of enhanced coagulation process: Water and energy nexus. *Water-Energy Nexus* **2019**, *2*, 25–36. [CrossRef]
56. Abbas, M.J.; Mohamed, R.; Al-Sahari, M.; Al-Gheethi, A.; Mat Daud, A.M. Optimizing FeCl 3 in coagulation-flocculation treatment of dye wastes. *Songklanakarin J. Sci. Technol.* **2021**, *43*, 1094–1102.
57. Zhang, C.; Chen, X.; Chen, M.; Ding, N.; Liu, H. Response Surface Optimization on Ferrate-Assisted Coagulation Pretreatment of SDBS-Containing Strengthened Organic Wastewater. *Int. J. Environ. Res. Public Health* **2023**, *20*, 5008. [CrossRef]
58. Dayarathne, H.; Angove, M.J.; Jeong, S.; Aryal, R.; Paudel, S.R.; Mainali, B. Effect of temperature on turbidity removal by coagulation: Sludge recirculation for rapid settling. *J. Water Process Eng.* **2022**, *46*, 102559. [CrossRef]
59. Amran, A.H.; Zaidi, N.S.; Muda, K.; Loan, L.W. Effectiveness of natural coagulant in coagulation process: A review. *Int. J. Eng. Technol.* **2018**, *7*, 34–37. [CrossRef]

60. Choudhary, M.; Ray, M.B.; Neogi, S. Evaluation of the potential application of cactus (*Opuntia ficus-indica*) as a bio-coagulant for pre-treatment of oil sands process-affected water. *Sep. Purif. Technol.* **2019**, *209*, 714–724. [CrossRef]
61. Ejimofor, M.; Ezemagu, I.; Menkiti, M. RSM and ANN-GA modeling of colloidal particles removal from paint wastewater via coagulation method using modified Aguleri montmorillonite clay. *Curr. Res. Green Sustain. Chem.* **2021**, *4*, 100164. [CrossRef]
62. Kusuma, H.S.; Amenaghawon, A.N.; Darmokoesoemo, H.; Neolaka, Y.A.; Widyaningrum, B.A.; Anyalewechi, C.L.; Orukpe, P.I. Evaluation of extract of Ipomoea batatas leaves as a green coagulant–flocculant for turbid water treatment: Parametric modelling and optimization using response surface methodology and artificial neural networks. *Environ. Technol. Innov.* **2021**, *24*, 102005. [CrossRef]
63. Meerburg, F.A.; Boon, N.; Van Winckel, T.; Pauwels, K.T.; Vlaeminck, S.E. Live fast, die young: Optimizing retention times in high-rate contact stabilization for maximal recovery of organics from wastewater. *Environ. Sci. Technol.* **2016**, *50*, 9781–9790. [CrossRef]
64. Cabirol, N.; Barragán, E.; Durán, A.; Noyola, A. Effect of aluminium and sulphate on anaerobic digestion of sludge from wastewater enhanced primary treatment. *Water Sci. Technol.* **2003**, *48*, 235–240. [CrossRef] [PubMed]
65. Flaten, T.P. Aluminium as a risk factor in Alzheimer's disease, with emphasis on drinking water. *Brain Res. Bull.* **2001**, *55*, 187–196. [CrossRef] [PubMed]
66. EPA, E.P.A. National primary drinking water regulations: Long term 1 enhanced surface water treatment rule. Final rule. *Fed. Regist.* **2002**, *67*, 1811–1844.
67. Chen, H.; Xu, H.; Zhong, C.; Liu, M.; Yang, L.; He, J.; Sun, Y.; Zhao, C.; Wang, D. Treatment of landfill leachate by coagulation: A review. *Sci. Total Environ.* **2023**, 169294. [CrossRef]
68. Chaturvedi, S.; Dave, P.N. Removal of iron for safe drinking water. *Desalination* **2012**, *303*, 1–11. [CrossRef]
69. Rostern, N.T. The effects of some metals in acidified waters on aquatic organisms. *Fish Ocean OPJ* **2017**, *4*, 555–645. [CrossRef]
70. Xu, B.; Chi, Y.; Chi, Y.; Zhao, J.; Fu, C.; Wang, X.; Tian, S.; Ding, Y. Response surface optimization and floc structure analysis of magnetic flocculation technology for anaerobic digestion reject water. *Water* **2023**, *15*, 707. [CrossRef]
71. Shi, Y.; Fan, M.; Brown, R.C.; Sung, S.; Van Leeuwen, J.H. Comparison of corrosivity of polymeric sulfate ferric and ferric chloride as coagulants in water treatment. *Chem. Eng. Process. Process Intensif.* **2004**, *43*, 955–964. [CrossRef]
72. Asfaha, Y.G.; Tekile, A.K.; Zewge, F. Hybrid process of electrocoagulation and electrooxidation system for wastewater treatment: A review. *Clean. Eng. Technol.* **2021**, *4*, 100261. [CrossRef]
73. Sahu, O.; Mazumdar, B.; Chaudhari, P. Treatment of wastewater by electrocoagulation: A review. *Environ. Sci. Pollut. Res.* **2014**, *21*, 2397–2413. [CrossRef]
74. Zaied, B.; Rashid, M.; Nasrullah, M.; Zularisam, A.; Pant, D.; Singh, L. A comprehensive review on contaminants removal from pharmaceutical wastewater by electrocoagulation process. *Sci. Total Environ.* **2020**, *726*, 138095. [CrossRef] [PubMed]
75. Vik, E.A.; Carlson, D.A.; Eikum, A.S.; Gjessing, E.T. Electrocoagulation of potable water. *Water Res.* **1984**, *18*, 1355–1360. [CrossRef]
76. Naje, A.S.; Abbas, S.A. Electrocoagulation technology in wastewater treatment: A review of methods and applications. *Civ. Environ. Res.* **2013**, *3*, 29–42.
77. Boinpally, S.; Kolla, A.; Kainthola, J.; Kodali, R.; Vemuri, J. A state-of-the-art review of the electrocoagulation technology for wastewater treatment. *Water Cycle* **2023**, *4*, 26–36. [CrossRef]
78. Islam, S.D.-U. Electrocoagulation (EC) technology for wastewater treatment and pollutants removal. *Sustain. Water Resour. Manag.* **2019**, *5*, 359–380. [CrossRef]
79. Syam Babu, D.; Anantha Singh, T.; Nidheesh, P.; Suresh Kumar, M. Industrial wastewater treatment by electrocoagulation process. *Sep. Sci. Technol.* **2020**, *55*, 3195–3227. [CrossRef]
80. Tahreen, A.; Jami, M.S.; Ali, F. Role of electrocoagulation in wastewater treatment: A developmental review. *J. Water Process Eng.* **2020**, *37*, 101440. [CrossRef]
81. Ahmadzadeh, S.; Asadipour, A.; Pournamdari, M.; Behnam, B.; Rahimi, H.R.; Dolatabadi, M. Removal of ciprofloxacin from hospital wastewater using electrocoagulation technique by aluminum electrode: Optimization and modelling through response surface methodology. *Process Saf. Environ. Prot.* **2017**, *109*, 538–547. [CrossRef]
82. Shen, M.; Zhang, Y.; Almatrafi, E.; Hu, T.; Zhou, C.; Song, B.; Zeng, Z.; Zeng, G. Efficient removal of microplastics from wastewater by an electrocoagulation process. *Chem. Eng. J.* **2022**, *428*, 131161. [CrossRef]
83. Moussa, D.T.; El-Naas, M.H.; Nasser, M.; Al-Marri, M.J. A comprehensive review of electrocoagulation for water treatment: Potentials and challenges. *J. Environ. Manag.* **2017**, *186*, 24–41. [CrossRef]
84. Babu, R.R.; Bhadrinarayana, N.; Begum, K.; Anantharaman, N. Treatment of tannery wastewater by electrocoagulation. *J. Univ. Chem. Technol. Metall.* **2007**, *42*, 201–206.
85. Kobya, M.; Gengec, E.; Sensoy, M.T.; Demirbas, E. Treatment of textile dyeing wastewater by electrocoagulation using Fe and Al electrodes: Optimisation of operating parameters using central composite design. *Color. Technol.* **2014**, *130*, 226–235. [CrossRef]
86. Hansen, H.K.; Peña, S.F.; Gutiérrez, C.; Lazo, A.; Lazo, P.; Ottosen, L.M. Selenium removal from petroleum refinery wastewater using an electrocoagulation technique. *J. Hazard. Mater.* **2019**, *364*, 78–81. [CrossRef] [PubMed]
87. Tir, M.; Moulai-Mostefa, N. Optimization of oil removal from oily wastewater by electrocoagulation using response surface method. *J. Hazard. Mater.* **2008**, *158*, 107–115. [CrossRef]
88. Bashir, M.J.; Mau Han, T.; Jun Wei, L.; Choon Aun, N.; Abu Amr, S.S. Polishing of treated palm oil mill effluent (POME) from ponding system by electrocoagulation process. *Water Sci. Technol.* **2016**, *73*, 2704–2712. [CrossRef]

89. Varank, G.; Yazici Guvenc, S.; Demir, A. A comparative study of electrocoagulation and electro-Fenton for food industry wastewater treatment: Multiple response optimization and cost analysis. *Sep. Sci. Technol.* **2018**, *53*, 2727–2740. [CrossRef]
90. Singh, S.; Singh, S.; Lo, S.L.; Kumar, N. Electrochemical treatment of Ayurveda pharmaceuticals wastewater: Optimization and characterization of sludge residue. *J. Taiwan Inst. Chem. Eng.* **2016**, *67*, 385–396. [CrossRef]
91. Bukhari, A.A. Investigation of the electro-coagulation treatment process for the removal of total suspended solids and turbidity from municipal wastewater. *Bioresour. Technol.* **2008**, *99*, 914–921. [CrossRef]
92. Casillas, H.A.M.; Cocke, D.L.; Gomes, J.A.; Morkovsky, P.; Parga, J.R.; Peterson, E.; Garcia, C. Electrochemistry behind electrocoagulation using iron electrodes. *ECS Trans.* **2007**, *6*, 1. [CrossRef]
93. Rajaniemi, K.; Raulio, M.; Tuomikoski, S.; Lassi, U. Comparison of batch and novel continuous electrocoagulation processes in the treatment of paint industry wash water. *Desalination Water Treat.* **2019**, *170*, 394–404. [CrossRef]
94. Sadaf, S.; Roy, H.; Fariha, A.; Rahman, T.U.; Tasnim, N.; Jahan, N.; Sokan-Adeaga, A.A.; Safwat, S.M.; Islam, M.S. Electrocoagulation-based wastewater treatment process and significance of anode materials for the overall improvement of the process: A critical review. *J. Water Process Eng.* **2024**, *62*, 105409. [CrossRef]
95. Nasrullah, M.; Singh, L.; Krishnan, S.; Sakinah, M.; Mahapatra, D.; Zularisam, A. Electrocoagulation treatment of raw palm oil mill effluent: Effect of operating parameters on floc growth and structure. *J. Water Process Eng.* **2020**, *33*, 101114. [CrossRef]
96. Bazrafshan, E.; Kord Mostafapour, F.; Farzadkia, M.; Ownagh, K.A.; Mahvi, A.H. Slaughterhouse wastewater treatment by combined chemical coagulation and electrocoagulation process. *PLoS ONE* **2012**, *7*, e40108. [CrossRef] [PubMed]
97. Swain, K.; Abbassi, B.; Kinsley, C. Combined electrocoagulation and chemical coagulation in treating brewery wastewater. *Water* **2020**, *12*, 726. [CrossRef]
98. Meng, X.; Zeng, P.; Lin, S.; Wu, M.; Yang, L.; Bao, H.; Kang, J.; Han, H.; Zhang, C.; Sun, W. Deep removal of fluoride from tungsten smelting wastewater by combined chemical coagulation-electrocoagulation treatment: From laboratory test to pilot test. *J. Clean. Prod.* **2023**, *416*, 137914. [CrossRef]
99. Kundu, A.; Gupta, N.; Kalamdhad, A.S. Optimization and kinetic analysis of electrocoagulation-assisted adsorption for treatment of young landfill leachate. *J. Environ. Manag.* **2024**, *366*, 121779. [CrossRef]
100. Faraj, H.; Jamrah, A.; Al-Omari, S.; Al-Zghoul, T.M. Optimization of an electrocoagulation-assisted adsorption treatment system for dairy wastewater. *Case Stud. Chem. Environ. Eng.* **2024**, *9*, 100574. [CrossRef]
101. Al-Qodah, Z.; Al-Zghoul, T.M.; Jamrah, A. The performance of pharmaceutical wastewater treatment system of electrocoagulation assisted adsorption using perforated electrodes to reduce passivation. *Environ. Sci. Pollut. Res.* **2024**, *31*, 20434–20448. [CrossRef]
102. Deveci, E.Ü.; Akarsu, C.; Gönen, Ç.; Özay, Y. Enhancing treatability of tannery wastewater by integrated process of electrocoagulation and fungal via using RSM in an economic perspective. *Process Biochem.* **2019**, *84*, 124–133. [CrossRef]
103. Pérez, L.S.; Rodriguez, O.M.; Reyna, S.; Sánchez-Salas, J.L.; Lozada, J.D.; Quiroz, M.A.; Bandala, E.R. Oil refinery wastewater treatment using coupled electrocoagulation and fixed film biological processes. *Phys. Chem. Earth Parts A/B/C* **2016**, *91*, 53–60. [CrossRef]
104. Al-Othman, A.A.; Kaur, P.; Imteaz, M.A.; Ibrahim, M.E.H.; Sillanpää, M.; Kamal, M.A.M. Modified bio-electrocoagulation system to treat the municipal wastewater for irrigation purposes. *Chemosphere* **2022**, *307*, 135746. [CrossRef] [PubMed]
105. Djajasasmita, D.; Lubis, A.B.; Ma'mur, I.D.; Pratiwi, S.T.; Rusgiyarto, F.; Nugroho, F.A.; Aryanti, P.T.P. High-efficiency contaminant removal from hospital wastewater by integrated electrocoagulation-membrane process. *Process Saf. Environ. Prot.* **2022**, *164*, 177–188. [CrossRef]
106. Aryanti, P.T.P.; Nugroho, F.A.; Anwar, N.; Rusgiyarto, F.; Phalakornkule, C.; Kadier, A. Integrated bipolar electrocoagulation and PVC-based ultrafiltration membrane process for palm oil mill effluent (POME) treatment. *Chemosphere* **2024**, *347*, 140637. [CrossRef] [PubMed]
107. Al Hawli, B.; Benamor, A.; Hawari, A.A. A hybrid electro-coagulation/forward osmosis system for treatment of produced water. *Chem. Eng. Process.-Process Intensif.* **2019**, *143*, 107621. [CrossRef]
108. Cao, Y.; Malmali, M.; Qian, X.; Wickramasinghe, S.R. Continuous electrocoagulation-membrane distillation unit for treating hydraulic fracturing produced water. *J. Water Process Eng.* **2022**, *50*, 103219. [CrossRef]
109. Deghles, A.; Kurt, U. Treatment of tannery wastewater by a hybrid electrocoagulation/electrodialysis process. *Chem. Eng. Process. Process Intensif.* **2016**, *104*, 43–50. [CrossRef]
110. Bani-Melhem, K.; Elektorowicz, M.; Tawalbeh, M.; Al Bsoul, A.; El Gendy, A.; Kamyab, H.; Yusuf, M. Integrating of electrocoagulation process with submerged membrane bioreactor for wastewater treatment under low voltage gradients. *Chemosphere* **2023**, *339*, 139693. [CrossRef]
111. Changmai, M.; Das, P.; Mondal, P.; Pasawan, M.; Sinha, A.; Biswas, P.; Sarkar, S.; Purkait, M. Hybrid electrocoagulation–microfiltration technique for treatment of nanofiltration rejected steel industry effluent. *Int. J. Environ. Anal. Chem.* **2022**, *102*, 62–83. [CrossRef]
112. Amosa, M.K.; Jami, M.S.; Alkhatib, M.a.F.R.; Majozi, T. Studies on pore blocking mechanism and technical feasibility of a hybrid PAC-MF process for reclamation of irrigation water from biotreated POME. *Sep. Sci. Technol.* **2016**, *51*, 2047–2061. [CrossRef]
113. Martínez-Huitle, C.A.; Panizza, M. Electrochemical oxidation of organic pollutants for wastewater treatment. *Curr. Opin. Electrochem.* **2018**, *11*, 62–71. [CrossRef]
114. Ganiyu, S.O.; Martinez-Huitle, C.A.; Rodrigo, M.A. Renewable energies driven electrochemical wastewater/soil decontamination technologies: A critical review of fundamental concepts and applications. *Appl. Catal. B: Environ.* **2020**, *270*, 118857. [CrossRef]

115. Linares-Hernández, I.; Barrera-Díaz, C.; Bilyeu, B.; Juárez-GarcíaRojas, P.; Campos-Medina, E. A combined electrocoagulation–electrooxidation treatment for industrial wastewater. *J. Hazard. Mater.* **2010**, *175*, 688–694. [CrossRef] [PubMed]
116. Ryan, D.R.; Maher, E.K.; Heffron, J.; Mayer, B.K.; McNamara, P.J. Electrocoagulation-electrooxidation for mitigating trace organic compounds in model drinking water sources. *Chemosphere* **2021**, *273*, 129377. [CrossRef] [PubMed]
117. Sanni, I.; Estahbanati, M.K.; Carabin, A.; Drogui, P. Coupling electrocoagulation with electro-oxidation for COD and phosphorus removal from industrial container wash water. *Sep. Purif. Technol.* **2022**, *282*, 119992. [CrossRef]
118. Barzegar, G.; Wu, J.; Ghanbari, F. Enhanced treatment of greywater using electrocoagulation/ozonation: Investigation of process parameters. *Process Saf. Environ. Prot.* **2019**, *121*, 125–132. [CrossRef]
119. Wagh, M.P.; Nemade, P.D.; Naik, U.; Sengupta, A. Enhancing color and chemical oxygen demand degradation in distillery spent wash by electrocoagulation and ozone assisted electrocoagulation. *Desalin. Water Treat.* **2020**, *197*, 213–223. [CrossRef]
120. Navas-Cárdenas, C.; Rajendran, S.; Ramírez, T.; Muñoz, F. Combined sedimentation, electrocoagulation and ozone processes for the wastewater treatment in an Ecuadorian MDF industry. *Int. J. Environ. Sci. Technol.* **2022**, *19*, 11699–11710. [CrossRef]
121. Ding, J.; Jiang, M.; Zhao, G.; Wei, L.; Wang, S.; Zhao, Q. Treatment of leachate concentrate by electrocoagulation coupled with electro-Fenton-like process: Efficacy and mechanism. *Sep. Purif. Technol.* **2021**, *255*, 117668. [CrossRef]
122. Rezgui, S.; Ghazouani, M.; Bousselmi, L.; Akrout, H. Efficient treatment for tannery wastewater through sequential electro-Fenton and electrocoagulation processes. *J. Environ. Chem. Eng.* **2022**, *10*, 107424. [CrossRef]
123. Sun, C.; Wang, J.; Gu, C.; Wang, C.; Sun, S.; Song, P. MOF-derived N-Co/Fe-PC composite as heterogeneous electro-Fenton catalysis combined with electrocoagulation process for enhanced degradation of Cu-CIP complexes from wastewater. *Chem. Eng. J.* **2023**, *452*, 139592. [CrossRef]

energies

MDPI

Review

A Review of Renewable Energy Technologies in Municipal Wastewater Treatment Plants (WWTPs)

Derick Lima [1], Li Li [1,*] and Gregory Appleby [2]

[1] School of Electrical and Data Engineering, University of Technology Sydney, Sydney, NSW 2007, Australia; 13699638@student.uts.edu.au
[2] Sydney Water Corporation, Sydney, NSW 2124, Australia; gregappleby70@gmail.com
* Correspondence: li.li@uts.edu.au

Abstract: The global trend towards sustainable development has included the implementation of renewable energy recovery technologies in municipal wastewater treatment plants (WWTPs). WWTPs are energy-intensive consumers with high operational costs and often are dependent from the electricity supplied by the main grid. In this context, the integration of renewable energy recovery technologies into WWTPs emerges as an environment-friendly strategy that enhances energy efficiency, sustainability and reduces energy operating costs. Renewable energy recovery technologies, such as anaerobic digestion, microbial fuel cells, and sludge gasification, can offer multiple benefits for a WWTP. Anaerobic digestion is the most widely adopted technology due to its efficiency in treating sewage sludge and its ability to generate biogas—a valuable renewable energy source. The use of biogas can offset the energy demands of the wastewater treatment process, potentially leading to energy self-sufficiency for the WWTP and a reduction in reliance from the electricity supply from the main grid. Similarly, microbial fuel cells harness the electrochemical activity of bacteria to produce electricity directly from wastewater, presenting a promising alternative for low-energy processes for sustainable power generation. Gasification of sewage sludge is a promising technology for managing municipal sewage sludge, offering key advantages, especially by generating a renewable energy production (sludge is converted into syngas), which further decreases the sludge volume and operating costs with sludge management, helps to eliminate odour associated with sewage sludge, and effectively destroys the pathogens. Adoption of renewable energy sources in WWTPs can be a great alternative to overcome issues of high operating costs and high dependency of electricity from the main grid, but their successful integration requires addressing challenges such as technological maturity, economic feasibility, and regulatory frameworks. This study aims to comprehensively explore the significance of different renewable energy technologies in municipal WWTPs, including site-specific and non-site-specific sources, evaluating their impact on sustainability, energy efficiency, and overall operational effectiveness. This review also highlights some studies in which different strategies were adopted to generate extra revenue and/or reduce operating costs. Through a comprehensive review of current practices and emerging technologies, this study underscores the transformative potential of these innovations in advancing low-emission wastewater management.

Keywords: wastewater treatment plants (WWTP); renewable energy resources; sewage sludge

1. Introduction

Wastewater treatment plants (WWTPs) are an essential component of a municipal system. These facilities receive raw municipal wastewater, treat it and send the treated effluent back to the environment [1]. WWTPs are large energy consumers, requiring a significant amount of electricity for their operation. Treating municipal wastewater is a complex and costly process; for example, the energy costs can range from 2% to 60% of the total operating costs, depending on the level of treatment used [2]. On average, the energy required to treat 1 million liters of sewage is between 150 to 1400 kWh. The reason for this energy range may be associated with several factors, such as treatment levels, facility capacity, energy efficiency, operational processes, and regulatory requirements [3]. In addition, due to more stringent effluent standards and the increase in the volume of wastewater generated globally, energy demand for WWTPs has been increasing [4].

With rising electricity costs, sustainability targets, and efficiency goals, more and more plants are looking for alternatives to minimise their operating costs and reduce carbon emissions while increasing their energy efficiency [1]. A commonly used technology for treating wastewater in WWTPs is anaerobic digestion (AD), and through this process, biogas is generated from the sludge treatment. Although energy recovery from AD is a very mature and common alternative, other emerging renewable energy technologies are arising as complementary sources to improve the potential of energy production in WWTPs [3]. Renewable energy technologies can significantly decrease a plant's overall energy demand and, in some cases, help energy-intensive consumers move toward energy self-sufficiency or even energy positivity. For example, several WWTPs in the USA (i.e., East Bay Municipal Utility District, Gresham, Gloversville-Johnstown Joint, Point Loma, and Sheboygan Regional) and Europe (Aquaviva, Greves-muhlen, Strassim, Wolfgangsee-Ischl, and Zillertal) have implemented different strategies to reach self-sufficiency, including the adoption of renewable energy technologies and co-digestion [5]. Another possibility to reduce energy costs is to participate in demand-response programs and demand-side management [6].

A large number of studies have been conducted on this topic of renewable energy generation in WWTPs. Some studies have focused on a particular technology, including anaerobic digestion [7], fuel cells [8], combined heat and power (CHP) technologies [9], supercritical water gasification [10], hydrothermal liquefaction [11], thermochemical conversion processes [12], and hybrid systems [13]. Some reviews explored new technologies not widely implemented in large-scale plants, including dark fermentation (DF), photocatalysis, photo-fermentation (PF), microbial photoelectrochemical cells (MPEC) [14], Microbial Electrolysis Cells (MECs) [15], microbial fuel cells (MFCs) [16], and biodiesel [17]. Other studies explored different strategies and market opportunities for WWTPs, including demand-response and demand-side management [6] and energy spot markets [18]. Although review papers covering different types of renewable energy technologies adopted in WWTPs have been conducted, few articles have considered and summarised multiple technologies in one comprehensive review. In this context, this review paper aims to comprehensively review the different types of energy technologies that can be used in WWTPs, including site-specific and non-site-specific sources. These technologies can lead WWTPs to be less dependent on importing electricity from the main grid, become more energy-sustainable and reduce operating costs. In addition, this review also included some alternatives that these facilities can explore to generate extra revenue and/or reduce operating costs, including providing grid services. The main contribution of this paper is to summarise the several types of renewable energy technologies that can be used in WWTPs in one review paper which can provide a comprehensive, balanced, and forward-thinking overview on this topic.

This literature review was conducted by searching related works in the academic literature and government and industry reports. To conduct this study, two research questions were included: Which types of site-specific and non-site-specific renewable energy technologies can be used in WWTPs? Which strategies and opportunities can be explored by WWTPs to increase revenues and benefits? Based on the research questions, the following research objectives were developed: (1) review different renewable energy technologies, including site-specific and non-site-specific that can be adopted by WWTPs, and (2) investigate strategies and opportunities WWTPs can consider to increase revenues and/or reduce operating costs. As the first step, the literature search was conducted primarily using three online databases—Scopus, IEEE Xplore, and ScienceDirect—for academic papers published in journals and conference proceedings. Despite limitations, this work aims to provide a rigorous and comprehensive review of energy generation in WWTPs. Therefore, it is expected that the insights and results from this study will be useful in generating a clear, structured, and comprehensive understanding of the current state-of-the-art of renewable energy generation opportunities in WWTPs.

This paper is organised as follows. Section 2 presents a short WWTP overview. Section 3 explores different energy technologies that can be adopted in WWTPs for energy recovery and power generation. Section 4 explores some market strategies that were used by WWTPs to generate revenues, including providing grid services, demand-response participation, and exposure to the spot electricity market. Challenges and opportunities for WWTPs in adopting renewable energy sources (RES) and providing grid services Section 4. This review is concluded in Section 5.

2. WWTP Overview

The main objective of a WWTP is to accelerate the natural process of purifying wastewater and return it back to the environment. The wastewater treated by WWTPs, also known as effluent, must meet quality standards and regulatory requirements (i.e., solids, carbon, phosphorus and nitrogen limits). The effluent's discharge standards can be defined by the state, territory, or on a national level. They can also vary from plant to plant depending on different factors, such as age, impact on waterways, and treatment capacity. WWTPs may differ in size, capacity, efficiency, effluent quality requirements, and type of treatment processes. In terms of treatment process, wastewater plants can be composed of one (primary), two (secondary), or three (tertiary) stages. Figure 1 shows a schematic diagram of a facility with two stages, primary and secondary [19].

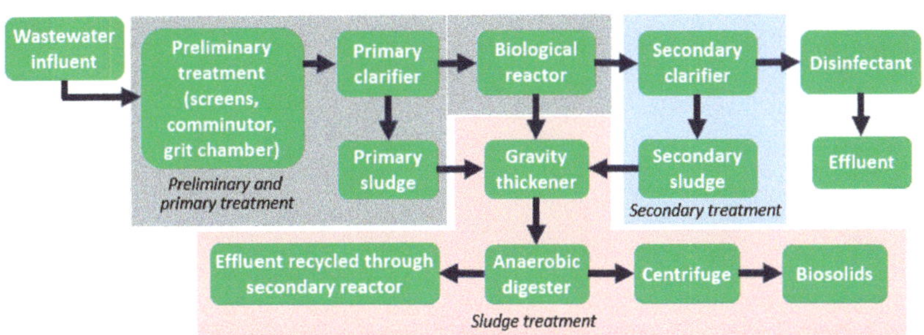

Figure 1. Basic schematic of a WWTP that provides primary and secondary treatments [19].

Preliminary and primary treatments are associated with the physical processes that are used to clean the wastewater. The objective of this stage is to remove solids from wastewater. In the preliminary stage, pollutants and large particles (i.e., rags, twigs, and stones), which can cause operational damage in treatment processes, especially to pumps, are removed from the wastewater. This stage usually uses screens, grit chambers, comminutors, and flotation. Primary treatment, also known as sedimentation, is where the scum, suspended, and settled solids are removed by sedimentation on the primary clarifiers. In most facilities, the preliminary stage is considered as a component of the primary treatment [19]. Secondary treatment consumes the organic matter that was not removed during the primary treatment. In this stage, biological mechanisms are able to remove up to 85% of the organic material contained in the wastewater. Usually, secondary treatment relies on biological reactors followed by a second clarifier with an activated sludge process. Some WWTPs are required to remove nutrients (i.e., nitrogen and phosphorus), and, in this case, additional tanks may be needed to accomplish the effluent quality standards in a process called nitrification–denitrification. Nutrient removal can be achieved by chemical or biological methods or a combination of both. Tertiary treatment aims to remove specific pollutants that were not removed from the previous treatment stages. It usually relies on chemical processes and filtration to provide the required polishing before the effluent is disposed. Additional treatment can be provided to extract nutrients and remove pathogens if required [19,20]. Choosing the most appropriate effluent treatment technology for a WWTP involves considering various technical factors (i.e., effluent volume, water demand, and plant size), and some commercial software tools, such as Winflows, IMSDesign, and WAVE, can support the planning and design processes of a WWTP.

Sewage Sludge

Sewage sludge is the solid material (organic and inorganic) removed from the raw wastewater during the treatment process. The amount of sludge produced in a WWTP is proportional to the quantity of wastewater treated by the plant, and can be expressed in mass or volume. On average, the total sludge produced may range from 0.2 to 0.3 kg/m^3 of wastewater treated [19]. It contains a mixture of organic and inorganic matter and a high concentration of several pathogenic organisms that have to be treated before disposal. Sewage sludge disposal is costly, and most WWTPs must follow regulatory requirements for waste management [21]. The energy recovery potential from sludge is directly related to three main components: (i) water content, (ii) organic (volatile) and (iii) inorganic (inert) matter. Several methods can be found on both literature and industry to treat sewage sludge, including anaerobic digestion (most common), co-digestion, composting, incineration, disposal in landfills, or being used in agriculture as a soil conditioner. However, due to the high costs of conventional treatment methods and the aim to improve sustainability targets and reduce operating costs of sludge disposal, WWTPs can recover energy from utilise sewage sludge using different methods [12], as shown in Figure 2. The energy content of dried sewage sludge, with 21–48% volatile matter, can range between 11.1 and 22.1 MJ/kg, and the calorific values for primary, secondary, and digested sludges were reported to be 16.2 MJ/kg, 12.2 MJ/kg, and 11.4 MJ/kg, respectively [22].

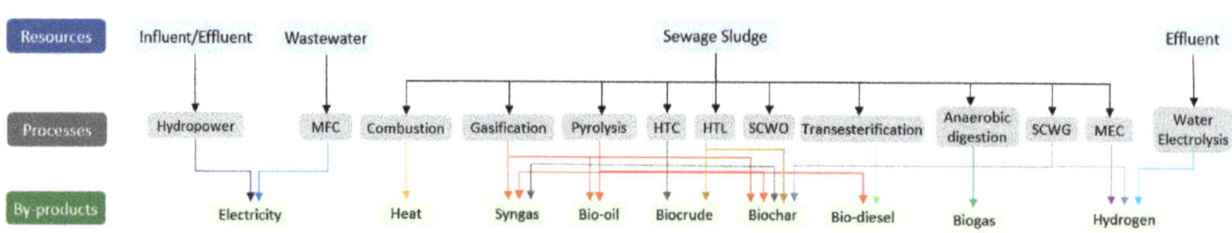

Figure 2. Site-specific sources of renewable energy in a WWTP.

3. Energy Resources in WWTP

Renewable energy technologies have become an essential component of the energy transition and one of the most effective solutions for reducing costs and emissions from WWTPs. Several types of renewable energy sources can be adopted by a WWTP. Some options can take advantage of the specific plant's sources, including energy recovery from wastewater and sewage sludge, an influent/effluent flow, and treated water. In contrast, others can be non-site-specific, such as solar and wind. Although many technologies can be implemented in WWTPs, some are still in the research and development stages and have not been widely used in large-scale facilities [23].

3.1. Site-Specific Sources

Site-specific sources refer to the energy recovery embedded in the WWTP's resources, including wastewater, sewage sludge, an influent/effluent flow, and treated water. In theory, the energy contained in a municipal wastewater and sewage sludge can be about five to ten times more than the energy required for their treatment processes, and this energy is mainly found in three forms: chemical, hydraulic/kinetic, and thermal. Based on this huge potential, according to the Water Environment Federation, WWTPs have the potential to become energy-neutral facilities or even net-energy producers [24]. Although many energy recovery methods can be used in WWTPs, each technology has its advantages, disadvantages, limitations, and potential benefits. Some of them are very widely used in large-scale plants (i.e., anaerobic digestion), and others are not as common (i.e., water electrolysis, MFC and MEC). Figure 2 illustrates the energy resources that can be recovered from a WWTP and the conversion processes associated.

3.1.1. Anaerobic Digestion

Anaerobic digestion (AD) is the most common technology used to treat sewage sludge in WWTPs (48% of the total wastewater is treated through AD in the USA). In this process, microorganisms consume the organic matter in the sludge and generate biogas as a by-product of the digested sludge. This process occurs in the absence of oxygen and is divided in four stages: hydrolysis, acidogenesis, acetogenesis, and methanogenesis. Biogas is a renewable gas that contains mostly methane and carbon dioxide, which can be used on-site to generate both electricity and heating or can be upgraded and injected into gas networks [25]. Several factors can influence the AD process, such as feedstock characteristics (i.e., total solids, suspended solids, volatile solids, chemical oxygen demand, and the carbon/nitrogen ratio) and operational parameters (i.e., temperature, retention time, organic load rate, pH, alkalinity, and reactor volume) [7].

Sewage sludge is known for its low degradability and low carbon/nitrogen ratio (two main parameters for anaerobic digestion efficiency). An effective alternative to overcome these challenges is to use co-digestion. It combines two or more organic feedstocks aiming to enhance the overall efficiency and performance of the digestion process, including biogas production and methane concentration. The average biogas generation from sewage sludge only can vary around 0.9–1.1 $m^3/day.m^3$ digester volume, whereas using co-digestion, it can reach up to 2.5–4.0 $m^3/day.m^3$ digester volume. Depending on the feedstock type, biogas production with co-digestion may increase by 25 to 400% compared to single-feedstock. Examples of feedstocks that can be mixed with sewage sludge include agricultural residues, beverage/dairy/food processing waste fats, food waste/scraps, oils, and grease (FOG), microalgae, an organic fraction of municipal solid waste (OFMSW), and high-strength waste (HSW). Some full-scale WWTPs have successfully adopted co-digestion of sewage sludge and other feedstocks to increase biogas production, such as Grevesmuhlen WWTP and Köhlbrandhöft WWTP, both in Germany [26] and East Bay Municipal Utility District WWTP and Des Moines Metropolitan Wastewater Reclamation Authority WWTP, both in the USA [25].

Pre-treatment techniques can also be used to improve biogas production and AD process; however, its main objective is to enhance sewage sludge digestibility and dewaterability, reduce sludge volume, and minimise environmental impacts. Improving sludge biodegradability and solubility may allow higher load ratings and shorter retention times, which can result in a higher processing efficiency [7]. Table 1 summarises some methods used in sewage sludge treatment, whereas Tables 2 and 3 show some studies found in the literature related to co-digestion and pre-treatment techniques.

Other methods, such as lime and alkaline stabilisation, are primarily used to reduce pathogens in the SS as a pre-treatment for land applications and aerobic digestion. Although the several benefits which can be provided by co-digestion and pre-treatment methods, the successful implementation of these processes in large-scale facilities may be challenging, especially in terms of re-designing the operation process and physical upgrades/updates. Additionally, new operating costs can be added when adopting these new alternatives [7].

Table 1. Pre-treatment methods that can be used to treat sewage sludge.

Type	Process	Advantages	Disadvantages	Applications
Th	SS heated (150–200 °C) under pressure, breaking down complex OC and making SS more digestible.	- Enhance biodegradability - Reduce pathogens and OC - Increase biogas production	- High CAPEX and OPEX - High energy consumption	- Used before AD - Improve biogas yield
M	SS is exposed to microwave radiation to rupture cell structures of the solid particles within SS and disrupt OM, improving digestion.	- Increases CH_4 production - Reduces pathogens - Improves biogas yield - Enhances dewaterability	- High CAPEX and OPEX - High energy consumption - Require controlled temperature to avoid OM degradation	- Used before AD - SS with high OC - Co-digestion processes
Uc	High-frequency sound waves (ultrasonication) are used to disrupt SS particles, improving degradability.	- Improve dewaterability - Increase CH_4 production - Reduce need for chemicals	- High CAPEX and OPEX - Energy-intensive process - Need of special equipment	- Increase biogas yield - SS with high MC - Co-digestion processes
Ud	High-frequency sound waves (20 kHz–1 MHz) to disrupt SS particles to cause cavitation, improving digestibility.	- Enhance digestibility - Increase dewaterability - Reduce pathogens	- High energy consumption - Expensive equipment - Need special setup and O&M	- SS dewatering - SS with high OC - Co-digestion processes
O	Ozone gas is bubbled through SS, breaking down OC and killing pathogens.	- Strong pathogen control - Effective in reducing odors - Increases biodegradability	- High operational cost - Requires energy-intensive ozone generation	- Odor control - Enhance degradability - Before AD
FA	Apply high pressure and high temperature to SS in the presence of water or organic additives.	- Enhance biodegradability - Increase CH_4 production - Reduce need for chemicals	- High CAPEX and OPEX - High energy consumption	- Enhance AD efficiency - Improve biogas yield - SS with high OC

Th: Thermal, M: Microwave, Uc: Ultrasonic; Ud: Ultrasound; O: Ozonation; FA: Free ammonia; SS: Sewage sludge: OC: Organic content; MC: Moisture content; O&M: Operation and maintenance.

Table 2. Co-digestion of sewage sludge with different feedstocks.

Ref.	Substrate	Parameters and System Design	OLR (gVS/L.d)	VS Removal (%)	Methane Production (L/kgVS)		Methane Concentration (%)	
[27]	SS and FW	6 × 5 L reactor, 37 °C, 22 days HRT, 10–50% FW:SS ratios	Min: 2.8 | Max: 4.2	Min: 66 | Max: 76	SS: 230–280 | FW: 290–330	Co-d: 318–385	SS: 45–57 | FW: 49–57	Co-d: 53–55.7
[28]	SS and FW	2 L reactor, 30–38 °C, 22 days HRT, ratios (1:1, 1:2 and 1:3)	NI	Min: 51.47 | Max: 60.36	SS: 356–478 | FW: 511	Co-d: 453–609	SS: 53 | FW: 50.4	Co-d: 52–70.3
[29]	SS and FW	5 L reactor, 30–38 °C, 22 days HRT, different mixing ratios (1:1, 1.5:1, 2:1, 1:1.5 and 1:2)	0.5–7 g VS/L	Min: 82.8 Max: 87.7	SS: 625.4 FW: 385.9–507.5	Co-d: 384.6–492	SS: 55.9–58.6 FW: 58.8	Co-d: 53–60.4
[30]	SS and FW	6 L reactor volume, 35 °C, 8–30 days HRT and three different SS:FW mixing ratio (2.4:1, 0.9:1, 0.4:1)	Min: 4.6 Max: 18.5	Min: 39.7 Max: 70	SS: 157–237 FW: 377–465	Co-d: 215–400	SS: 63–65 FW: 50–54	Co-d: 53–61

Table 2. *Cont.*

Ref.	Substrate	Parameters and System Design	OLR (gVS/L.d)	VS Removal (%)	Methane Production (L/kgVS)		Methane Concentration (%)	
[31]	SS and FVW	100 L reactor, 35 °C, 11–14 days HRT	Min: 1.46 Max: 2.8	Min: 35 Max: 43	VW: 335 SS: 84	Co-d: 90–430	NI	NI
[32]	SS, GTS, OFMSW	2 × 6 L reactor, 38 °C, 20 days HRT and 5–30% mixing ratio	Min: 1.15 Max: 2.17	Min: 52 Max: 65	SS: 300 | GTS: NI OFMSW: NI	Co-d: 456–547	SS: 66 | GTS: NI OFMSW: NI	Co-d: 66–69
[33]	TWAS and RS	0.25 L reactor, 37 °C, 50 days HRT, mixing ratio of 1:1 and 1:3 on volume basis	5% TS	Min: 34.5 Max: 69.1	RS: 216.3 TWAS: 184.6	Co-d: 304	NI	NI
[34]	SS and RS	1.2 L reactor, 35 °C and 55 °C, between 25 (55 °C) and 75 days (35 °C) HRT, 4:1 (weight basis) mixing ratio	20% TS	Min: 61 (35 °C) Min: 70.2 (55 °C)	RS: 222 (35 °C), 248 (55 °C) SS: 308 (35 °C), 344 (55 °C)	Co-d: 518 (35 °C), 602 (55 °C)	NI	Co-d: 36–60
[35]	SS and WS	0.5 L reactor, 37 °C, 30 days, 1.15:1.94 (VS basis) ratio	7.73	Min: 56.38 | Max: 63.59	SS: 136.8 | WS: 243	Co-d: 176.7–333.9	NI	NI
[36]	WAS, WS, and RS	0.15 L reactor, 30 days HRT, 1:1 mixing ratio	4 g VS/L	NI	RS: 95 | WS: 103 WAS: 87	Co-d: 36–223	RS: 50 | WS: 48 WAS: 50	Co-d: 30–58
[37]	SS and GTW	0.25 L reactor, 35 °C, 10–31 days HRT, 4 mixing ratios (14%, 24%, 43% and 39% GTW)	NI	Min: 31 Max: 35	SS: 223 GTW: 606	Co-d: 214–517	NI	NI
[38]	SS and GTW	6 L reactor, 35 °C, 15 days HRT, 1:1 mixing ration (PS and WAS) + GTW 5% VS.	2.93	Min: 47 Max: 59	SS: 384 GTW: NI	Co-d: 641	SS: 61 GTW: NI	Co-d: 69
[39]	SS and FOG	For 55 °C: 10 L reactor, 20 days HRT. For 70 °C: 2 L reactor, and 2 days HRT. Mixing ratios: 20, 40, 60 and 80% FOG	NI	Min: 46.8 Max: 82	SS: 138.3 FOG: NI	Co-d: 102–673	SS: 61.6–62.8 FOG: NI	Co-d: 49.7–67.3
[40]	SS and FOG	1 L reactor, 35 °C, 15 days HRT, 4 mixing ratios (14%, 24%, 43% and 39% GTW)	4 g COD/L.d	Min: 43.1 Max: 54.6	SS: 114–128 FOG: 143–290	Co-d: 453–609	SS: 62.8–71.2 FOG: 63.3	Co-d: 66.1–68.4
[41]	SS, FW and G	0.25 L reactor, 37 °C, 20 days HRT, 1:1 mixing ratio (SS:FW) with 1% and 3% glycerol.	0.66–1.1	Min: 8.6 Max: 17.4	SS: 138.3 FW: NI | G: NI	Co-d: 236–526	SS: 85.9% FW: NI | G: NI	Co-d: 77.4–79.4
[42]	SS and G	5.5 L reactor, 35 and 55 °C, 20 days HRT, 2 SS:G mixing ratios (99:1, 98.8:1.2 *v*/*v*%)	1.2–1.6 (35 °C); 1.1–1.3 (55 °C)	Min: 36 (35 °C), 45 (55 °C) Max: 64 (35 °C), 73 (55 °C)	SS: 296 (35 °C), 354 (55 °C) G: 277–475 (35 °C), 349–490 (55 °C)	Co-d: 250–660 (35 °C) 230–530 (55 °C)	NI	Co-d: 63–72 (35 °C) 57–66 (55 °C)
[43]	SS and PM	0.16 L reactor, 35 °C, 47 days HRT, different SS:PM mixing ratios (21:1, 14:1, 7:1 VS)	0.54 (average)	Min: 56.3 Max: 71.4	SS: 182 PM: 239	Co-d: 190–200	SS: 52–58 PM: 52–58	Co-d: 52–58
[44]	SS, SM PoM	1 L (b) and 3 L (s-c) reactor, 35 °C, 15–30 days HRT, different mixing ratios	1.27–2.86 (s-c)	37.9–45.8 (b) 23.9–38 (s-c)	SS: 184 (b) PoM: NI | SM: NI	Co-d: 198–290 (b); 186–273 (s-c)	SS: 67–68 PoM: NI | SM: NI	Co-d: 67–68
[45]	SS, WV PoM	5 L reactor, 35 °C, 6–20 days HRT, 50:50 *w*/*w*% SS:WV mixing ratio plus 10 g/L of PoM	NI	Min: 38 Max: 58	SS: 130 (10 days HRT) PM: NI | WV: NI	Co-d: 210–261	NI	NI

Table 2. *Cont.*

Ref.	Substrate	Parameters and System Design	OLR (gVS/L.d)	VS Removal (%)	Methane Production (L/kgVS)		Methane Concentration (%)	
[46]	SS and CM	1 L reactor, 37 °C, 20 days HRT, 3 CM:SS mixing ratio (30:70, 50:50, 70:30 *w*/*w*%)	NI	Min: 78.4 Max: 97.3	NI	Co-d: 335–511	NI	NI
[47]	SS and M	70 L reactor, 35–37 °C, 30 days, SS:M ratio of 0.2–1.8% VS	NI	Min: 4 | Max: 35	NI	Co-d: 226	NI	Co-d: 79–85.5
[48]	WAS and M	0.13 L reactor, 35 °C, 25–30 days HRT, 4 WAS:M mixing ratio (3:1, 1:1, 1:3)	NI	NI	WAS: 362 M: 318	Co-d: 354–442	NI	NI

Co-d: Co-digestion; NI: Not informed; FVW: Fruit and Vegetable waste; RS: Rice straw; WS: Wheat straw; GTW: Grease trap waste; G: Glycerol; PM: Pig manure; PoM: Poultry manure; SM: Swine manure; WV: Wine vinasse; CM: Cow manure; M: Microalgae; b: batch mode; s-c: semi-continuous mode; FW: Food waste; OLR: Organic loading rate; SS: Sewage sludge.

Table 3. Pre-treatment techniques applied to sewage sludge.

Ref.	Type	Sludge Collection	Operating Parameters	Biogas Production/Methane Yield
[49]	Th	WWTP in Valladolid, Spain	System include steam boiler, 1.5 L hydrolysis reactor connected to a 5 L flash tank. Reactor loaded with 0.75 L sludge, under 110–200 °C, and 10–50 min.	CH_4 yield increased up to 50% (thermal) under 30 min under 180–200 °C. CH_4 production improve up to 45% using thermal compared to conventional process.
[50]	Th	WWTP, in Korea.	The hydrolysis thermal plant had the capacity to treat 1 ton SS/cycle. SS was dewatered before the treatment and the thermal hydrolysis process operated at 75–225 °C and 15–105 min reaction time.	Optimal parameters included 76 min reaction and 180 °C. At 150 °C and 1 h reaction, the maximum CH_4 yield was 273.2 mL/g COD (40% increase compared to control). The minimum CH_4 yield was 221.7 mL/g COD under 30 min and 200 °C.
[51]	Th and FA	WAS obtained from a WWTP in Changsha, China.	Three temperatures (35, 50 and 70 °C) were used. Reactors fed with 0.4 L of WAS, and different concentrations of ammonium stock solution was used (220–450 mg). FA concentrations between 79.7 and 163.1 mg $NH_3 - N/L$ and operate at 25 °C.	Without any treatment (control), the biochemical methane potential (BMP) was 183.4–200.6 mL/g VSS. With FA only and thermal only, the BMP was 188.1–201.8 mL/g VSS, and 184.1–196.9 mL/g VSS, respectively. With combined methods, BMP varied between 195.7 and 229.4 mL/g VSS.
[52]	Th and Uc	Ulu Pandan municipal WWTP, in Singapore.	Batch mode and operating temperature of 35 °C. Sewage sludge treated with ultrasonic at 5 MJ/kg TS and optimal temperature of 65 °C for thermal treatment.	Combining thermal and ultrasonic (30 s, 5000 kJ/kg TS) resulted in high COD solubilisation (760–10,200 mg/L), proteins (115–2900 mg/L) and carbohydrates (60–660 mg/L) than a single treatment. Biogas production improved by 20%.
[53]	Th, M, and Uc	Copero urban WWTP in Seville, Spain.	Batch mode and operating temperature of 35 °C. Thermal parameters included 75 L volume autoclave, 120 °C, 2 atm for 15 min. Microwave parameters included 100–900 W at 80 °C for 1.4 min. Ultrasonic parameters: 6 L volume 25 °C, atm pressure, 150 W power generator for 45 min.	CH_4 production improved by 20%, 29%, and 95% based on microwave, thermal and sonication, respectively. The specific energy applied for the thermal, sonication and microwave was 36, 102 and 20 kJ/g TS, respectively. Sludge solubility increased by 19.2% using thermal and 83.4% using microwave.
[54]	M	WWTP located in the city of Leon, Spain.	Microwave oven (2450 MHz frequency) was used. SS samples were irradiated with a power output of 650–900 W. Reaction was carried out at 34 °C in 2 × 3 L reactors with mechanical stirrers. A 30:70 mixing ratio of PS:WAS was used. SS samples irradiated at energy values of 975 kJ/L.	CH_4 yield without treatment was 166, 209, 213, and 226 mL/g VS considering 5, 10, 20, and 25 days, respectively. With microwave treatment, the CH_4 yield increased to 214, 295, 308, and 324 mL/g VS for a 5, 10, 20, and 25 days of retention times respectively. CH_4 yield increased between 29% and 43%.
[55]	M	Copero urban WWTP in Spain.	Pre-treatment applied (400 W and 700 W), and the specific energy used was up to 30 kJ/g TS and maximum temperature of 100 °C.	CH_4 yield for raw SS was 111 mL/g VS. Using microwave pre-treatment, CH_4 yield increased to 118–130 mL/g VS.

Table 3. *Cont.*

Ref.	Type	Sludge Collection	Operating Parameters	Biogas Production/Methane Yield
[56]	M	Qinghe WWTP in Beijing, China	2 L lab-scale reactors with 112 rpm and 20 days HRT. OLR of 2.92 g VS/L.d at 37 °C. Microwave-H_2O_2 pre-treatment was used. A 2-stage reactor was set up with a 0.65 L bottle (the first stage reactor) in series with a 2 L reactor.	CH_4 production improved from 215.5 mL/g VS (control) to 258.4 mL/g VS (microwaved) for a 1-stage reactor. For 2-stage, CH_4 yield increased from 258.4 mL/g VS (control) to 288.3 mL/g VS (microwaved). The 2-stage was 11.6% higher than 1-stage.
[57]	Uc	Full-scale WWTP located in Antwerp-South, Belgium.	Ultrasound reactor at 10 L/min fixed flow rate and 25 kHz frequency applied. Power up to 1 kW was used. The AD happened at 37 °C and retention time of 3 weeks in a 1 L reactor filled with 0.5 L WAS.	CH_4 production increased up to 20% using the pre-treatment. For a 23 day reaction, CH_4 production was 126 mL/g DS, and 1.014 kWh/kg DS energy value. Energetic content of the surplus biogas by ultrasonic was 0.195 kWh/kg DS.
[58]	Th and Ud	Biobío WWTP, in Concepción, Chile.	Ud and low-thermal (55 °C) pre-treatment were conducted. Ud using specific energies of 0.5, 15.5 and 30.5 MJ/kg TS. Th treatment with retention times of 3, 8 and 13 h were also performed.	CH_4 production ranged between 472 and 611 mL/g V based on 0.5–30.5 MJ/kg TS (Ultrasound) at for 3–13 h reaction, The CH_4 yield increased between 16% and 50% using the methods.
[59]	M and Uc	WWTP in of Mechelen-Noord, Belgium.	0.6 L reactor used with 500 g SS. Applied 100 W power for 8 min, and 800 W power for 1 min. Total energy of 96 kJ/kg SS to both treatments.	Biogas production increased to 0.26–0.28 L/g VS compared to 0.22 L/g VS (control case). Biogas improved by 27% and 20% based on Uc and M treatments, respectively. Both methods were considered not cost feasible.
[60]	U and O	Municipal WWTP, in Singapore.	130 W and 20 kHz (ultrasound) and 180 W (ozonation) applied to a 0.2 L of sewage sludge.	Biogas production improved by 11–15.4%. The maximum CH_4 production rate increased from 3.53 (control) to 4.32–4.54 mL CH_4/d.

Th: Thermal, M: Microwave, Uc: Ultrasonic; Ud: Ultrasound; O: Ozonation; FA: Free ammonia; WAS: Waste activated sludge; PS: Primary sludge; DS: Dry solids; OLR: Organic Loading Rate; SS: Sewage sludge.

Biogas can be used to generate electricity from different CHP technologies, including fuel cells (FCs), gas turbines (GTs), and Internal Combustion Engines (ICEs). Traditionally, an ICE is the most widely technology used in WWTPs due to its low capital and operational costs, technology maturity, flexibility, robustness, and simple maintenance. GTs are versatile due to their multi-fuel capabilities, relative low maintenance costs, and high-power density, but compared to ICEs, they are still a more complex technology, requiring more maintenance and higher investment costs [9]. FCs have grown rapidly worldwide, especially because of their flexibility, which can be widely used in cogeneration applications. Different from ICEs and GTs, FC generates electricity directly from chemical energy without the need for an intermediate mechanical conversion.

Multiple criteria can be considered for selecting the most suitable technology, such as quantity of biogas available, biogas purity, system capacity, and investment/operation costs [7]. Table 4 shows the main characteristics of FCs, GTs, and ICEs.

Table 4. Characteristics of fuel cell, gas turbines, and CHP systems.

	FC	Micro GT	GT	ICE
Technology status	Emerging	Mature	Mature	Mature
Capacity (kW)	200–1200	30–250	1200–4700	110–3700
Electrical Efficiency (%)	36–45	26–30	26–37	30–42
Thermal Efficiency (%)	30–40	30–37	30–52	35–49
CH_4 minimum level (%)	85	40	30	60
Emissions	Extremely low	Very low	Low	Medium/High
Capital costs (USD/kW)	3800~5280	800~1650	1100~2000	465~1600
O&M costs (USD/kWh)	0.004~0.019	0.012~0.025	0.008~0.014	0.01~0.025

The biogas generated from WWTPs can also be upgraded into biomethane and used as renewable natural gas (RNG). For example, the Janesville WWTP used the biomethane generated on-site to power eight biogas-based fuel vehicles in 2012, with a goal to refuel up to 40 vehicles in the future. The plant has a treatment capacity of over 68,000 m^3/d and a maximum fuel capacity of up to 33,357 MJ/d. From the total biogas generated (225 m^3/h), about 35.7% (80.4 m^3/h) was used in the biogas upgrade to refuel the vehicles, and the remaining part was used in the CHP system [61]. Point Loma WWTP secured a long-term clean biogas power purchase agreement (PPA) with two customers: University of California (2.8 MW) and San Diego South Bay Water Reclamation Plant (1.4 MW). The project cost USD 45 mi, and 75% of the project was subsidised. The plant has capacity to generate around 45,300 m^3/d, and it was the first project in California (USA) to inject biomethane into the natural gas network [62]. San Antonio WWTP, located in Texas, USA, partnered with Ameresco under a PPA to develop a biogas project. The system treats and upgrades the biogas generated during the sewage treatment process and injects it into a commercial gas pipeline. The system has the capacity to process more than 42,475 m^3/d of biogas, and about 60% is injected into the gas network, providing an annual reduction of over 19.700 tons of CO_2 [63]. Newtown Creek WWTP, located in New York (USA), aims to provide enough RNG to supply 5200 homes, saving up to 90,000 metric tons of GHG emissions [64]. Many studies have been conducted on the use of biogas-fuelled gas turbine cogeneration systems in WWTPs. Table 5 summarises some studies that investigated the energy generation potential using FCs, GTs, and ICEs in WWTPs.

Table 5. Studies found in the literature based on different generation technologies, including GT, FC and ICE, in WWTPs.

Ref.	Type	Study Aim	Operating Parameters
[5]	FC	Study the Castiglione WWTP in Italy to achieve self-sufficiency based on co-digestion and SOFC system.	• Castiglione WWTP (5th largest WWTP in Europe) serves 25 million PE, and the co-digestion of SS:OFMSW ratio of 76:24, in terms of mass. • The biogas yield increased from 402 to 500 L/kg VS and biogas production increased from 18.800 to 47,100 m^3/d after co-digestion. • With SOFC and co-digestion, thermal and electrical load coverage was 95.4% and 155.8% (the current system supplies about 49–54.6% of the loads).
[65]	FC	Assess the economic benefits of using SOFC in a WWTP.	• The investment cost of a SOFC system was found to be 7000 EUR/kW, which is currently competitive against CHP technology. • SOFC system can be a feasible option if the cost is around 5600 EUR/kW (with subsidies) or between 2900 and 3400 EUR/kW (without subsidies).
[66]	FC	Study a SOFC system in a WWTP in the UK. WWTP serves 750,000 PE and 105,000 m^3/day inflow.	• WWTP's electricity demand and generation were 60 and 40 MWh/day. Thermal and biomethane production was 48 MWh/day, and 8230 kg/day. • 159 kg/h of H_2 could be generated from biomethane using the SMR system. • For a 120 kW system, the WWTP could generate up to 58 MWh/day (increase by 45%), and the WWTP could achieve 100% electricity self-sufficiency.
[67]	FC	Integration of a SOFC in Parand WWTP to eva-luate economic benefits. Two objective functions were optimised.	• For the single-objective optimisation (payback period), the SOFC electrical power output was 51.56 GWh/year, the investment cost was USD 36.9 mi, the annual benefit was about USD 5.6 mi, the payback period was 1.68 years, and overall exergy efficiency of 19.47%. • For the two-objective optimisation (payback period and exergy efficiency), the SOFC electrical power output was 55.6 GWh/year (29.2% was sold to the grid), the investment cost was USD 39.3 mi and the annual benefit about USD 6.6 mi, payback period was 1.74 years, and overall exergy efficiency of 20.94%.
[68]	GT	Investigate the optimum size of a gas turbine system in a sewage Wastewater plant which serves about 100,000 people.	• The WWTP average monthly wastewater treated volume, biogas production and electricity demand are 1564 mL, 130,000 m^3 and 638 kW, respectively. • Three GT sizes (30, 65 and 200 kW), and three ambient temperatures (continental, temperate, and tropical) were considered. • The annual average power generation for the 30, 65 and 200 kW GT systems were around 1.4–2.4 GWh, 1.5–2.6 GWh, and 1.5–3 GWh, respectively. • The NPV of the 30, 65 and 200 kW GT systems were around USD 2.64–4.73 mi/year, USD 2.96–5.51 mi/year, and 3.1–6.74 mi/year, respectively.
[69]	GT	Study the feasibility of a GT system to supply energy in a real WWTP, in Iran, which treats around 74.2 mL of sewage daily.	• Two scenarios were considered: scenario 1 (local energy prices with no tax on emissions) and scenario 2(international prices with tax on emissions) • For scenario 1, the LCOE, annual electricity generation and the payback period was USD 0.1373/kWh, 10.34 GWh and 14.8 years, respectively, whereas when the total cost is minimised, it was USD 0.0395/kWh, 10.73 GWh and 18.24 years, respectively. For scenario 2, the LCOE, annual electricity generation and payback period were USD 0.3166/kWh, 11.73 GWh and 2.38 years, and when the total cost is optimised, it was USD 0.067/kWh, 2.17 GWh and 2.52 years.
[70]	GT	Study a GT system in the Bali WWTP in Taiwan that has a daily treatment capacity of 1320 mL.	• Daily biogas production is around 4424 m^3/d; around 25.8% is used to supply heating, and the remaining can be used for electricity generation. • The GT's model (30 kW) selected was the CR-30 from Capstone Turbine, and the yearly net electricity generation for the system is about 172 MWh. • The actual power generated from the system is 629 kWh/d based on the system's efficiency of 23.4% and the available biogas of 433 m^3/d.

Table 5. *Cont.*

Ref.	Type	Study Aim	Operating Parameters
[71]	GT	Investigates the injection of hydrogen into biogas to generate electricity in a WWTP.	• Generation system included a 10,000 m^2 PV system to power PEM electrolysers, and 2 × GTs (C65 and C200 Capstone turbines) to generate electricity. • Yearly electrical power and thermal energy varied between 70–400 kW and 120–500 kW, respectively, and available biogas is around 94 m^3/h. • The LCOH can vary between 8.3 and 27.5 USD/kg with a payback period of 6.5–11.6 years.
[72]	ICE	Investigate the combined utilisation of syngas and biogas to power a CHP system in a WWTP.	• Two CHP systems considered: compression ignition (CI) and natural gas (NG) engines. Biogas production estimated at 25 m^3/h, and syngas production for undigested and digested sludges were 0.05 m^3/s and 0.02 m^3/s. Total investment costs for the CI and NG systems were around EUR 1.60–1.66 mi and EUR 0.63–0.99 mi, respectively. The total annual cost savings for the CI and NG systems were around EUR 171,000–235,000 and EUR 110,000–174,000, respectively.
[73]	ICE	Study the potential of a biogas-fuelled system to supply electricity to 5 WWTPs, in Spain.	• Five WWTPs which supply 121,500–451,250 PE and flow between 20 and 30 mL/d. • CHP systems selected based on biogas production: 300 kW (plant 1), 500 kW (plant 2), 500 kW (plant 3), 700 kW (plant 4) and 800 kW (plant 5). • The payback period for all 5 plants was between 2 and 3 years. The CHP systems could supply electricity between 39 and 76%, and 100% thermal energy.
[74]	ICE	Investigated the biogas energy potential in 4 WWTPs in Brazil.	• The 4 WWTPs serve between 41,000 and 133,000 PE. The theoretical net power estimated was 71–329 kW, total investment costs were USD 205,000–315,000, and total generation potential for all plants could be up to 23.6 GWh/year. • The NPV for a 15 years lifetime could vary from USD 1.31–4.74 mi.
[75]	FC and GT	Investigate a system of SOFC and GT in Collegno WWTP, in Italy, which serves around 270,000 PE.	• A 180 kW system generates 1.58 GWh/y with an 11.6-year payback time. The LCOE was 0.134 EUR/kWh. • A 390 kW SOFC-GT generates 2.54 GWh/y with a 7.7-year payback time, and the LCOE was 0.116 EUR/kWh. • For a 275 kW SOFC-GT system, the system could generate 2.2–2.4 GWh/y with an 8–8.5-year payback time and LCOE of 0.118–0.125 EUR/kWh, whereas for a 310 kW SOFC-GT system, the system could generate 2.3–2.7 GWh/y with a 7.9–8.3 years payback period and almost the same LCOE.
[76]	GT, ORC, chiller	Investigate the benefits of a CHP system in WWTPs.	• For 70,000 PE and 400,000 PE plants, the total investment cost was USD 0.25–0.4 mi, and USD 0.75–1.27 mi, respectively. For a 70,000 PE, the payback period for the GT, ORC and chiller were 2.6–3.3, 10–15, and 21–25 years, respectively. For a 400,000 PE, it could be reduced to 1.6–2.1, 5.5–7, and 16–18 years.
[77]	FC, ICE and GT	Study the opportunities and challenges of using FC, ICE and GT on 6 WWTPs.	• 2 WWTPs used: 100,000 PE (12.35 mL/d flow and 62.5 m^3/h biogas production) and 500,000 PE (61.5 mL/d flow and 312.5 m^3/h biogas production). • For the 100,000 PE plant, the system consisted of 143 kW ICE, 2 × 60 kW GT, 300 kW MCFC and 150 kW SOFC. For a 500,000 PE plant, it was 844 kW ICE and 1.2 MW MCFC. The LCOE for the ICE, GT, MCFC and SOFC systems were 4.6–10.4, 6.7–11.5, 16.6–19.9, and 15.2–19.4 EUR/kWh, respectively.
[78]	SCWG and CC	Study an integrated system for H_2 production and power generation.	• Syngas yield was between 1.48 kg/kg_{feed} (500 °C) and 1.78 kg/kg_{feed} (700 °C), and H_2 was 0.02 kg/kg_{feed} (500 °C) and 0.09 kg/kg_{feed} (700 °C). • At 500 °C, total net power was 62.1–90.7 kW with 23.3% efficiency. At 700 °C, the net power and system efficiency were 56.6–88.4 kW, and 63.48%.

3.1.2. Thermochemical Processes

Thermochemical Processes aim to recover the sludge energy content by changing its composition, including biological, chemical, and physical characteristics, based on thermal treatment to produce a valuable fuel, which can be solid, liquid, and/or gas renewable fuels. Additionally, it can reduce the operating costs related to sludge disposal and its negative impacts on the environment. These processes have been receiving significant attention over the last few decades, and they can be classified into incineration (also known as combustion), gasification, pyrolysis, and hydrothermal treatments [79]. Figure 3 illustrates different thermochemical methods that can used to treat sewage sludge, producing valuable by-products, including hydrogen-rich gas, syngas, bio-oil, and biochar.

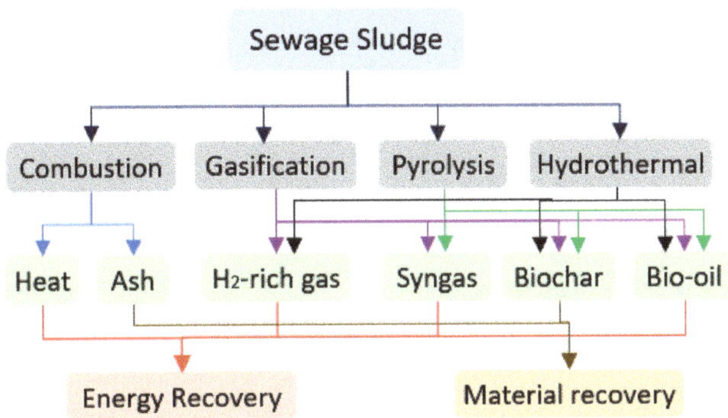

Figure 3. Thermochemical processes applied to sewage sludge for energy recovery.

(a) Combustion

Also known as incineration, it occurs under high temperatures (between 800 °C and 1000 °C) and an excess of air. It is a very mature technology and one of the most traditional methods to convert biomass into energy. Its applications on sewage sludge have been widely studied, and the main benefits of using combustion includes achieve complete oxidation of the organic elements in the biosolids, reduce the sludge volume, and ensure pathogen destruction. In this process, heat and electricity can be recovered, and other co-products, such as ash (inert material), flue gas, and water, are generated [80]. Its main advantages include the ability to significantly reduce the sludge volume (up to 90%), destroy pathogens and toxic elements, mature technology, lower costs compared to other methods, odour control, and potential for energy recovery. However, this method also results in low energy efficiency, emission of toxic compounds, low economic viability, and usually requires additional fuel. Due to its high ash and water content, sewage sludge is typically mixed with other feedstocks (i.e., wood pallets) to achieve a better performance (process known as co-combustion) [81]. Some commercial sludge treatment facilities have been reported in the literature, including plants in Belgium, China, Germany, Japan, the USA, the UK, and the Netherlands, which have used incineration to treat municipal sewage sludge [80]. For instance, a sludge treatment facility (STF) was constructed and started its operation in 2015 in Hong Kong, China, to process 2000 tons/day of sewage sludge from the Stonecutter's Island WWTP. The STF is supplied by two generators capable of producing all the electricity needed to operate the plant. The STF is not only energy self-sufficient but also exports the electricity surplus (up to 2 MW) to the main grid [82].

(b) Gasification

Gasification is a method where the biomass (i.e., dried sludge) is exposed to a high temperature (about 1000 °C) in a reduced air environment, generating syngas, heat, and ash. The syngas (synthetic gas) composition may include CH_4, CO, CO_2, H_2, O_2, and N_2, which can be cleaned to produce a higher quality fuel, whereas the heat can be further recovered to generate electricity. The high heating value (HHV) of syngas from sewage sludge gasification depends on different parameters, including gasification technology, operating conditions (i.e., temperature, pressure, and air-to-fuel ratio), and sewage sludge characteristics. On average, it may vary from 5.8 to 18 MJ/m^3. The benefits of gasification are related to the ability to deal with the inorganic elements, generation of syngas and heat, low energy needs, mature technology, and combined feedstocks without quality/efficiency problems. However, issues related to low energy efficiency, emissions, economic viability, and additional fuel requirements are the main drawbacks [83].

(c) Pyrolysis

Sludge pyrolysis (also known as liquefaction) operates under a temperature between 350 and 500 °C in a restricted oxygen environment. Some parameters, including temperature, heating rate, residence time, pressure, sludge characteristics, reactor type, and process design, can influence pyrolysis efficiency [79]. The main co-product generated from this process, which can be used as a renewable energy source, is bio-oil, which has a calorific value of 30–37 MJ/kg [84]. Syngas and biochar are also other co-products generated that can be used for electricity generation and carbon sequestration or soil amendments, respectively [85]. Different types of pyrolysis can be found in the literature, and the most common are listed in Table 6.

Table 6. Types of pyrolysis and their characteristics.

Pyrolysis Type	Residence Time	Temperature	Heating Rate	Pressure
Slow	>60 min	300–700 °C	0.1–1 °C/s	0.1 MPa
Intermediate	~10 min	500–650 °C	1–10 °C/s	
Fast	0.5–20 s	550–1250 °C	10–300 °C/s	
Flash	<0.5 s	800–1300 °C	>1000 °C/s	
Microwave	30 min	500–1000 °C	<5 °C/s	-
Vacuum	30–120 min	300–600 °C	0.1–1 °C/s	0.01–0.02 MPa
Hydro	240 min	350–600 °C	10–300 °C/s	5–20 MPa

Reactor type is a very important parameter in the pyrolysis that directly impacts the conversion efficiency of organic matter into useful energy (i.e., solid, liquid or gas fuels). Examples of pyrolysis reactors include fixed bed, fluidised bed, rotary kiln, auger (screw), microwave, vacuum, conical spouted bed, bubbling fluidised, circulating fluidised, plasma, and solar, as illustrated in Table 7. Among them, the most common are the fixed and fluidised bed reactors due to their low cost, simplicity, efficiency, flexibility, heat uniformity, and controllability [85].

Table 7. Types of reactors used in pyrolysis.

Type	Details	Efficiency	Op. Costs	Scale	Benefits	Limitations
FiBR	Stationary bed, batch or continuous operation	Low to moderate (15–25%)	Moderate (USD 30~80/ton SS)	Small	- Simple design - Low capital cost - Easy to operate - Well-established	- Lower throughput - Low heat transfer - Energy efficiency
FBR	Granular material, heated by upward gas flow, continuous operation	Moderate to high (25–40%)	Moderate (USD 50~120/ton SS)	Large	- High heat transfer efficiency - High throughput - Uniform heating - Well-established	- Complex design - Requires filtration - Higher capital cost
RKR	Rotating drum, externally heated, continuous operation	Moderate (20–35%)	High (USD 70~150/ton SS)	Large	- High throughput - Suitable for high moisture content - Well-established	- High energy consumption - Expensive maintenance

Table 7. *Cont.*

Type	Details	Efficiency	Op. Costs	Scale	Benefits	Limitations
AR	Screw conveyor moves SS through a heated zone continuously	Moderate to high (20–35%)	Moderate (USD 40~90/ ton SS)	Small/ Medium	- Continuous operation - Easy to control - Low maintenance	- Limited scalability - Needs specialized equipment - Less mature
MR	Direct heating using microwave radiation	High (50–70%)	High (USD 90~150/ ton SS)	Small	- Fast heating - High energy efficiency - Smaller footprint - Uniform heating	- Limited scalability - High capital cost - Complex control systems - Emerging (pilot-scale)
VR	Pyrolysis under reduced pressure, improving bio-oil quality and volatile product yields.	Moderate to high (30–50%)	High (USD 70~130/ ton of SS)	Small/ Medium	- High-quality bio-oil - Improved energy recovery - Better volatile product yields	- Requires vacuum pumps - High capital cost - Complex design - Emerging (limited use)
CSR	Conical reactor with particles suspended by gas flow. SS is fed in from the top and heated through gas flow.	Moderate (25–35%)	Moderate (USD 50~100/ ton of SS)	Medium/ Large	- Efficient heat transfer - Can handle particles of varying sizes - Uniform heating	- Complex design - Requires precise flow control - Less scalable than fluidized beds - Less mature
BFBR	Reactor with a bubbling action caused by gas flow and typically used for lower-density feedstocks.	Moderate (20–35%)	Moderate (USD 50~120/ ton of SS)	Small/ Medium	- Simple design - Moderate heat transfer - Suitable for moderate-sized sludge feedstocks	- Lower efficient heat transfer - Requires gas–solid separation - Mature (widely used)
CFBR	Fluidised bed with a continuous loop of particles in motion and continuous operation	High (30–50%)	High (USD 60~150/ ton of SS)	Large	- High heat transfer efficiency - Suitable for large-scale use - Uniform heating - Well-established	- Complex design - High energy consumption - Expensive maintenance
PR	Pyrolysis using high-temperature plasma arc	Very high (60–80%)	Very high (USD 150~300/ ton of SS)	Small/ Medium	- Very high temperatures (up to 10,000 °C) - Capable of decomposing almost all feedstocks	- Extremely high capital and operating cost - Need of special equipment - High energy use - Emerging (high-tech)
SR	Uses concentrated solar energy to heat the sludge for pyrolysis.	Moderate to high (20–40%)	Low (USD 80~150/ ton of SS)	Small/ Medium	- Renewable energy source - Sustainable	- Dependent on sunlight - Requires large infrastructure - Expensive setup

SS: Sewage sludge; FiBR: Fixed-bed reactor; FBR: Fluidized-bed reactor; RKR: Rotary kiln reactor; AR: Auger reactor; MR: Microwave reactor; VR: Vacuum reactor; CSR: Conical spouted bed reactor; BFBR: Bubbling fluidized bed reactor; CFBR: Circulating fluidized bed reactor; PR: Plasma reactor; SR: Solar reactor.

(d) Hydrothermal treatment

Hydrothermal treatment applies high pressure and a high temperature the sewage sludge in the presence of water in order to generate useful products, including biogas, biofuels, and solid biochar. Hydrothermal treatment methods (see Table 8) can classified in three groups: Hydrothermal Carbonisation (HTC), Hydrothermal Liquefaction (HTL), and Hydrothermal Gasification (HTG).

Table 8. Characteristics of HTC, HTL and HTG.

Type	Temperature (°C)	Pressure (MPa)	Water State	Generated Products	Advantages
HTC	150–280	0.1–11	Subcritical	Solid (char), small amount of liquid (biocrude) and gas	Sludge stabilisation, volume reduction, fertiliser
HTL	280–375	8–22	Sub/near critical	Liquid (biocrude), small amount of solid (char), water-soluble fractions	Use wet sludge without the need for drying
HTG	>375	>22.1	Super-critical	Gas (Syngas), small solid amount (char), and water-soluble fractions	Produce high concentrations of H_2

HTC focuses on producing carbon-rich solids (hydrochar) and offers benefits in carbon sequestration and soil enhancement. However, it has lower energy recovery compared to other processes. HTL generates bio-crude oil with high energy density. It requires further upgrading, and involves complex operating conditions. HTG efficiently converts organic materials into syngas, and it is highly effective for waste-to-energy conversion, but there is high operational costs and material challenges [79]. Gasification of sewage sludge has been reported in different studies, as illustrated in Table 9. Some researchers have investigated the energy potential from thermochemical processes applied to sewage sludge, as shown in Table 10. Table 11 highlights studies that investigate the generated products and the potential for applying HTC and HTL on sewage sludge.

(e) Supercritical water

There are two main methods to treat sewage sludge via supercritical water technologies: supercritical water gasification (SCWG) and supercritical water oxidation (SCWO). SCWO and SCWG (also known as HTG) are considered promising technologies to treat organic wastes, including sewage sludge, because this technology not only generates fuel gases (including hydrogen), but also has low environmental and social impact. In supercritical water processes, the feedstock (i.e., sewage sludge) is subjected to a supercritical water environment (temperature and pressure higher than 374 °C and 22.1 MPa), and the organic compounds in the feedstock are dissolved. The by-products of SCWO are typically water vapour, carbon dioxide, and hydrogen (can be generated as a secondary product depending on the composition of the feedstock and reactor operation conditions), whereas the SCWG method generates syngas, which is mostly composed primarily of hydrogen but also smaller amount of carbon dioxide and methane. The composition of the gases generated from SCWO and SCWG may vary depending on different factors, including the feedstock composition, temperature, pressure, residence time, and the presence of catalysts [10,86,87]. Some companies, such as General Atomics, EcoWaste Technologies, Chematur, and Supercritical Fluids International, have been developing both small and large-scale SCWO systems that can be applied to a range of areas, including wastewater treatment [86].

Ref. [87] studied different SCWO systems (small and large scale systems), which were implemented in different locations across several years and based on different feedstocks, including sewage sludge. The authors summarised the main information of different systems, highlighting their location, year of construction, company name, system capacity, and feedstock type. They also highlighted the common operation challenges, which include (i) corrosion due to both a high temperature and pressure, (ii) corrosion due to acid formation, (iii) salt precipitation or plugging, and (iv) high energy consumption and operating costs. Some of the plants stopped operations due to technical and/or economic reasons.

Table 9. Studies focused on sewage sludge gasification from small-scale and commercial plants.

Ref.	Sewage Sludge	Feedstock Characteristics	Operation Parameters	Generated Products
[88]	DSS	HHV 13.5 MJ/kg dry, 8.6% MC, 51.8% VM, 43.1% ash. UA: 51% C, 7.3% H, 7.9% N, 2.05% S, and 31.8% O.	100 kW_{th} FBR, DSS feeding rate of 9–16 kg/h, temperature 790–815 °C, air flow 11–15 m^3/h,	Syngas LHV 3.6–4.5 MJ/m^3 dry, gas composition includes 4.6–6.6% CH_4, 8.4–10.2% H_2, 7.3–9.9% CO, 15.2–16.5% CO and other gases.
[89]	DSS + sp (mix ST)	HHV 13.2 MJ/kg dry, 29.9% MC, 67.7% VM, 9.8% ash. UA: 45.2% C, 5.8% H, 1.2% N, 0.4% S, and 37.4% O.	100 kW_{th} LTCFB reactor, 45 kg of dry SS gasified for 28 h, temperature 500–750 °C, Air eq. ratio of 0.2–0.3	Syngas LHV and HHV as 1.76 and 1.88 MJ/m^3 dry. Gas composition includes 2.5% CH_4, 1.6% H_2, 8% CO, 18% CO_2 and other gases.
	Dry SS + sp (BJ)	HHV 14.8 MJ/kg dry, 12.5% MC, 66.5% VM, 13.3% ash. UA: 43.5% C, 4.2% H, 1.2% N, 0.4% S, and 35.7% O.	6 MW_{th} LTCFB reactor, 7.5 ton of dry SS gasified for 30 h under 500–750 °C, Air eq. ratio of 0.3	Syngas LHV and HHV as 3.84 and 4.2 MJ/m^3 dry. Gas composition includes 4.2% CH_4, 9% H_2, 11.8% CO, 19.4% CO_2 and other gases.
	Dry SS (SLU BJ)	HHV 11.4 MJ/kg dry, 12.5% MC, 42.7% VM, 43.1% ash. UA: 27.6% C, 4.3% H, 3.9% N, 0.8% S, and 19.8% O.	100 kW_{th} LTCFB reactor, 240 kg of dry SS gasified for 14 h, temperature 500–750 °C, Air eq. ratio of 0.2–0.3	Syngas LHV and HHV as 1.88 and 2.05 MJ/m^3 dry. Gas composition includes 2.6% CH_4, 5% of H_2, 4.5% CO, 18.5% CO_2 and other gases.
	Dry SS (SLU RA)	HHV 12.2 MJ/kg dry, 4.6% MC, 43.1% VM, 43.8% ash. UA: 28.7% C, 4.2% H, 3.9% N, 1.3% S, and 18.1% O.	100 kW_{th} LTCFB reactor, 800 kg of dry SS gasified for 40 h under 600–750 °C, Air eq. ratio of 0.2–0.3	Syngas LHV and HHV as 1.96 and 2.16 MJ/m^3 dry. Gas composition includes 2.4% CH_4, 6.6% H_2, 4.2% CO, 17% CO_2 and other gases.
[90]	SS	SS: HHV 2.84 MJ/kg dry, 80.1% MC, 9.78% VM, 8.3% ash. UA: 6.27% C, 1.1% H, 0.77% N, 0.3% S, 3.2% O.	Fluidised-bed gasifier, temperature of 784–822 °C with 0.3 air:fuel ratio. SS feeding rate of 25 kg/min,	Syngas LHV 2.2–2.6 MJ/m^3. Gas composition: 2.9–3.4% CH_4, 3.3–3.5% H_2, 5.6–6.9% CO, 23.2–27.8% CO_2. Gas yield: 28.5 m^3/min, 46.6–53% CGE.
	SS and PMS	PMS characteristics: HHV 4.25 MJ/kg dry, 51.4% MC, 29.9% VM, 18.33% ash. UA: 13.9% C, 2.03% H, 0.42% N, 0.15% S, and 13.8% O.	Fluidized-bed gasifier, 1:1/1:2 SS:PMS mix ratios at 816–858 °C and 0.3 air:fuel ratio. SS feeding rate of 8.3–12.5 kg/min, and 16.7–12.5 kg/min for PMS.	Syngas LHV 1.7–2.42 MJ/m^3 dry. Gas composition: 2.6–3.65% of CH_4, 2.6–4.13% of H_2, 3.95–4.7% of CO, 17.4–25.42% of CO_2 and other gases. Gas yield of 29.4 m^3/min and CGE around 45.6–61.6%.
[91]	Dry SS	6.5% MC, 48.1% VM, 47.6% ash, 4.3% fc. UA: 51% C, 6.9% H, 7.5% N, 2.4% S, and 32% O.	20 kW bubbling fluidised bed reactor, temperature of 850 °C, air ratio of 0.25, fuel mass flow of 7 kg/h.	Syngas yield of 1.3 m^3/kg. Gas composition in fraction mole: 6% of CH_4, 40% of H_2, 20% of CO, 32% of CO_2 and other gases.
[92]	Dry SS	6.8% MC, 38% VM, 50% ash, 12% fc. UA: 29.7% C, 4.3% H, 4.6% N, 1.4% S, 59.9% O. 11.73 MJ/kg HHV.	20 kW reactor operating for 12 h/day at 20 kg/h feeding rate (max of 240 kg/day) at 1000–1150 °C.	Gas composition in fraction mole: 1.2% CH_4, 23.3% H_2, 18.5% CO, 13.4% CO_2 and 43.6% N_2. 1.2 kg SS could produce up to 1 kWh
[93]	Dry SS	LHV 9.2 MJ/kg dry, 7.8% MC, 47.8% VM, 47.5% ash. UA: 48.8% C, 7.4% H, 7.1% N, 2.3% S, and 34.2% O.	20 kW fluidised bed reactor at 800 °C. Parameters: 2.7 kg/h steam, 3.6–5.4 kg/h fuel, 1.5 mol/mol steam:carbon.	Gas yield: 0.85 m^3/kg (800 °C), 0.4 m^3/kg (710 °C). CGE: 60% (800 °C), 45% (710 °C). Gas composition in m^3/m^3: 0.1 CH_4, 0.46 H_2, 0.12 CO, 0.3 CO_2.
[94]	Dry SS	Raw SS: HHV 10.98 MJ/kg dry, 3.2% fc, 55% VM, 41.8% ash. UA: 21.86% C, 3.4% H, 3.8% N, 0.64% S, 28.5% O.	Reactor at 700–900 °C and 1–60 min reaction. SS:SD mix ratios of 1:0, 1:3, 1:1, 3:1, and 0:1. SS upgraded via HTC.	At 900 °C, syngas yield was 0.93 m^3/kg, LHV 5.62 MJ/kg dry and 67% efficiency. Gas composition: 2% CH_4, 5% H_2, 35% CO and 58% CO_2.
	Dry SS and SD	SD-SS: HHV 12–17 MJ/kg, 13–21% fc, 46–65% VM, 13–41% ash. A total of 33–49% C, 3–5% H, 2.5% N, 0.5% S, 19–30% O.	130 W and 20 kHz (ultrasound) and 180 W (ozonation) applied to a 0.2 L of sewage sludge.	At 900 °C and 3:1 SS:SD ratio, syngas yield was 1.04 m^3/kg, LHV 8.15 MJ/kg dry and 77.7% efficiency. Gas content: 7% CH_4, 8% H_2, 45% CO, 40% CO_2.
[95]	SS	64.2% VM. UA: 41.2% C, 5.2% H, 3.2% N, 20.7% O, and calorific value of 14.1 MJ/kg.	Temperature around 750–850 °C, SS mass rate around 170–260 g/h.	Gas yield was 0.45–1.13 m^3/kg (20–23% H_2 and 4–13% CH_4), LHV was 6.5–9.7 MJ/m^3. Biochar HHV up to 15 MJ/kg
[96]	Dry SS	LHV 13.2 MJ/kg, 7.1% MC, 51.53% VM, 4% fc, 37.4% ash. UA: 31.3% C, 4.56% H, 4.72% N, 1.3% S, 20.7% O.	Fluidized bed reactor at 813–817 °C, 60–260 min reaction time, and usage of additives (AC, Ni, Fe)	Syngas LHV 5.04–6.52 MJ/m^3 dry, 67.5–88.87% CGE. Gas content: 4–6.5% CH_4, 13.4–28.13% H_2, 10–13.4% CO, 9.4–14.5% CO_2, 44–51.7% N_2 and others.

Table 9. *Cont.*

Ref.	Sewage Sludge	Feedstock Characteristics	Operation Parameters	Generated Products
[97]	Dry SS	LHV 13.1 MJ/kg dry, 8.7% MC, 51.3% OM, 41.7% ash. UA: 29.5% C, 4.9% H, 4.1% N, 1.6% S, and 15.0% O.	Fluidised bed gasified with 0.3 air: fuel ratio and 800 °C operating temperature. A total of 1.2–3.7 g/min SS feeding rate, and use Do as catalyst (10% mix with sludge)	Without De: Syngas LHV 2.9–3.6 MJ/m^3, 35.7–43.2% CGE of, gas yield 2.7–3 m^3/kg. Gas content: 2.3–3.3% CH_4, 8.5–10.3% H_2, 6.1–9% CO, 13.2–13.9% CO_2. With De: Syngas LHV 2.9–3.9 MJ/m^3, 35–49% CGE, gas yield 2.8–3.2 m^3/kg. Gas content: 2.2–3.2% CH_4, 9.6–14% H_2, 5.3–9.7% CO, 12.6–15% CO_2.
[98]	Dry SS	LHV 17 MJ/kg dry, 5.3% MC, 61.56% VM, 26.14% ash, 7% fc. UA: 39.5% C, 5.8% H, 5.3% N, 0.9% S, 24.4% O.	2-stage gasifier in series. A total of 2 kg/h feeding rate of, and use additives (AC, CaO). A 0.25 eq. ratio and 785–820 °C.	52.5–66% syngas, 19.2–23% char, 14.3–20.3% condensate liquid and tar. Syngas LHV was 9.2–11.7 MJ/m^3 dry and 51.7–80.1% CGE. Gas content: 7.2–8.5% CH_4, 28–52.2% H_2, 15.9–19.3% CO, 20–32.4% CO_2 and 3.8–8.9% N_2.
[99]	DSS	SS Characteristics: LHV 14.26 MJ/kg dry, 18.75% MC, 33.4% VM, 11.6% ash, 36.3% fc. UA: 45.55% C, 6.6% H, 1.1% N, 1.2% S, and 33.9% O.	2-stage gasifier (fluidized bed and tar-cracking reactors) connected in series. Operating temperature of 800 °C, air flow rate 13–17 L/min, operating time 60 min.	65.5–75.3% syngas, 15–22.4% char, 9% condensate liquid and tar. Syngas LHV of 5.35–6.1 MJ/m^3 dry and 67–92.4% CGE. Gas composition: 4.5–5.4% CH_4, 14.6–22.6% H_2, 9.7–12.4% CO, 10.64–10.9 CO_2 and 51.6–54.5% N_2.
	DSS and coal	Coal Characteristics: 1% MC, 22.5% VM, 17.8% ash, 64.95% fc. UA: 78.5% C, 0.6% H, 0.43% N, 0.4% S, and 2.32% O. Coal:DSS mix ratio of 70:30, 50:50, and 30:70.		74% syngas, 12.5–13.7% char, 11.8–13.3% condensate liquid and tar. Syngas LHV of 5.1–5.4 MJ/m^3 dry, 80.6–84.3% CGE. Gas composition: 3.8–4% CH_4, 24–26.6% H_2, 9.46–10.4% CO, 11.12–12.5% CO_2 and 46.2–50.8% N_2.
[100]	Dry SS	LHV 15.1 MJ/kg dry, 8.2% MC, 56.9% VM, 30.3% ash, 4.6% fc. UA: 39.8% C, 6.4% H, 5.6% N, 1.2% S, and 24.7% O.	2-stage gasifier (fluidized bed and tar-cracking reactors) connected in series, 760–815 °C temperature, 90–100 min reaction, and use of AC as additive.	68.6–76.9% syngas, 12–19.5% char, 6.2–20% condensate liquid and tar. Syngas LHV of 5.4–7.5 MJ/m^3 dry. Gas composition: 2.7–7% CH_4, 9.6–34.1% H_2, 9.2–17.2% CO, 6.5–14.6 CO_2, 38.4–64.4% N_2 and other elements.
[101]	Dry SS	LHV 14.1 MJ/kg dry, 4.7% MC, 51.3% VM, 24.1% ash. UA: 39.46% C, 5.8% H, 5.35% N, 0.9% S, and 24.4% O.	2-stage gasifier, 13 L/min ratio, and 785–810 °C. A 10.4–16.6 g/min feeding rate for 75–220 min, and used additives.	Syngas LHV was 5.65–7.1 MJ/m^3 dry. Gas content: 3.2–5.8% CH_4, 11.8–31.3% H_2, 9.1–18.4% CO, 7.6–14.7% of CO_2, 39.5–54.4% of N_2 and other elements.

SS: Sewage sludge; DSS: Dewatered sewage sludge; PMS: Paper-mill sludge; SD: Sawdust; LTCFB: Low Temperature Circulating Fluidised Bed; De: Dolomite, CGE: Cold gas efficiency; HHV: Higher heating value, LHV: Lower hearing value; AC: Active carbon; OM: Organic matter; MC: Moisture content; UA: Ultimate analysis; fc: Fixed carbon; sp: Straw pellets; FBR: Fluidized-bed reactor.

Table 10. Studies focused on sewage sludge pyrolysis.

Ref.	Sludge Collection	Feedstock Characteristics	Operation Parameters	Generated Products
[102]	WWTP in Medellin, Colombia	6.56 MJ/kg dry HHV, 6.1% MC, 27.24% VM, 3.3% fc, 63.4% ash. UA: 12.8% C, 1.74% H, 1.2% N, 0.55% S, 16.22% O.	Fluidised bed reactor operated at an atm pressure. Operating temperature between 300 °C and 800 °C.	At 500 °C and 600 °C, the biochar, bio-oil and gas yields were 14–27%, 28–39%, and 45–47%, respectively. At 800 °C, gas content included 50% H_2, 20% CO_2, 20% CO, and 10% CH_4. At 600 °C, it was 43% H_2, 37% CO_2, 12% CO, and 8% CH_4.
[103]	Municipal WWTP in Dalian, China	13.6 MJ/kg dry HHV, 76.6% MC, 27.24% VM, 3.3% fc, 63.4% ash. UA: 12.8% C, 1.74% H, 1.2% N, 0.55% S, and 16.22% O.	Quartz cylindrical reactor. Sludge pyrolised for 4 h with heating rate of 3 °C/min at temperatures ranging from 500 °C to 800 °C.	The char, bio-oil and gas yields were around 17.5–25%, 38–43.5%, and 37–42%, respectively. Highest yields for char, bio-oil and gas were at 800 °C, 600 °C, and 500 °C. The lowest value was at 600 °C, 500 °C/800 °C, and 800 °C, respectively.

Table 10. *Cont.*

Ref.	Sludge Collection	Feedstock Characteristics	Operation Parameters	Generated Products
[104]	Carter's Creek WWTP in Texas, USA	18.5 MJ/kg dry HHV, 1.9% MC, 68.1% VM, 14.1% fc, 17.8% ash. UA: 39.4% C, 5.6% H, 7.8% N, 0.8% S, and 28.6% O.	Fast pyrolysis using a bench-scale bubbling fluidized bed reactor. Temperature between 425 °C and 550 °C.	Highest bio-oil yield was 35.7%, biochar was 28.7% and biogas 23.5%, and 11.8% losses. The bio-oil HHV ranged between 24.3 MJ/kg (at 425 °C) and 37.6 MJ/kg (at 550 °C). The generated biochar HHV was 7.4 MJ/kg.
[105]	Urban WWTP in Barcelona, Spain.	11.1 MJ/kg dry HHV, 5.6% MC, 54.2% VM, 8.6% fc, 37.2% ash. UA: 25.5% C, 4.5% H, 4.9% N, 2.1% S, and 25.8% O.	Flash pyrolysis using a conical spouted bed reactor (CSBR) at between 450 °C and 600 °C.	Char and bio-oil yields at 450 °C, 500 °C, 550 °C, 600 °C were 51%, 46%, 44%, 43%, and 45%, 48.5%, 48.5%, 46%, respectively. Gas yield of 4–11%.
[106]			Fast co-pyrolysis with SS and lignocellulosic biomass in a conical spouted bed reactor.	At 500 °C, the gas, bio-oil and biochar yields were 12%, 55%, and 33%, respectively,
[107]	WWTP in Minnesota, USA.	4.5% MC, 68.6% VM, 0.3% fc, 16.4% ash. UA: 53.24% C, 7.4% H, 6.12% N, and 33.25% O. A total of 24.42 MJ/kg HHV.	A cfMAP with 1 kW power and 2450 MHz was used. Temperature of 450–600 °C. Used ZSM-5 as catalyst (activated at 550 °C for 4 h)	The energy value for bio-oil, biochar and gas varied between 2.2 and 7.7 MJ/kg, 2.14 and 5.4 MJ/kg, and 5.6 and 9.4 MJ/kg, respectively. The char, bio-oil and gas yields were between 33 and 62.5%, 16 and 40%, and 21.5 and 40%, respectively.
[108]	WWTP in Shaanxi, China	4.57% MC, 63.13% VM, 5.47% fc, 26.83% ash. UA: 29.12% C, 5.98%, 3.98% N, 1.64% S, 21% O. A total of 16.7 MJ/kg HHV.	Fast pyrolysis using a fixed bed quartz tube reactor at operating temperature between 500 °C and 900 °C.	Char, bio-oil and gas yields ranged between 14 and 35%, 37.5 and 71%, and 15 and 27.5%, respectively. The generated gas included 8–36% H_2 and below 7% CH_4. CO content was 35%, at 900 °C, and below 10% for the other temperatures
[109]	urban WWTP in Madrid, Spain.	7% MC, 50% VM, 3% fc, 40% ash. UA: 27.9% C, 4.7% H, 4.5% N, 1.4% S, and 34.6% O. A total of 12.5 MJ/kg HHV.	Bench-scale stirred batch reactor at 525 °C under N_2 atmosphere.	At 525 °C, the biochar, bio-oil and gas yields were approx. 50%, 41%, and 9%, respectively. LHV of the pyrolised SS was 10 MJ/kg.
[110]	WWTP in Beijing, China.	SS: 2.25% MC, 61.52% VM, 6.7% fc, 29.5% ash. UA: 53.2% C, 7.5% H, 6.4% N, 2% S, and 30.9% O. Cb: 4.6% MC, 78.1% VM, 15.64% fc, 1.7% ash. UA: 49.2% C, 6.3% H, 0.5% N, 0.3% S, and 43.7% O.	Tube furnace reactor under 400–800 °C. Pyrolysis rate of 20 °C/min under N_2 protection (99.99%, flow rate = 25 mL/min). Mix ratio of SS:Corncob of 1:1	The yields of char, bio-oil and gas varied from 30.4% (800 °C) to 75.1% (400 °C), 10.2% (400 °C) to 51.8% (800 °C), and 14.7% (400 °C) to 20% (700 °C), respectively. Gas content: CH_4 and CO below 10%, H_2 ranged from 22% to 59% and CO_2 from 35% to 59%.
[111]	Nanjing urban WWTP in Nanjing, Jiangsu province, China.	SS: 79% MC, 31.52% VM, 5.25% fc, 63.23% ash. UA: 20.9% C, 8.7% H, 3.5% N, 0.9% S, 2.8% O. HHV 12.5 MJ/kg. SD: 6.3% MC, 73.6% VM, 14.4% fc, 5.7% ash. UA: 49.5% C, 7% H, 0.3% N, 0.4% S, 42.7% O.	Co-pyrolysis conducted in a screw moving bed reactor. Operating parameters included 900 °C at 20 °C/min rate and kept for 30 min based on the 16 g/min feeding rate. SD content mixed with SS were 0, 20%, 40%, 60% and 80%.	Generated gas had an HHV of 14 MJ/kg, 13.4 MJ/kg, 13.45 MJ/kg, 13.41 MJ/kg and 13 MJ/kg for a mix of SD about 0, 20%, 40%, 60% and 80%, respectively. Dry gas yield for a mix of SD of 0, 20%, 40%, 60% and 80% was 0.24 m^3/kg, 0.36 m^3/kg, 0.55 m^3/kg, 0.66 m^3/kg, 0.74 m^3/kg, respectively. The H_2, CH_4, CO_2 and CO were 29.13–42.35%, 10.5–18.9%, 14.6–24.1%, and 26.8–31.4%, respectively.
[112]	WWTP in Australia.	Biosolids: 20–80% MC. UA: 35.7% C, 5.2% H, 3.5% N, 25.4% O, 0.7% S on wt%.	450–850 °C, 1 atm and feed rate of 265 kg/h. Costs included AUD 1000/kW (reactor and turbine), AUD 600/m^3 (N_2) plant, AUD 3/m^3 for water, and AUD 0.03/kW for electricity.	The biochar, bio-oil and gas composition was 43.2–53%, 37.7–40.4% and 9.3–17.2%, respectively. From 650 °C to 850 °C, the gas increases (HHV decreased from 23.2 to 20 MJ/kg) and oil decreases. For a 30-year plant operation, the NPV was AUD 2.3–2.6.

HHV: Higher heating value; LHV: Lower hearing value; VM: Volatile Matter; UA: Ultimate Analysis; MC: Moisture; fc: Fixed Carbon wt: wet basis; cfMAP: continuous fast microwave-assisted pyrolysis; SD: sawdust; SS: sewage sludge; Cb: Corncob.

Table 11. Studies that applied hydrothermal treatments using sewage sludge.

Ref.	Type	Sludge Collection	Feedstock Characteristics	Parameters	Generated Products
[113]	HTC	Jiangxinzhou WWTP in China.	82.5% MC. UA: 39.88% C, 6.20% H, 6.04% N, 20.5% O, 5.62% S. A total of 17.97 MJ/kg HHV.	1 L batch reactor with maximum pressure and temperature of 35 MPa and 500 °C.	Gas production of 1.59 L at 200 °C under 2 MPa and 2.86 L at 360 °C under 19.4 MPa. At 380 °C, H_2 and CH_4 yields were 0.14 and 0.24 mol/kg, respectively.
[114]		WWTP in Nanjing, China.	89.2% MC, 56.9% VM. UA: 25.6% C, 4.4% H, 4.6% N, 22% O, 0.2% S. 11 MJ/kg HHV.	5 reactors operation under 200–250 °C and 2–26 MPa.	H_2 yield reached 0.7 mol/kg at 450 °C, accounting for 11.2 *v*/*v*% of the syngas. At 400 °C, H_2 yield was below 0.15 mol/kg.
[115]		Il-San municipal WWTP, in Korea	66.9% VM. UA: 38.55% C, 6.46% H, 8.05% N, 46.50% O, 0.44% S. HHV of 16.5 MJ/kg.	1 L autoclave reactor under 180–280 °C and 30 min reaction time.	The HHV of the solid fuel was around 17.3–22.4 MJ/kg. Energy recovery efficiency decreased (from 92.2% to 89.6%) with the increase in temperature.
[116]		WWTP in Japan.	85.94% MC. UA: 51.20% C, 6.64% H, 8.85% N, 31.94% O, 1.37% S on db. 18.81 MJ/kg HHV	Operating temperature around 180–240 °C and 15–45 min reaction time.	Lowest HHV was 18.30 MJ/kg at 180 °C and 15 min. Highest value was 20.17 MJ/kg at 240 °C and 45 min. Recovery efficiency of 40% (>200 °C).
[117]		WWTP in Changsha, China.	89.3% MC, 47.5% VM. UA: 25% C, 4.2% H, 4.8% N, 15.3% O, 0.74% S, 11 MJ/kg HHV	0.5 L 316 stainless steel reactor (180–300 °C temperature and 30–480 min reaction time.	Maximum HVV was 12.06 MJ/kg at 260 °C and 1 h reaction. Lowest value (9.8 MJ/kg) was for 3 h reaction and 260 °C (high reaction time does not increase HHV).
[118]		WWTP in Ratmal-na, Sri Lanka	81% MC. UA: 34.41% C, 5.21% H, 4.75% N, 23.3% O, 1.73% S on db. 15.2 MJ/kg HHV	Temperature between 100 and 200 °C and reaction up to 1 h.	Maximum (89%) and minimum (73.7%) char yields. Highest HHV (16.17 MJ/kg) at 150 °C and lowest HHV (13.57 MJ/kg) at 100 °C.
[119]		WWTP in Kar-miel city, Israel.	81% MC. UA: 40.3% C, 5.8% H, 4.7% N, 19.4% O on db. A total of 18.0 MJ/kg HHV	0.5 L stainless steel stirred reactor, 200–300 °C and 0.5, 1, and 2 h retention times.	Hydrochar HHV ranged from 18.2 MJ/kg (200 °C, 0.5 h) to 21.5 MJ/kg (300 °C, 0.5 h). Highest and lowest BMP yield were 236.0 and 25.7 mL CH_4/gCOD.
[120]		WWTP in Shimodate, Japan	79% VM. UA: 48.94% C, 7.09% H, 2.51% N, 33.4% O, 0.65% S on db. 21.0 MJ/kg HHV	0.2 L reactor for 0.5 h and 180 °C, at 6:1, 4:1, 3:1 and 2:1 mixing ratios	SS hydrochar HHV was 21.59 MJ/kg (180 °C and 30 min) and HTC input energy was 115.96–117.7 MJ. Hydrochar HHV was about 23.46–25.34 MJ/kg.
[121]		WWTP in Gwangju, Korea.	72.33% VM. UA: 52.29% C, 7.89% H, 6.39% N, 32.62% O, 0.81% S on db. A total of 20.6 MJ/kg	1 L reactor operating under 180–270 °C and 30 reaction time.	Hydrochar HHV was around 18.66–23.44 MJ/kg. Maximum (93.13%) and minimum (40.78%) char yields were achieved at 180 °C and 270 °C.
[122]	HTL	Adelaide plant, Ontario, Canada.	62.2% VM, 96.1% MC. UA: 38.0% C, 5.23% H, 7.2% N, 25.2% O, 0.75% S. 16.0 MJ/kg HHV	0.1 L stirred reactor under 2 MPa, 200–350 °C, 10–60 min reaction.	Composition of oil, solid and WSP were 11.3–33.6 wt%, 9.9–61.2 wt%, and 27.3–62.3 wt%. From WSP, biogas was recovered (up to 0.8 L in 31 days).
[123]		WWTP at Viborg, Denmark.	Dry matter of 4 wt%. UA: 46.5% C, 6.1% H, 3.3% N, 31.2% O, 0.4% S. A total of 19.8 MJ/kg HHV	Reactor with 1.66 L/min feed capacity (20 L vol.), 350 °C temperature and 5 h reaction.	Bio-crude average yield of 24.5 wt%, chemical energy recovery of about 33.6%, and an average of HHV of around 26.9 MJ/kg.
[124]		Marselisborg WWTP, Denmark	Not informed	HTC batch reactor (20 mL) at 340 °C and 20 min reaction time.	With K_2CO_3 catalyst, the HHV and chemical energy recovery increased from 36.1 to 38 MJ/kg and 56 to 67%, respectively.
[125]		Daugavgriva plant in Riga, Latvia.	80.5% MC, 56.8% VM. UA: 52.0% C, 7.6% H, 7.5% N, 30.4% O, 2.6% S. A total of 15.3 MJ/kg HHV	Batch stainless steel autoclave reactor at 200–300 °C and 10–100 min reaction time.	Highest bio-oil yield (47.8%) and 70% recovery achieved under 5.0 MPa, 300 °C, and 40 min time. Lowest bio-oil yield (34.5%) with 36.2 MJ/kg HHV at 200 °C.
[126]		WWTP in Aalborg Forsyning, Denmark.	73.4% MC, 50.5% VM. UA: 50.9% C, 7.4% H, 6.9% N, 34.8% O, 0.8% S. A total of 22.15 MJ/kg HHV	0.1 L stainless steel reactor at 350–400 °C, 10 MPa, and 15 min reaction time.	At 350 °C and no catalyst, bio-crude HHV was 35.3 MJ/kg and 64% energy recovery. With catalyst, HHV was 36.6 MJ/kg and 74.6% energy recovery.

Table 11. *Cont.*

Ref.	Type	Sludge Collection	Feedstock Characteristics	Parameters	Generated Products
[127]	HTL	Beishiqiao plant in Xi'an, China	90% MC. UA: 33.9% C, 5.1% H, 5.8% N, 16.5% O, 3.2% S. A total of 16.1 MJ/kg HHV	4.4 mL mini-batch reactors at 18 MPa with temperature ranging 260–350 °C.	Highest biocrude yield (23 wt.%), 35.4 MJ/kg HHV and energy recovery (50.2%) at 340 °C. Lowest energy recovery (32.6%) with 34.84 MJ/kg HHV at 260 °C.
[128]		WWTP in State College, PA, USA.	97.8% MC. UA: 46.5% C, 7.0% H, 2.1% N, 33.3% O, 0.8% S. A total of 19.9 MJ/kg HHV	4.1 mL reactor at isothermal HTL (673 K, 1 h) and non-isothermal HTL (773 K, 1 min).	With 10 wt% and 50 wt% additives, biocrude yield ranged around 18.9–21.7% and 10.2–18.6% (isothermal), and 24.8–29.1% and 10.9–27.5% (non-isothermal).
[129]		Qinghe WWTP in Beijing, China	84.5% MC. UA: 46.7% C, 6.8% H, 8.1% N, 37.6% O, 0.8% S.	0.6 L batch stainless 316 reactor, at 210–330 °C, 30 MPa and 0.5 h.	At 210 °C, biocrude yield was 39.9% and 86.3% HTL conversion. At 270 °C, it increased to 47.5% and 97.7%, and at 330 °C, it decreased to 41% and 90.1%.
[130]		Municipal WWTP in Changsha, China.	UA: 28.9% C, 4.5% H, 4.2% N, 13.9% O, 0.6% S.	0.5 L autoclave reactor (316 stainless steel) at 350–400 °C, 35 MPa (max), and 0.5 h.	Bio-crude HHV about 37.8–39 MJ/kg at 350 °C and 0.5 h reaction time. For SS with no pyrolysis, bio-crude HHV was 37.35 MJ/kg at 350 °C and 30 min reaction.
[131]		WWTP in Doha, Qatar.	83.6% MC. UA: 30.5% C, 6.2% H, 5.5% N. A total of 16.9 MJ/kg HHV	0.1 L reactor under 275–400 °C and 30–120 min reaction time.	For a 0.5h reaction time and 350 °C, maximum biocrude yield was 44.8%, whereas at 275 °C, the lowest biocrude yield was reached (less than 20%).
[132]		H.C. Morgan Water Facility, in Alabama, USA.	82.4% MC, 52.9% VM. UA: 33.1% C, 5.5% H, 5.0% N, 25.9% O, 0.7% S. A total of 14.1 MJ/kg HHV	1.8 L reactor at 350 °C, and 1 h. Used red mud (RM) catalyst: calcined RM, reduced RM at 500 °C, and reduced RM at 700 °C	At 25 wt% of CRM ethylene, RM at 500 °C ethylene and RM at 700 °C ethylene, the biocrude yield reached 27.1%, 31.3% and 38.3%, respectively, and HHV was 30.43 MJ/kg, 28.29 MJ/kg and 28.44 MJ/kg, respectively.
[133]		WWTP in Shenyang, China	84.9% MC. UA: 40.6% C, 4.7% H, 3.7% N, 49.6% O, 1.2% S. A total of 14.3 MJ/kg HHV	Batch-type 0.5 L reactor at 340 °C and 40 min reaction time.	Bio-oil HHV was 32.2 MJ/kg. After using treatment methods, it increased to 33.5–35.3 MJ/kg, and maximum HHV was reached (37.2 MJ/kg).
[134]		Qinghe WWTP, in Beijing, China.	53.5% OM. UA: 44.7% C, 7.6% H, 7.2% N, 39.6% O, 1.0% S. A total of 21.3 MJ/kg HHV	1 L stainless 316 steel reactor with 400 °C and 20 MPa.	Bio-oil HHV was 29.05 MJ/k (control). With treatment, Bio-oil HHV was 25.3–41.5 MJ/kg. Bio-oil HHV treated with HCl reached up to 45.2 MJ/kg.

BMP: Biomethane Potential; MC: Moisture; UA: Ultimate Analysis; OM: Organic Matter; db: dry weight basis; WSP: water-soluble products.

Ref. [135] reviewed supercritical technologies exploring their influences on the hydrogen production based on different parameters (i.e., sludge properties, moisture, and temperature). They presented some real-world projects, including small and large-scale plants, including the pilot-scale systems developed by Duke University, University of Missouri, University of Valladolid, and Xi'an Jiaotong University). Commercial projects were also highlighted as follows:

- HydroProcessing: An SCWO system implemented in the Harlingen WWTP in Texas, USA, in 2001. Some operating parameters of this system included a 150 ton/day capacity, 592 °C temperature, 24.5 MPa pressure, 20–90 s reaction time, and 6–9% solid content. The consumption of heater, oxygen, and pumps were 4.1 MWh/dry ton sludge, 1.5 ton/dry ton sludge, and 0.55 MWh/dry ton sludge, respectively. The capital and operating costs of the project were about 3 million USD and 100 USD/dry ton of sludge, respectively.
- Chematur AB: Two SCWO systems were developed: (i) first system included a 250 kg/h capacity for demonstration purposes, and (ii) second had a capacity of 1.1 ton/h, built with a plan to treat the sewage sludge of Kobe, Japan. The operational parameters of the second system included a 25 MPa pressure, 30–90 s reaction time, 400–600 °C, and operation at 15% dry solids. The consumption of natural gas, oxygen, cooling water, and electricity was around 21.9 Nm^3/dry ton sludge, 1.05 ton/dry ton sludge, 100 m^3/dry ton sludge, and 228.6 kWh/dry ton sludge, respectively. The capital and operating cost of the project were about 5 million GBP and 70 GBP/dry ton sludge, respectively.
- SuperWater solution: It was implemented in the Iron Bridge Regional Water Reclamation facility in Orlando, USA. Tested between 2009 and 2011, the SCWO system had a capacity of 5 t/d. The capital and operating costs were around 268 USD/dry ton sludge and 33.7 million USD, respectively. System's parameters included 35 dry ton sludge/d capacity, a 600 °C operating temperature, 26 MPa pressure, a 30–60 s reaction time, and 10% dry sludge.

Table 12 presents some studies that used supercritical water technologies to investigate the potential of hydrogen production using sewage sludge as feedstock.

3.1.3. Transesterification

The transesterification of sewage sludge is a chemical process in which the triglycerides/lipids (i.e., fats, oils, and greases) present in the feedstock are converted into biodiesel and glycerol by reacting them with an alcohol (usually methanol or ethanol) in the presence of a catalyst (typically an alkaline). This reaction produces methyl esters (biodiesel) and glycerol as by-products. The biodiesel produced is a fatty acid methyl ester (FAME) that can be used as a renewable fuel in diesel engines. It has properties similar to conventional diesel fuel but is derived from biological sources [136]. Two main techniques, lipid extraction and esterification/transesterification or direct in situ esterification/transesterification, are used to produce biodiesel from sewage sludge. Biodiesel production from sewage sludge can be divided into five main steps: (i) sludge pre-treatment (improve lipid extraction), (ii) lipid extraction from sewage sludge, (iii) use of extracted lipids to generate biodiesel, (iv) use of catalysts (optimise process), and (v) extraction of valuable by-products. This process provides a great alternative to managing sewage sludge by reducing its volume and converting waste into a useful product [17].

Pyrolysis also generates biodiesel from sewage sludge, but not directly. The by-products of pyrolysis include gaseous products, solid char and bio-oil. The bio-oil can be further processed and refined to be converted into biodiesel. Pyrolysis can be adjusted to increase the desired bio-oil fraction and, consequently, biodiesel production [137]. Table 13 summarises some works based on converting sewage sludge into biodiesel using transesterification and pyrolysis.

Table 12. Studies related to utilising supercritical water gasification on sewage sludge for hydrogen production.

Ref.	Sludge Collection	Sewage Sludge Properties	Parameters	Gas Production
[138]	WWTP in Jiangsu, China.	73.5–88.5 wt% MC. UA: 7.6–31% C, 2.3–6.2% H, 0.4–3.5% N, 20.2–34.1% O, 0.9–3.3% S. Up to 13.7 MJ/kg HHV	316 L batch reactor under 400 °C, 1 h reaction time.	Total gas production ranged from 10.7 to 43.3 mol/kg (mean gas production of 18.9 mol/kg).
[139]	WWTP in Shaanxi, China.	84 wt% MC. UA: 38.2% C, 2.4% H, 4.7% N, 23.7% O, 1% S. A total of 14.6 MJ/kg LHV	Heating rate reactor (70 °C/min) under 550–750 °C and 30 MPa	H_2 production up 18.9 mol/kg under 750 °C and 20 min
[140]	Municipal WWTP in Zhengzhou, China.	79 wt% MC, 65 wt% OM. UA: 7.4% C, 15.5% H, 1.3% N, 55.7% O, 3.8% P, TOC of 0.88 g/L	0.6 L reactor, 27 MPa, 6 min retention time and 380–460 °C temperature	H_2 production ranged from 2.5 mol/kg (380 °C) to 19.9 mol/kg (460 °C), and CH_4 production ranged from 1.8 mol/kg (380 °C) to 8.2 mol/kg (460 °C).
[141]	Municipal WWTP in Jiangsu, China.	77–88.5 wt% MC. UA: 7.6–27.5% C, 2.1–5.2% H, 0.4–3.8% N, 12.9–34.1% O, 0.9–2.5% S. Up to 12.6 MJ/kg HHV.	316 L batch reactor under 400 °C, 10 min reaction time and 24 Mpa.	H_2 yield was 1.06 mol/kg (control). With 5 wt% of NaOH and Ni, it was 2.7 and 3.6 mol/kg, respectively. For a mix of 3.3 wt% Ni and 1.67 wt% NaOH, 4.8 mol/kg,
[142]	Municipal WWTP in Nanjing, China.	83.2 wt% MC, organic matter of 45.1 wt%. UA: 19.5% C, 3.7% H, 3.2% N, 18.5% O, 0.2% S, 8.7 MJ/kg HHV	316 L batch reactor under 400 °C, 1 h reaction time and 22.1 MPa.	With Ni of 2–10 wt%, H_2 yield was 0.6–1.05 mol/kg. With H_2O_2, H_2 yield decreased from 0.31 mol/kg (2 wt% H_2O_2) to almost 0 mol/kg (10 wt% H_2O_2).
[143]	WWTP in Jiangsu, China.	73.9 wt% MC, 26.2 wt% OM. UA: 12.9% C, 2.1% H, 1.93% N, 4.24% O, 1.01% S. 4.8 MJ/kg HHV.	316 L batch reactor under 400 °C, 0.5 h reaction.	With no FA, H_2 yield was 0.16 mol/kg. With FA of 1, 2, 4 and 6 wt%, H_2 yield was 0.52, 1.2, 3.47 and 10.07 mol/kg, respectively.
[144]	WWTP in Higahashi-Hiroshima, Japan.	79.16 wt% MC. UA: 43.1% C, 6.6% H, 4.4% N, 25.9% O, 2.4% S on db.	316 L reactor under 500–600 °C, 25 MPa and 5–60 s reaction time.	At 550 °C and 600 °C, H_2 composition was about 38.5–39.4 vol%. At 500 °C, CO_2 composition was 49.5 vol%.
[145]	Domestic WWTP in Japan.	78.8 wt% VM. UA: 38.3% C, 5.9% H, 7.9% N, 33% O, 1% S on db.	Bench-scale reactor under 600 °C, 23 MPa and 1 h reaction time.	Total gas yield of 9.8 mol/kg with a composition of 60% of H_2.
[146]	WWTP in Nanjing, China.	75 wt% MC, 40.8 wt% OM. UA: 19.5% C, 3.7% H, 3.18% N, 14.25% O, 0.17% S on db. A total of 9.45 MJ/kg HHV.	Batch reactor under 23 MPa pressure, 400 °C for 10 min.	H_2 yield was 0.12 mol/kg without Ni catalyst, and 0.47 mol/kg with 1 cycle of Ni. H_2 yield decreased with more Ni cycles.
[147]	WWTP in Hangzhou, China.	35.14 wt% VM, 57.4 wt% OM. UA: 18.94% C, 2.21% H, 2.89% N, 12.79% O, 0.6% S on db. A total of 5.89 MJ/kg LHV.	0.5 L batch reactor under 26–28 MPa, 1 h reaction and 380–460 °C temperature.	H_2 yield was 6.44 mol/kg with 38.4% of H_2 composition.
[148]	Paşaköy WWTP in Istanbul, Turkey	57.4 wt% VM. UA: 29.4% C, 4.4% H, 17.9% O, 5.29% N, 0.47% S on db. pH of 5.84.	3.12 L reaction (20 L feed tank), operated with a 25 mL/min flow.	H_2 production increase from 3.4 L/h at 500 °C to almost 4.5 L/h at 650 °C.
[149]	Xi'an WWTP in Shaanxi, China.	87 wt% MC, 51.12 wt% VM. UA: 37.58% C, 4.4% H, 5.72% N, 24.44% O, 0.84% S. 9.64 MJ/kg HHV.	Batch reactor under 25 MPa pressure and 20 min reaction time.	The H_2 yield was 0.66, 1.93, 3.95, 7.44 and 11.81 mol/kg for 400 °C, 450 °C, 500 °C, 550 °C and 600 °C, respectively.

TOC: Total organic carbon; UA: Ultimate Analysis; OM: Organic Matter; MC: Moisture Content; db: dry weight basis.

Table 13. Studies on biodiesel production from sewage sludge.

Ref.	Type	Sludge Collection	Parameters	Gas Production
[136]		Municipal WWTP located in Tamil Nadu, India.	Lipid extraction performed for 6 h using 50 mL of chloroform methanol (2:1 ratio), diethyl ether, n-hexane and ethanol. Samples heated (50 °C for 0.5 h), and removed solvent	Total lipids extracted from PSS and SSS were between 3 and 6.5 g and 3.3 and 4.9 g using different solvents with concentration of 32.5% (PSS) and 24.5% (SSS). The generated biodiesel included: 89.2–91.2%, ester content, of 40.6–42.9 MJ/kg HHV, 65–72.6 cetane and saponification value of 131–162 mg of NaOH.
[150]	T	SS collected from two WWTPs in Beijing, China.	SS heated (45–75 °C), stirred at 300 rpm, added to methanol, H_2SO_4 and hexane solutions for 8 h. NaCl and hexane added, and the final mixture was centrifuged (3000 rpm) for 5 min, and filtrated.	Average biodiesel yield was 14.9% and 3.7% for A^2/O and MBR processes. Maximum biodiesel yield of 16.6% for A^2/O treatment was obtained using methanol:SS ratio of 10, 60 °C, and 5% H_2SO_4 con-centration. For the MBR process, the maximum value of 4.2% used ratio of 8, 50 °C, and 5% H_2SO_4.
[151]		WWTP at Universi-dad Rey Juan Car-los, Madrid, Spain.	Used methanol and n-hexane for extraction and reaction, and Zr-SBA-15 as catalyst. FAME producted in a 25 mL reactor at 209 °C, 2000 rpm, 50:1 methanol ratio and 12.5 wt% catalyst.	2 approaches used. (i) 1-step direct conversion: Overall weight FAME yield for PSS and SSS were about 15.5 wt% and 10 wt% (based on dried sludge mass), respectively. (ii) 2-step process: FAME yield was lower than 6% for PSS and almost negligible for SSS.
[152]		WWTP in Osong City, Korea.	1 L flask using a mixture of dewatered SS, methanol and n-hexane, and H_2SO_4 as catalyst, and stirred at 100 rpm. Reaction time varied from 1 to 8 h, and temperature of 55 °C (n-hexane) and 105 °C (xylene).	Maximum biodiesel yield generated from in situ transesterification were between 8.04% and 9.68% using n-hexane (solvent) and 10 mL/g methanol:SS ratio, and for a mix of PSS and SSS using 2 mL/g methanol:sludge ratio, it was 3.28% (n-hexane) and 8.12% (xylene)
[153]		WWTP in Villapérez-Oviedo (Asturias, Spain).	Hexane and methanol as solvents and extraction used 1:2 SS:hexane ratio. Solvent removed at 70 °C, and samples stored after 1 h drying at 105 °C. For solid liquid extraction, 1:10 SS:hexane ratio for 4 h.	Maximum production of 0.4 g FAME/100 g dry SS was achieved (26.8% of the total lipid extracted) for 24 h reaction. Total lipids extraction was 9% (1.75 g lipids/100 g dry SS) using hexane. For methanol (4% *v*/*v*), a 2.1% FAME content was achieved, whereas 0.4% was obtained for solid–liquid procedure.
[154]		Municipal WWTP in Oviedo, Spain	Reactor at 60 °C for 24 h, and methanol. NaOH (catalyst) at 4 ratios of methanol: 4%, 30%, 50% and 70%. SS:methanol ratio of 1:10. Used also H_2SO_4 in methanol (4% *v*/*v*) and mixed with SS in a 20:1 ratio. Microwave, sonication and particle sieving used for lipid extraction.	Maximum biodiesel yield of 14.3% (mass of FAME/lipid content). After 5 h reaction, biodiesel yield using 0.4% H_2SO_4 transesterification reached 22.2%, whereas the biodiesel yield using between 30% to 70% NaOH reached between 7 and 14.5%. Using microwave applied to dried SS for 4 min increased the FAMEs production by 110% (from 22.1 to 46.7%), whereas sonication improved by up to 42%.
[155]	T	PSS and SS from the Gangneung WWTP, in Korea.	Parameters used: 0.08% (*w*/*w*) alkaline/acidic catalysts with 40 mL/g-lipid methanol, 20 mL/g-lipid n-hexane at 50 °C for 24 h. Biodiesel extracted with n-hexane, centrifuged, separated, and dried for 24 h.	The contents of carbohydrate, crude lipid, ash, crude protein, and other elements were 8.9%, 14.5%, 17.8%, 42.8% and 16%, respectively. Lipid content organic solvents ranged from 2.9% to 5.7%, and using treatment methods (i.e., BDM, microwave, autoclave and ultrasonication), was about 10–14.5%.
[156]		WWTP in Beijing, China.	KOH, KOH/AC, and KOH/CaO as catalysts. Optimised reaction for KOH/AC used 1:10 SS:methanol ratio at 60 °C, 300 rpm, and 8 h extraction. For KOH and KOH/CaO, extraction time was 5 h.	The biodiesel yield was 1.2%, 6% and 6.8% using KOH, KOH/CaO and KOHAC as catalysts, respectively. The results showed that these are not good catalysts for biodiesel production from SS when compared with H_2SO_4.
[157]		WWTP in Beijing, China.	$SO_4^{2-}/Al_2O_3 - SnO_2$ catalysts prepared by 79 wt% of H_2SO_4. Dry SS (10 g) added to n-hexane and ethanol (0.2 L each), and extraction at 80 °C for 10 h. Catalysts (0.4–1.2 g), n-hexane (50 mL) and methanol (128 mL) used to extract crude fat and reacted for 0.5–6 h, 130–170 °C.	Biodiesel yields of 57%, 50.3% and 50.5% were achieved for temperatures of 130 °C, 150 °C and 170 °C, respectively. The biodiesel yield increased from 33.7% to 73.3% from 0.5 h to 4 h reaction time, but for 6 h reaction time, the biodiesel yield was 72.1%. The highest FAME yield (73.3%) was achieved at 130 °C, 4 h reaction time, using 10 g of dry SS and 0.8 g catalyst.

Table 13. *Cont.*

Ref.	Type	Sludge Collection	Parameters	Gas Production
[137]	P	WWTP in Pavia, Italy.	Microwave system at different operating conditions, including temperature of 180–650 °C and reaction times of about 1–28 min.	Highest value of oil to sludge was about 25% under 280 °C and 8 min reaction time, and the lowest value (7%) was at 180 °C at 50 min reaction
[158]		WWTP in Texas, USA.	Bench-scale fluidized bed reactor, 150–300 °C, ethanol:bio-oil ratio (*w*/*w*) between 1 and 3, 2–4 hs reaction, and Ni/HZSM5 as catalyst.	Generated bio-oil upgraded to biodiesel. Bio-oil and biodiesel HHV were 36.43 MJ/kg and 39.97 MJ/kg. SS-derived biodiesel yield was 89.33% (max) at 150 °C, 3 h reaction time, and ethanol:oil ratio of 2.
[159]	T and P	Jungnang WWTP in Seongdong-Gu, Seoul city, Korea.	Dried SS (10 g) 0.2 L of solvent (hexane) used to extract the lipids at 80 °C for 24 h. To separate solvent, used an evaporator at 80 °C for 3 h. SSRB produced by pyrolysis. A total of 20 g of SSR used at 600 °C for 4 h.	Highest biodiesel yield (33.5 wt.%) at 305 °C via thermally induced transesterification in 1 min using the SS biochar. Biodiesel yield from (trans)esterification was less than 1% with 5 wt.% H_2SO_4. Kinetics (<1 min) of thermally induced transesterification was faster than normal transesterification (3–24 h).

P: Pyrolysis; T: Transesterification; SSRB: Solid residue biochar; BDM: Bligh and Dyer method.; PSS: Primary sludge; SSS: Secondary sludge; FAME: Fatty acid methyl esters.

3.1.4. Microbial Electrolysis Cell

A Microbial Electrolysis Cell (MEC) is an anaerobic process that converts organic matter contained in the wastewater into renewable fuel (mostly methane or hydrogen). MECs can utilise either wastewater or sludge as feedstocks, and this technology has the potential to reduce the costs associated with wastewater treatment and sewage sludge disposal, which are often the two main operating costs in a WWTP. In an MEC, the organic matter is oxidised through bacteria in the anode electrode, and hydrogen is produced as a by-product in the cathode by using a small electric voltage. An MEC's performance, can be measured in terms of organic removal and hydrogen production, is directly dependent on the electric voltage magnitude. Usually, when an MEC is fed with municipal wastewater, the electric current generated is low which is expected since municipal wastewater is known for its low organic matter content when compared to other feedstocks. This is one of the key challenges of applying this technology when used to treat municipal wastewater. The first commercial MEC reactor (Ecovolt) was developed by Cambrian Innovation to treat high-strength wastewater [160]. Small-scale experiments using MEC and domestic waste-water as feedstock have been reported in the literature, as shown in Table 14.

3.1.5. Microbial Fuel Cell

Microbial fuel cells (MFC) is similar to MEC, but it converts the organic matter contained in the municipal wastewater into electricity directly. In this process, the MFC converts chemical energy contained in the organic compounds to electrical energy under anaerobic conditions through the catalytic reactions of microorganisms. Bacteria oxidise the organic matter contained in the wastewater at the anode, and the reduction reaction occurs on the cathode. During the oxidation, electrons are released and transferred to the anode, generating the electrical current that drives the MFC. When the electrons reach the cathode, they are combined with electron acceptors (i.e., oxygen) and protons (i.e., hydrogen) to generate water and close the electrochemical circuit. The main parameters that influence the performance of MFCs include the pH, temperature, substrate characteristics, bacteria activity, electrodes material, and internal resistance. MFCs vary in design, size, and power density. For example, small-scale cells can have a high-power density (i.e., above 500 W/m^3), while large ones may have lower density (i.e., 30 W/m^3). The MFC scale up usually occurs via enlarging the single reactor or combining multiple reactors in one system. MFC technology has been receiving lots of attention in research, but it is still in the early stages of development [161,162].

The main drawbacks of MFCs include the process efficiency, power densitiy, longer start-up times, performance variability, and sensitivity. However, the main advantages of MFCs include energy savings (i.e., aeration and sludge treatment processes) and less sludge production. MFCs can operate in either batch or continuous mode, which usually depends on the wastewater characteristics, cell design, microorganisms group, and electrode materials [162]. Table 15 highlights some studies that have investigated the MFC technology applied to municipal and domestic wastewater treatment. As shown in Table 15, the coulombic efficiency may vary significantly due to several factors, including microbial metabolism, electrode material and configuration, system performance, substrate characteristics and concentration, and operating conditions. As these parameters are different across the different experiments, they lead to different values.

Table 14. Studies on MEC using wastewater from different WWTPs.

Ref.	Feedstock	Parameters	Gas Production
[163]	Wastewater collected prior to primary clarification from the Howdon WWTP, in Newcastle, UK.	100 L dual-chamber MEC with 6-cell cassettes (88 L working volume). Anode and cathode electrodes surface area of 16.4 and 3.4 m^2/m^3 (anode-to-cathode ratio of 5:1), respectively.	H_2 continuously generated for 1 year with average production rate of 7 L/m^3.d. Average energy recovery and coulombic efficient were 48.7% and 41.2%, respectively.
[164]	Different types of wastewater, including urban wastewater from Rubi WWTP, in Barcelona, Spain.	130 L dual-chamber MEC with 10-cell cassettes. Anode-to-cathode ratio volume of 3.5:1, and used an anionic exchange membrane (AEM) to separate the chambers.	The average H_2 production was 32 L/m^3.d with 95% of purity (5% was methane). The OM removal efficiency was around 25% for a 2-day retention time and OLR of 0.25 gCOD/L.d.
[165]	Wastewater obtained from a domestic WWTP in England.	175 L dual-chamber MEC with 13 m^2/m^3 cathode specific area and 34 m^2/m^3 anode surface area-to-volume ratio under 5 h HRT.	The average H_2 production was 5.2 L/m^3.d with 93% purity, and the COD removal was 63.5%.
[166]	Primary sludge collected from the Gold Bar WWTP in Alberta, Canada.	Dual-chamber MEC. Anode and cathode about 0.42 and 0.17 L, respectively. A 40 cm^2 membrane to separate the chambers. Semi- continuous fed mode (45 mL/d) and residence time of 8 days.	The H_2 production rate found was 145 L/m^3.d, and the COD removal efficiency was up to 73%.
[167]	Raw sludge obtained from a WWTP in Jinju, Republic of Korea.	2.5 L reactor with 16 mm anode-cathode electrode. Reactors under 30, 35 and 40 °C for 6 days and stirred at 100 rpm. Operated in fed-batch mode. Raw and seed sludge was mixed under 7:3 ratio.	Maximum CH_4 production was 111 L/m^3 under 35 °C, whereas under 30 °C and 40 °C, the CH_4 production was 85 L/m^3 and 98 L/m^3, respectively. The CH_4 yield at 30 °C and 40 °C was 82.1 L/kgCOD and 77.1 L/kgCOD, respectively.
[168]	Wastewater collected from the municipal WWTP of Leon, in Spain.	Single-chamber 3 L membrane less MEC operated under batch and continuous mode at 21 °C with 4 h, 8 h, 12 h and 24 h HRTs	At batch mode, CH_4 production rate was 1.3–1.4 L/m^3.d and 54% COD removal efficiency. The energy and net-energy consumptions were 2.92 and 2.14 kWh/kgCOD.

Table 15. Studies related to MFC technology using municipal or domestic wastewater.

Ref.	Feedstock	COD on Wastewater	MFC System	COD Removal Rate	Maximum Power Density	Max Energy Recovery	Coulombic Efficiency	Operation Time/HRT
[161]	Wastewater collected from the Pepper's Ferry Regional WWTP, in the USA.	155 mg/L.	96 tubular MFC modules (2 L liquid each)	76.8%	-	0.006 kWh/m^3	-	1 year and 18 h HRT.
[162]	Effluent from the primary clarifier of a WWTP, in Switzerland.	Up to 130 mg/L.	45 L (4 units with 11.2 L each) used in a full-scale WWTP.	13.5–67%	73 [4], 82 [5], 80 [6] mW/m^2	0.012 kWh/m^3	24.8%	9 months and 12–44 h HRT
[169]	Municipal wastewater from 2 WWTPs (Xiao Jiahe and Yong Feng), in China.	60–100 mg/L (Xiao Jiahe), 200–400 mg/L (Yong Feng).	1000 L system (50 stacked modules, 20 L each).	70–80%	60 W/m^3	0.033 kWh/m^3	41–75%	100 days to 1 year, 2 h HRT.

Table 15. *Cont.*

Ref.	Feedstock	COD on Wastewater	MFC System	COD Removal Rate	Maximum Power Density	Max Energy Recovery	Coulombic Efficiency	Operation Time/HRT
[170]	Primary effluent wastewater from Pennsylvania State University WWTP, in US.	376–428 mg/L.	3 cell sizes: 0.028 [2] L, 0.22 [3] L and 85 [1] L. CSA: 15 m^2/m^3 (small) and 7.3 m^2/m^3 (big)	75–80%	83 [1] W/m^3/ 304[2] W/m^3	-	13–27%	-
[171]	Primary effluent wastewater from the Pennsylvania State University WWTP, in US.	480–1010 mg/L	2 L reactor (1.4 L liquid vol.). A total of 0.86 L anode volume and CSA of 29 m^2/m^3.	57%	22 W/m^3	-	4.4% (min) and 42% (max)	8 h HRT
[172]	Effluent from the primary clarifier from the Haeundae domestic WWTP, in Korea.	144 mg/L.	5 MFC units (150 mL each) in series. CSA: 400 m^2/m^3	34%	16.7 W/m^3	-	12%	8 months 2.5 h HRT
[173]	Wastewater from primary clarifier from a municipal WWTP, in Switzerland.	200–450 mg/L	1000 L system (64 MFC × 16.25 L each) into 4-quadruple stacks.	34.4–95.4%	-	0.015–0.060 kWh/m^3	4.7–14.9% (25% max)	18 months
[174]	Effluent from the primary clarifier of the Mumbai municipal WWTP, in India.	1650 mg/L	0.7 L system.	68%	621 mW/m^2	-	47–48%.	-
[175]	Wastewater collected from Al Gabal Al Asfar WWTP, in Egypt.	92–350 mg/L	Double chamber MFC with 2 × 0.3 L (anode and cathode)	Up to 72.85%.	209 [7] mW/m^2/ 117 [8] mW/m^2	-	-	24 h HRT
[176]	Effluent from primary clarifier from the Taiping municipal WWTP, in China	200–350 mg/L	1.5 m^3 system	63% (92% max)	406 mW/m^3	0.0034 kWh/m^3	-	5 h HRT

[1]—large MFC; [2]—small MFC; [3]—medium MFC; [4]—44 h HRT; [5]—22 h HRT; [6]—12 h HRT; [7]—in summer; [8]—in winter; CSA: Cathode surface area.

3.1.6. Hydropower

Hydroelectric power is known as one of the most economical and popular energy resources. Hydropower is an affordable source of electricity, and compared to other sources, it has a relatively low cost during the project lifespan in terms of maintenance and operations costs. It is a flexible type of renewable energy resource that can be widely applied in WWTPs by using the water flow potential to generate renewable energy [177]. In a WWTP, the possible locations for installing a hydropower system typically include the upstream (raw/untreated wastewater) or downstream (treated effluent). Usually, it is classified based on size, including micro (1–100 kW), mini (0.1–1 MW), small (1–10 MW), and large (above 10 MW). The power output from a hydropower turbine is directly dependent on the water flow rate and available head. There are different types of hydropower turbines available on the industry, including an Archimedes screw, Crossflow, Francis, Kaplan, Pelton, and Propeller [178]. Some researchers have studied the application of hydropower in WWTPs, and some works are summarised in Table 16.

Ref. [178] listed 46 WWTPs that had installed hydropower systems (17 micro, 22 mini, and 7 small). Among them, 12 plants used Pelton turbines, 10 used Kaplan, 2 used Propeller type, 2 used Francis, 2 used pumps working as turbines (PATs), and the remaining were not specified. The authors noticed that the adoption of hydropower technology was not very high when compared to other types of renewable energy technologies. Some real hydropower plants which were installed at different WWTPs are listed in Table 17.

3.1.7. Water Electrolysis

Water electrolysis is a technology where an electric current is passed through water, splitting it into hydrogen and oxygen. This method is valued for its potential to generate clean hydrogen fuel (also known as green hydrogen), when the electricity used in the process comes from a renewable source. In WWTPs, hydrogen can be used to power CHP technologies, including FC, GT, or even modified ICEs, or can be injected into the gas network following the standards regarding the hydrogen-blending constraints. The oxygen part is a valuable element and can also be used in the wastewater treatment process. Oxygen in a WWTP can be used is in the aeration process, which is vital for the biological treatment of wastewater. In the activated sludge process, oxygen is supplied to the aeration tanks to support the growth of aerobic microorganisms that break down organic pollutants. Additionally, in advanced treatment stages such as membrane bioreactors (MBRs) or in systems focusing on nutrient removal, controlled oxygen levels can improve the performance of both nitrification and denitrification processes. Water electrolysis is a promising technology, and the possibility of using treated effluents from municipal WWTPs as the source of water for electrolysis can potentially enhance sustainable hydrogen production in these facilities. However, to utilise the treated effluent from a WWTP as a water supply, it must meet specific requirements, including being purified and demineralised and exhibiting a conductivity lower than 5 μS/cm [179,180].

Additionally, depending on the quality of the treated effluent, some further tertiary treatments may be required, including chemical treatments, accelerated filtration, ultrafiltration, nanofiltration, coagulation, reverse osmosis, ion exchange, and electrodeionization. Water electrolysis can be a competitive technology in terms of costs. For example, hydrogen production via water electrolysis can range between USD 2.05 and 10.5/kg H_2, whereas the costs for sewage sludge pyrolysis can be around USD 1.2–2.2/kg H_2, and the steam methane reform method is about USD 1.14/kg H_2 [180].

Table 16. Studies that considered hydropower technologies applied in WWTPs.

Ref.	Study Aims	Main Findings
[177]	Investigate the hydro potential in a WWTP in Wisconsin, in the USA.	• WWTP parameters: 190 MGD effluent flow, 3 m head, 6.5–9 m^3/s flow rate. • Hydro turbine parameters: Kaplan design, 271 kW nominal capacity, output between 150 kW and 207 kW, 100–200 rpm and inlet mass flow between 5000 and 10,000 kg/s. Expected savings was estimated to be around 1.56 GWh/year
[178]	Assess the potential of hydropower technology in WWTPs	• Listed 49 real cases studies of hydro application in WWTPs in different countries. • Details of the hydro projects included installed capacity, potential energy generation, capacity, available head and water flow.
[181]	Investigate the benefits/feasibility of a hydropower system in Zeekoegat WW-TP, in South Africa (60 mL/d capacity).	• 3 × 6.7 kW siphon turbines were considered in the study based on head net and flow discharge of 3.6 and 0.37 m^3/s (each unit). • Economic analysis results included 30 years of design life, 9 years pp, total investment, maintenance and annual costs were USD 69,000, USD 691, and USD 7520, respectively. • System could produce 181.3 GWh/year and for a 20-year project, the potential cost savings would be approx. USD 437,500.
[182]	Evaluate the energy recovery potential and economic viability of different WWTPs in Ireland and UK.	• In the UK, 5 out of 11 plants were economically viable: (a) Beckton (4 years pp, 2.6 m head, 14 m^3/s flow, 234 kW, EUR 816,000 c.c., EUR 203,000 a.s.), (b) Knostrop (3.1 years pp, 8.5 m head, 4.3 m^3/s flow, 184 kW, EUR 623,200 c.c., EUR 202,800 a.s.), (c) Crosness (4 years pp, 4 m head, 7.2 m^3/s flow, 184 kW, EUR 633,000 c.c., EUR 159,600 a.s.), (d) Minworth (4.2 years pp, 4 m head, 6.3 m^3/s flow, 32.5 kW, EUR 249,000 c.c., EUR 28,000 a.s.), (e) Long Reach (8.8 years pp, 1.7 m head, 3 m^3/s flow, 25.5 kW, EUR 222,800 c.c., EUR 22,000 a.s.). • In Ireland, 3 out of 14 plants were economically viable (<10 years pp): 2 largest plants: Ringsend (5.1 years pp, 3.7 m head, 4.4 m^3/s flow, 103 kW, EUR 443,000 c.c., and EUR 87,000 a.s.) and Carrageenan (7.7 years pp, 4 m head, 1.2 m^3/s flow, 30 kW, EUR 195,000 c.c., EUR 25,400 of a.s.).
[183]	Study the hydro potential of treated wastewater discharged from the Torun WWTP, in Poland.	• WWTP maximum treatment capacity of 90,000 m^3/day and 0.5 m^3/s average wastewater flow. • The highest turbine power (24.8 kW) was based on maximum wastewater flow of 0.56 m^3/s and system efficiency of 66.35% and maximum hydraulic head of 7.5 m, whereas the lowest turbine power (14.8 kW) based on system efficiency of 53.7% and flow of 0.39 m^3/s.
[184]	Study the implementation of micro hydropower turbines in 4 different WWTPs, in Ireland.	• The WWTPs included Ringsend (1640,000 PE, 377,600 m^3/d flow rate, and 3.7 m head), Carrigrennan (413,000 PE, 102,000 m^3/d flow rate, and 4 m head), Navan (50,000 PE, 10,000 m^3/d flow rate, and 16.3 m head), and Greystones (40,000 PE, 8000 m^3/d flow rate, and 6 m head). • For Ringsend, the power output, annual savings, project cost and pp were 103 kW, EUR 87,000/year, EUR 442,000 and 5.1 years. For Carrigrennan, it was 30 kW, EUR 25,400/year, EUR 195,000 and 7.7 years. For Navan, it was 11.8 kW, EUR 10,000/year, EUR 77,200 and 7.8 years, and for Greystones, 3.5 kW, EUR 3000/year, EUR 50,000 and 16.9 years.

Table 16. *Cont.*

Ref.	Study Aims	Main Findings
[185]	Feasibility study of hydropower in Kiheung Respia WWTP, South Korea.	• WWTP parameters: 0.35 m^3/s flow rate, and 4.3 m available head. System efficiency and electrical power output were 83%, and 12.3 kW. • Up to 96% of the total effluent treated by the facility could be used to generate electricity (able to produce 68 MWh/year).
[186]	Hydro system a WWTP, in Pakistan.	• Plant's parameters: 1.5 m net head and 0.24 m^3/s flow rate. Archimedes screw turbine used for the study for low heads. • Hydro system of 1.96 kW.
[187]	Investigate the performance of propeller turbines in recovering energy in the aeration tank of a WWTP.	• Data collected from 13 WWTPs were analysed in the study. • The proposed microturbine system could generate around 3.8 GWh/year and provide a potential savings up to 15.7% and USD 260,500 on the plant's annual electricity consumption and annual cost savings, respectively. • The cost of system's implementation varied from USD 191,400 to USD 528,00, and the average p.p. was considered as 15 years.
[188]	Find the optimal design and best hydropower technology for Tatlar WWTP, in Turkey.	• Tatlar WWTP had the capacity to treats 765,000 m^3/d (plan to be 1377, 000 m^3/d by 2025). • Plant's parameters: 12.5 m^3/s average flow. • 2 turbines studied: (i) Archimedean turbine, the payback period, energy generation, investment, O&M, and average energy revenue were 2.4 years, 7.88 GWh/year, EUR 709,000/year, EUR 13,600/year and EUR 2172,000, respectively. (ii) Kaplan model, the payback period, energy generation, investment, O&M, and average energy revenue were 3.2 years, 8.6 GWh/year, EUR 773,500/year, EUR 57,000/year and EUR 2279,000, respectively. The Archimedean screw type was selected as a better option.
[189]	System at Clarkson WWTP (Canada).	• WWTP the capacity to treat 350 mL/d of wastewater. Up to 1.1 GWh/year of electricity could be generated on-site.

pp: Payback period; c.c.: capital costs; a.s.: annual savings.

Table 17. Hydropower plants installed in WWTPs [190].

WWTP	Location	Turbine Type	Total Installed Power (kW)	Flow (m^3/s)	H (m)
North Head, Sydney	Australia	Kaplan	4500	3.5	60
Emmerich	Germany	Archimedes	13	0.4	3.8
As Samra	Jordan	Pelton	1600	3.2	104
		Francis	1680		41
Aïre, Geneva	Switzerland	Kaplan	200	3.2	5
Engelberg		Pelton	50	0.16	54.4
Grächen			262	0.09	365
La Douve I, Leysin			430	0.08	545
La Douve II, Leysin			75	0.108	83
Morgental, St. Gallen			1350	0.84	190
Profay, Le Chable			350	0.1	449
La Asse, Nyon		Pump as turbine	220	0.293	94.3
Elsholt	UK	Archimedes	180	2.6	-
Deer Island, Boston	USA	Kaplan	2000	13.1	8.8
Point Loma, San Diego		Francis	1350	7.6	27
Hsinchu	Taiwan	-	11	-	-
Taichung		-	68	-	-

Following the stoichiometry balance, to generate 1 kg of H_2 and 8 kg of O_2 is required 9 kg of deionized water. However, in real-world applications, electrolyser manufacturers recommend using a higher quantity of water (between 10 and 22.4 kg), considering water losses (i.e., 10%) and water used for equipment cleaning (around 25%) [179]. Few studies have investigated the utilisation of water electrolysis technology in WWTPs, as illustrated in Table 18.

Table 18. Studies exploring the electrolysis technology utilisation in WWTPs.

Ref.	Plant	Study Aim	Parameters/Methods	Gas Production and Utilisation
[191]	The WWTP Mainz, in Germany. Wastewater inflow up to 55 mL/d (peaks of 6.300 mL/h.	Studied a 1.25 MW Electrolyser system in a WWTP. PV and CHP system to power the electrolyser. H_2 could be injected into the gas network or power FC buses. O_2 could generate ozone to be used in the advanced WW treatment	WWTP power consumption of 8200 MW/y, PV generation of 227 MWh/y, CHP generation (using biogas) of 6173 MWh/y and 1800 MWh/y power bought from the grid.	H_2 and O_2 production was about 2975 MWh/year and 600 ton/year, respectively. Assuming 2% vol of H_2 feed-in limit, 1240 MWh of H_2 could be fed into the gas network per year. The remaining 1735 MWh of H_2 could be used in fuel cell buses in public transport.
[192]	RWW, PE, SE, TE, and SW collected from a WWTP in Gyeongsan city, South Korea.	Generate H_2 from low-grade wastewater using alkaline water splitting technology. Based on this investigation, the potential applicability for achieving energy independence in municipal WWTP is planned	WW filtered through UF membrane to produce ATW. After treatment, COD, TN, TDS in the treated effluent were around 2.8–37.9 mg/L, 0.9–28.7 mg/L and 44–377 mg/L, respectively.	H_2 production based on low-grade water was between 19.2 and 22.8 mL/h.L, whereas based on a UF treatment, it was between 20.4 and 23.4 mL/h.L for the different WW samples. For a deionised and tap waters (control), H_2 production was 23.6–26.6 mL/h.L.

WW: wastewater; RWW: raw wastewater; PE: primary effluent; SE: secondary effluent; TE: tertiary effluent; SW: Surface water; ATW: Advanced treated water; UF: ultrafiltration; COD: Chemical oxygen demand; TN: Total nitrogen: TDS: total dissolved solids.

3.2. Non-Site-Specific Sources

Non-site-specific sources do not rely on a specific geographical location or environment to be used. This group include renewable energy technologies that are not tailored to the particular characteristics or location, such as solar energy and wind [23].

3.2.1. Solar Energy

Solar energy has the least environmental impact compared to other types of renewable energy resources. It can be used in multiple pathways in WWTPs, and some researchers have studied its application, including solar thermal and PV generation, in those facilities. Solar thermal energy can also be used in several applications, such as heating, heat pumps, and sludge drying. Sludge drying can be used to dewater digested sludge, which is a very important stage of sewage sludge management in a WWTP. It not only reduces the amount of waste to a minimum but also helps eliminate bad odour and pathogen problems. Traditional thermal drying systems are very complex and require high investment and operational costs [193]. Solar PV is the most widely used type, especially due to its scalability, flexibility and lower costs. For example, in [23], from 105 WWTPs investigated in the USA, 41 plants adopted a PV system. It was also identified that solar PV was primarily installed in hybrid configuration combined with AD. For plants with a flow rate above 19 mL/day, PV system could supply around 8–30% of the plant's energy demand.

3.2.2. Wind

Although wind generation is one of the most widely used technologies, it is not commonly used in water facilities, mainly due to its initial costs and the complexity of building a small-scale plant. Most studies consider hybrid configurations consisting of PVs, batteries, and biogas systems. Few researchers have investigated the use of wind turbines alone in WWTPs.

3.2.3. Hybrid System

Due to the intermittency and uncertainty of renewable energy resources, a single technology may be unreliable and add challenges in terms of the dynamic load demandingness of a WWTP. A hybrid system combines two or more power generation resources to improve the generation potential which can overcome the issues of a single technology, providing cost-effective system with flexible capacity [23]. Some studies have considered non-site-specific power generation technologies in WWTPs, including solar energy and wind alone, and also hybrid systems, as shown in Table 19.

Several large-scale renewable energy projects have been implemented or are under development in WWTPs worldwide, as shown in Table 20. In Australia, WWTPs have been focused on becoming more sustainable and energy-efficient in the last few years. Table 21 shows some examples of sustainable renewable energy projects and future systems planned to be installed in WWTPs in Australia.

4. Grid Services and Energy Market Participation of WWTPs

With rising electricity costs, sustainability targets, and efficiency goals, more and more WWTPs are looking for alternatives to minimise their operating costs and reduce carbon emissions while increasing energy efficiency. The adoption of renewable energy sources can be considered one the best alternative for WWTPs; however, due to some potential constraints, such as high upfront investment costs and site limitations, their adoption may create some challenges. Other potential alternative to overcome these issues can be participating in demand-response programs, energy markets, and grid services. Looking for opportunities to take advantage of better energy pricing and incentives, some WWTPs are starting to consider the possibility of exploring these areas [23].

Energy utilities incentivize customers to reduce their energy usage at particular time periods of the day, especially during peak periods. Some wastewater facilities have demonstrated capabilities to shift their loads to assist the grid, which can range from a slightly reduction on energy consumption for a short period to larger amounts during longer periods. One of the most common incentives given by energy utilities to consumers is based on demand-response (DR) programs. Its main objective is to adjust the customer's consumption under energy price signals or grid conditions. Customers are required to reduce energy consumption during peak periods when the energy price and demand are higher. By participating in DR programs, customers receive energy discounts that can help to reduce their energy costs. Before participating in DR schemes, WWTPs should conduct technical evaluations to understand if they are capable to maintain its wastewater treatment efficiency and effluent discharge quality while providing these services [194].

DR are mainly based on two methods: load shedding (electricity consumption reduction during peak period without a change in the overall consumption behaviour) or load shifting (shift the energy demand from peak to off-peak). DR programs can also be incentive-based or price-based. Based on incentives, the customer receives fixed payments based on the DR duration event and the total number of hours provided annually. The customers receive a demand reduction signal from the system operator, which can be mandatory or voluntary. Based on price, the customer is subjected to variable electricity prices over time. Clients are motivated to adjust consumption following the energy tariff at a specific time [6]. Some WWTPs were reported participating in energy markets and providing grid services. For example, an EMWD facility located in California, USA, started to participate in three DR programs: (i) Southern California Edison's (SCE's) EnerNOC DR Program, (ii) Base Interruptible Program, and (iii) Price-Based program for day-ahead pricing. EnerNOC is an aggregator that is responsible for the connection between the WWTP and the energy provider. Participating in those programs, the overall savings of the facility was USD 555,000 annually in 2012. EMWD also plans to partner with Honeywell to convert some of its manual controls to automatic controls to increase its DR portfolio [195].

Table 19. Solar, wind and hybrid system generation in WWTPs.

Ref.	Type	Study Objective	Main Findings
[196]	Solar thermal	Optimise a solar dryer in a WWTP in Morocco.	• Solar thermal system included a 2.5 m^2 solar collector, drying chamber, centrifugal fan and thermo-regulator. • Drying system based on 3 temperatures (50, 70 and 90 °C) and air flow rate of 0.083 m^3/s. System able to reduce up to 60% of the sludge volume.
[197]	Solar thermal	Design a system for drying SS from Antalya Metropolitan Municipality WWTP, in Turkey.	• For the proposed system, the payback period, average collector efficiency, Levelised cost of heating (LCOH) and O&M costs were 2.9–3.5 years, 50.17%, 0.017–0.02 USD/kWh, and 6.41–7.86 USD/year. • Specific moisture extraction rate (SMER) and specific energy consumption (SEC) of about 0.77–1.34 kg/kWh and 1.77–2.86 kWh/kg, respectively.
[198]	Solar thermal	Proposed a thin layer sandwich-like chamber for SS drying	• The best result was achieved for a 5 mm sludge thickness layer with average drying rate of 6.72 g/h under 0.5 kW/m^2 solar radiation. • In 11 h operation, the water content of the sludge decreased from 79% to 5%, and the heat utilisation efficiency was around 24.3%
[199]	Solar thermal	Investigated the SS drying system for a WWTP in Beijing, China,	• Experiments conducted in a bench-scale convective dryer, SS exposed to temperatures around 100–200 °C and hot air speeds of 0.6, 1.4 and 2 m/s. • Average mass transfer and surface heat coefficients were between 1270–3460 m/s and 10.66–26.96 W m^2.K., respectively.
[200]	Solar thermal	Studied the potential of a solar air heater for drying SS.	• The average system's efficiency and SEC were around 70.12–81.70%, and 2.36–5.40 kWh/kg, respectively. • The average minimum and maximum temperatures were 82.1 °C and 86.4 °C, respectively.
[23]	Solar PV	Assess the status of solar PV in 105 WWTPs in California, USA, to evaluate its usage in WWTPs.	• 41 out of 105 WWTPs had on-site PV system with an average capacity of 0.86 MW (12 kW min., 4.2 MW max.). • 34% of the plants had a 1 MW PV solar system, and the total PV solar capacity installed in the WWTPs was 35.5 MW. • Solar PV supplied 8–30% of WWTPs with flow above 5 MGD (biogas supplied 25–65%), and facilities below 5 MGD flow, the solar PV supplied between 30% and100% of the load demand. • For Sacramento WWTP, the 4.2 MW plant supplied 8% of the plant's energy needs.
[201]	Solar PV	Study the benefits of a PV system in a WWTP, in Romania.	• The flow rate of the WWTP was about 745 L/s and 322 L/s under peak and off-peak period, respectively. • Proposed PV system could reduce up to 40% of the total energy demand and 12% of the carbon WWTP emission.
[202]	Solar PV	Assess the SBBGR in a WWTP using a solar system.	• Propose a 5.1 kW solar PV system to supply heating/cooling and replace the need to buy energy from the grid. • The thermal energy extracted from the SBBGR system could reach up to 14.5 kWh and operated for 4 months.

Table 19. *Cont.*

Ref.	Type	Study Objective	Main Findings
[203]	PV-Battery system	Integration of PV-battery system for 2 small-scale decentralised WWTPs, in the Netherlands.	• The average energy consumption of the BEVER III and MBR DWWTPs were 1.08 and 2.3 MWh/month, respectively. • Proposed PV system for BEVER III: 15 kWp PV and 20 kWh battery to supply 75% (winter) and 100% (summer) of plant's electricity demand. • Proposed PV system for MBR: 30 kWp PV and 50 kWh battery to supply 65% (winter) and 100% (summer) of plant's electricity demand.
[204]	Wind	Assess the benefits of the 100 kW wind turbine to supply electricity for a WWTP located in Texas, US.	• The WWTP treated about 189.3 m^3/d of wastewater which required approximately 236,000 kWh/year. • Wind project cost about USD 610,900 (in 2012), and in 3 years, the plant saved almost USD 16,000/year in electricity costs, generating about 155,700 kWh. • To be a positive NPV, the system would need to generate about 557 MWh/y (or economic benefit of USD 49,000).
[205]	PV and Biogas	Study the potential of anaerobic co-digestion and solar PV in a WWTP in Loures, Portugal.	• Co-digestion of SS and FW (40–60% SS:FW mixing ratio) and co-digested under 37 °C, OLR of 1.12 g TVS/L.d, and 15 days HRT. • The proposed PV system had an installed capacity of 730 kWp which could generate approx. 1250 MWh/y, 12.2% PV self-sufficiency ratio, and 12 y payback period. The cost of co-digestion implementation was calculated as EUR 524,000.
[206]	Solar thermal/biomass	Propose a 20 MW solar/biomass system to supply a WWTP in Spain.	• Proposed hybrid system could generate up to 148.9 GWh under a 0.85 capacitor factor. Total system investment of EUR 211 million. • The LCOE was EUR 0.25/kWh. System's total exergetic efficiency was between 15% (solar) and 34% (biomass).
[207]	PV, battery and diesel generator	Analyse the potential of using a hybrid system to reduce consum-ption in a WWTP, in Romania.	• WWTP serves 23,000 PE with 315 kW total installed power and 537,180 kWh/y electricity consumption. • The proposed PV system was designed as 310.5 kWp (which could provide around 370,00 kWh), and 862 kWh battery capacity. • The LCOE for PV-battery and PV-only systems were 0.154 EUR/kWh and 0.01 EUR/kWh, respectively. PV-battery system costs of around EUR 1 million.
[208]	PV, heat pump, water electrolyser)	Investigate the economic benefits for implementing a hybrid generation in a WWTP.	• WWTP details: 8.7 MW thermal load, 4.1 MW electrical demand, 40 t/d oxygen demand and biogas supply of about 404,750 GJ/year. • Five configurations were proposed: (a) PV only (10 MW), (b) PV + HP (15 MW + 2.175 MW), (c) PV + HP + TES (15 MW + 5 MW + 43.5 MW), PV + HP + TES + grid (same as c), and (d) PV + electrolyser (15 MW + 2.175 MW).
[209]	MHP and MEC	Integration of MEC and MHP to assess potential benefits for a WWTP.	• MHP was used to power the MEC, based on 4 scenarios: (i) S1: water flow of 0.1–1.1 m^3/s, (ii) S2: water flow of 0.1–1.1 m^3/s and 10 m head, (iii) S3: water flow of 0.1–0.55 m^3/s and 10 m head, and (iv) S4: water flow of 0.0345–0.55 m^3/s and 20 m head. • Maximum power for S1, S2, S3 and S4 were 91.6, 32.6, 13.1, and 38.8 kW. It could power an MEC system of 2.8, 1.4, 0.7, and 0.5 GW. • The MHP-MEC integrated system required a capital cost of 453.77, 45.38, 56.87 and 45.57 USD million for S1, S2, S3 and S4 systems, respectively. • The payback period for S1, S2, S3 and S4 would be about 19.7, 2, 5.2 and 4.1 years, respectively.

Table 19. *Cont.*

Ref.	Type	Study Objective	Main Findings
[210]	FC and Solar thermal	Conduct an economic analysis for a hybrid system in Collegno WWTP, in Italy, which serves serves 270,000 PE and treats 38,500 m^3/day.	• LCOE for the SOFC only was EUR 0.144/kWh and a payback period of 20 years, considering a SOFC system generating 1 MW/year. • LCOE for the hybrid system for 3 different ST system sizes, including 300 m^2, 700 m^2 and 1000 m^2, was 0.141, 0.134 and 0.123 EUR/kWh, respectively. • System's payback period considered SOFC and ST systems sizes of 300 m^2, 700 m^2 and 1000 m^2 was 19, 13.5 and 9 years, respectively. • If the annual production of SOFC increases to 2, 5 or 10 MW/year, the payback period could reduce to 7.5, 3.5 or 3 years, respectively.
[211]	SOFC, solar thermal, and GT	Evaluate the potential benefits of a hybrid system in a WWTP, in Italy.	• 4 scenarios considered: (i) SOFC only, (ii) SOFC and ST, (iii) GT and SOFC, and (iv) Trigen (SOFC, RC, and absorption chiller). • The total investment costs, operating costs, electrical load coverage, total energy savings and payback period for the considered scenarios were around EUR 1.2–1.33 mi, EUR 113,000–134,000, 28–39.4%, EUR 220,000–380,000 and 2.95–5 years. The LCOE was calculated to be 0.069–0.087 EUR/kWh.

TES: Thermal energy storage, MHP: Micro-Hydropower; MEC: Microbial Electrolysis Cell; ST: Solar Thermal; GT: Gas turbine; SOFC: Solid Oxide Fuel Cell; RC: Rankine Cycle; MGD: Mega gallons per day; SS: Sewage sludge; SBBGR: Sequencing Batch Biofilter Granular Reactor; FW: Food waste.

Table 20. Large-scale renewable energy projects in WWTPs.

Ref.	WWTP	Location	Technology/ System	Findings
[212]	ACUA WWTP	New Jersey, US	7.5 MW Wind	• The 7.5 MW installed capacity wind farm started operation in 2006 • The plant can supply around 60% of the WWTP's electricity needs, and the surplus can supply the main grid. • The cost of the project was USD 12.5 mil, and in the first year of operation, the plant was able to save 20% of the investment cost.
[213]	Field's Point WWTP	Providence, US.	4.5 MW Wind	• WWTP treats up to 246 million litres per day of wastewater. It supplies around 25% of the facilities' annual demand. • Wind project cost was USD 14 mil, and reduced the electricity costs by 40% (from USD 2.5 mil to USD 1.5 mil) per year.

Table 20. *Cont.*

Ref.	WWTP	Location	Technology/ System	Findings
[214]	JRDWRF	City of Pueblo, Colorado, US	309 kW solar PV	• The city received USD 1.5 mi to install a PV system. The system can cover around 40% of the facility's electricity demand. • Project was economically viable since it had support from deferral fund and tariff rebates from the local energy utility.
[195]	EMWD	California, US.	PV solar, FC, and turbines	• PV systems in 5 plants. Total PV system capacity of 21 MW (potential to produce 45 GWh/year and supply 30% of the electricity demand of each site, generating a revenue of USD 2 mi yearly). FC units installed in 2 facilities which can supply 25–40% of the energy needs for each site. • FC system uses biogas generated on-site and has the potential to save around USD 1 mi annually in electricity costs. • 8 micro-turbines of 60 kW capacity generate additional streams of USD 300,000 a year.
[215]	PLWWTP	San Diego, USA	Biogas upgrading	• Project cost USD 45 mi to implement of which the federal government provided 30% of rebates and USD 12 mi in tax-exempt bonds. • Plant generates more electricity than needed. It sells about 3.5 MW (plant produces 5.5 MW and consumes 2 MW) • The purified biogas powers 3 FC system (2.6–4.5 MW system). With the incentives, the 2.8 MW FC system received USD 7.65 mi in rebates.
[216]	Werdhölzli	Switzerland	Sludge incineration	• Plant uses the sludge from 70+ WWTPs to generate renewable energy. • The system can handle up to 100,000 metric tons of sludge yearly with a capacity of 875 kW of electrical power and 4450 kW of heat.
[217]	Deer Island WWTP	Boston, US	Solar PV, wind, and hydro	• WWTP treats 1300 MLD of wastewater. Hybrid system: 2 × 600 kW wind, 736 kW PV, and 2 MW hydro (up to 8.85 GWh/year generation). • Facilities' annual energy demand is 18 MW, costing USD 16 mi/year. After installing the hybrid system, electricity costs are reduced by 25%

EMWD: Eastern Municipal Water District; PLWWTP: Point Loma Wastewater Treatment Plant; JRDWRF: James R. Dilorio Water Reclamation Facility; MLD: Million litres per day.

The EMWD plant started participating in DR programs in 2007 by committing to reduce 1.5 MW of power capacity. The facility received USD 100,000 per year by providing grid services to relieve the SCE network. Based on the terms, EMWD could choose to contribute in different levels of responses, depending on its operation and needs, which could be achieved by running its capacity at a lower level or even shutting down completely [218].

Pennsylvania American Water identified that its Shire Oaks facility had the potential to provide grid balance services while fulfilling its technical obligations. The facility started by adding one pump, which was able to respond to the requests without impacting its operating processes. Providing the services of only one pump, the facility was able to offset approximately 2–3% of the total electricity bill, and a second pump was planned to start providing grid services [219].

A DR solution was implemented as a pilot program for two large customers, including the Kind Country South Treatment Plant, located in the USA. The local energy provider, Puget Sound Energy (PSE), and Generac Grid Services conducted a trial to evaluate the project. Generac was responsible for monitoring, controlling, and managing the capacity curtailment in real-time from both customers if any automatic event was needed. In total, the energy provider called four DR events in the winter 2017 peak capacity season, and it was able to curtail an average of 4.09 MW in the season out of the maximum capacity of 6 MW [220].

SA Water has been participating in the electricity spot market in Australia since 2013. Between 2013 and 2017, the company started to be exposed to the spot prices but under hedge contracts, which provided a fixed volume and prices. Since 2017, SA Water has been registered as an AEMO participant. As a market customer, the company began to be completely exposed to spot market prices and it was responsible for managing the risks. Since 2020, SA Water has become a registered self-retailer. The company also participates in the Frequency Control Ancillary Services (FCAS) market [18]. Although water and wastewater utilities can provide grid services and participate in DR programs, not many studies have explored these topics, as shown in Table 22.

Challenges and Opportunities of Using RES and Market Strategies in WWTPs

Most wastewater treatment facilities are not primarily designed to be cost-efficient systems. Almost 90% of the energy consumption in WWTPs is related to three stages: secondary treatment, pumping, and sludge treatment combined with dewatering [2]. The most common alternatives that these facilities adopt to reduce energy costs in WWTPs include (i) optimising energy consumption by improving efficiency, (ii) on-site power generation to reduce operating costs related to energy importation from the main grid, and (iii) energy demand management [3]. Planning and operation studies are essential to understand the potential risks and evaluate the opportunities for WWTPs. They can help to design the best power generation configuration and find the optimal operation strategy for the generation system. Power system planning is a critical techno-economic assessment that is used to investigate and plan future system expansion, feasibility, and generation potential. The main objective of power generation planning is to determine the necessary generating capacity to satisfy the load demand in real time. In the system's planning, it is ideal to have a generation capacity that can supply 100% of the load anytime. Still, it is very challenging to match the generation and demand, mainly if the system is composed of renewable energy sources [221].

Table 21. Renewable energy projects in WWTPs in Australia.

Ref.	Organisation	Technology/System	Details
[222]	Hunter Water	6 MW solar PV	• At Dungog, Tanilba Bay and Karuah facilities, there is a 2.96 MW total operating capacity. A 3.1 MW Balickera Park solar system is the biggest of HW
[223]	Icon Water	720 kW solar PV and 1.23 MW Hydro	• 2.29 MW generation capacity to be installed. At SWTP, the installed capacity of the mini hydro system is 630 kW (able to produce up to 3.4 GWh/y). • Googong Mini Hydro installed capacity of 600 kW with generation potential of 4.2 GWh/y (13% of total electricity demand of Icon Water in 2015–16)
[224]	QST, GoR, and Melb W	Hydrothermal liquefaction	• Technology converts biosolids into biocrude, and the project cost around AUD 11.8 mi. • The plant aims to produce 12–15 megalitres of biofuel per year which can be used as renewable diesel and aviation fuel.
[225]	Loganholme WWTP	Gasification	• Convert 34,000 tonnes of sewage sludge to produce synthetic gas, and the project's investment cost is around AUD 17.3 mi. • Before the project, biosolids were disposed of in landfills, costing around AUD 1.8 mi (30% of the facility's operating costs).
[226]	Melb W	25 MW biogas, 25 MW hydro and 24 MW Solar	• 86.2%, 0.02% and 93.6% of the electricity generated from biogas, hydropower and solar were consumed on-site, respectively. • In ETP, the biogas and new PV systems will provide up to 48% of plant's demand yearly. In WTP, the new PV system will supply up to 12.4 GWh/y. • New mini-hydro projects in three WWTPs (St Albans, Upper Yarra, and O'Shannassy) will generate about 7.1 GWh/year
[227]	SA Water	154 MW PV and 34 MWh battery	• Generate 242 GWh, supplying about 70% of the SA Water's electricity demand yearly, and the system's total investment cost was AUD 300 mi.
[228]	Sydney water	PV, biogas, hydro	• Sydney Water Corporation produces around 20% of its electricity needs through on-site generation based on biogas, hydropower, and PV systems.
[229]	WC	PV, wind, biogas, hydro	• 1.5 GWh of renewable projects. Plan to invest an extra AUD 30 mil in solar generation and participate in pilot programs to provide grid services.

ETP: Eastern Treatment Plant, WTP: Winneke Treatment Plant; SWTP: Mount Stromlo Water Treatment Plant; QST: Queensland sewage treatment; Melb W: Melbourne Water; GoR: Gladstone oil refinery: WC: Water Corporation.

Table 22. Grid services and market participation studies found in the literature.

Ref.	Study Aim	Main Findings
[18]	Investigate the electricity spot price behaviour for SA Water in Australia	• Studied the energy prices in the NEM. SA Water has taken advantage of electricity spot price to manage plant's operation, and could save up to AUD 400,000. • Before 2013, SA Water used energy retail with fixed negotiations; in 2013–2017, operated under spot price with fixed price and fixed volume hedge contracts; from 2017, managed market risks itself and since 2020, became a registered self-retailer.
[194]	Investigated the Laguna WWTP in California, USA, participating in a DR scheme.	• The plant serves around 230,000 people and treats about 66 mL/day of wastewater. • Plant's energy consumption around 3–5 MWh, when not shifting energy loads. • CHP system consists of 4 × 1.1 MW generators and 2 MW Tesla battery • 18 events were requested to evaluate the plant's capability from March to June 2019. • The WWTP could save up to USD 45,500 yearly from DR only and USD 68,340 if combined with other strategies, which represents around 4.8% of plant's energy costs.
[230]	WWTPs, aggregated as VPP, to provide services.	• Integration is needed since one plant has not enough capacity to provide service. • WWTP's potential control reserve in Germany was approximately 300 MW. • Theoretical energy production potential from biogas in WWTPs was 2.1–2.6 TWh/y.
[231]	Control strategy for WWTPs for a short-term demand side	• The proposed control strategy could provide a modest cost savings of 1% if the day-ahead market is considered • If it is applied to the regulating market, the potential savings could be up to 27%.

The planning process can be considered an optimisation problem in which the optimal solution consists of finding the optimal system configuration, including the best technological generation mix, optimal size, capacity, location, and construction time with minimum investment cost [221]. Due to the uncertainty and intermittency, adopting a hybrid configuration system is usually a good alternative for WWTPs. Hybrid systems can maximise the energy generation potential from different resources, and lead to a better cost-effectiveness, energy efficiency, modularity, and flexibility [232].

Although there are several technological solutions and strategies to increase economic benefits for a WWTP, their implementation in a full-scale plant is still very limited. The biggest challenge is commonly related to capital and operating costs. Because of the vast number of available technology resources, determining the optimal planning design and operation strategy of power generation in WWTPs is not an easy task and is no longer a simple technical problem but a complex and difficult challenge that requires an integrated and comprehensive approach to make a cost-effective solution [233].

5. Conclusions

This paper reviewed different renewable energy technologies that can be adopted by WWTPs to improve energy efficiency, reduce operating costs, and move toward sustainability, focusing on providing an overview of the current status of each technology by exploring their energy recovery potential, benefits and drawbacks.

AD is still the most widely used alternative to treat sludge and recover energy in a WWTP due to its maturity technology, sewage sludge treatment efficiency, low operating costs and biogas production as a co-product. In the USA, almost 50% of all the wastewater is treated through AD. Co-digestion and pre-treatment methods are some alternatives to improve biogas production and process efficiency, but the second one is primarily used to enhance sludge biodegradability, sludge management, and reduce environmental impact. Thermochemical processes, including pyrolysis and gasification, can also offer several benefits for treating sewage sludge by converting organic matter into valuable products while addressing the challenges associated with sludge management. Both gasification and pyrolysis are mature technologies that generate high-value co-products, including syngas and bio-oil, and can be used to generate electricity and heating. Additionally, hydropower and wind technologies offer significant benefits as renewable energy sources, but their application in WWTPs is limited due to several practical, technical, and economic challenges. For example, conventional hydropower systems require a significant and consistent flow of water to generate electricity, and, typically, WWTPs do not have the necessary flow rates or water head heights to support hydropower turbines effectively, and wind turbines require substantial space to operate efficiently, and many municipal facilities are located in urban areas or constrained sites where space is limited. On the other hand, other technologies, including Microbial Fuel Cells (MFCs) and Microbial Electrolysis Cells (MECs), hold significant potential for sustainable energy production and municipal wastewater treatment. Despite their promising laboratory-scale performance, large-scale implementation of MECs and MFCs can face some challenges related to scalability, energy efficiency, capital and operational costs, material durability, microbial stability, reactor design, and regulatory approval. Therefore, more academic research and technological development are needed to achieve both technical advancements and system integration. Many of these challenges can be overcome, opening up opportunities for bioelectrochemical technologies to contribute to sustainable energy production, waste treatment, and environmental management on larger-scale facilities. Overall, the successful adoption of renewable energy technologies involves overcoming several challenges that span technical, economic, regulatory, and operational domains. It is important to address these challenges and technology limitations to outline future prospects for achieving a more energy-efficient and sustainable wastewater treatment paradigm.

WWTPs also have a significant potential to contribute to grid services and participate in demand-response programs by leveraging their operational flexibility and energy consumption patterns. By shifting load consumption, optimising processes, integrating renewable energy, and investing in advanced control systems and energy storage, WWTPs can provide valuable services to the grid while benefiting from cost savings and incentives. Additionally, the effective implementation of demand response and the ability to provide grid services for a WWTP requires addressing operational, economic, and regulatory challenges to ensure that energy management strategies align with both treatment efficiency and grid requirements.

The main challenges and limitations of this study included (i) a review and compilation of several types of renewable energy technologies limited only to municipal wastewater and sewage sludge (some technologies can use other types of feedstocks, such as AD, thermochemical processes, MFC, MEC); (ii) some renewable sources are still under experimental stages and early stage development, and thus, their applications in large-scale WWTPs are very limited, making it difficult to quantify their effectiveness, efficiency, or financial viability; (iii) due to the wide range of WWTPs design, treatment processes and capacity, applications of these technologies can have significant differences in performance, including energy production and efficiency; and (iv) many of the studies on renewable energy in WWTPs are based on small-scale pilot projects or case studies and they are not widely implemented in large-scale facilities. Thus, it is hard to measure key parameters, including efficiency, system performance, costs, and operational control.

Therefore, the future research directions regarding the adoption of energy resources in WWTPs can be based on two main topics: (a) further research on emerging renewable energy sources (RES) and (b) optimising traditional RES. For traditional methods, such as AD, the focus should be on enhancing biogas production, improving process efficiency (i.e., co-digestion and pre-treatment), and integrating AD with other energy recovery systems. However, thermochemical processes are also a mature technology, and further research exploring better conversion efficiencies and the production of high-value co-products from sewage sludge while reducing emissions can be developed. In terms of hydropower and solar PV, investigations to overcome site-specific challenges, such as limited space or water flow, with compact systems and more efficient systems can be explored. On the other hand, emerging technologies, particularly MFCs and MECs, improving scalability, energy efficiency, and material durability, alongside overcoming regulatory barriers for large-scale deployment, are the main challenges to be overcome. Further research is needed to integrate these bioelectrochemical systems with other treatment processes to maximise energy recovery and treatment efficiency. Additionally, water electrolysis holds promise for hydrogen production from wastewater, which could contribute to energy storage and fuel cells. The future of WWTPs will also involve incorporating digitalization and smart grids to optimize energy management, enabling WWTPs to provide grid services and become key players in decentralized energy systems.

Author Contributions: Conceptualization, D.L., L.L. and G.A.; methodology, D.L., L.L. and G.A.; validation, D.L.; formal analysis, D.L.; investigation, D.L.; resources, D.L., L.L. and G.A.; data curation, D.L.; writing—original draft preparation, D.L.; writing—review and editing, L.L. and G.A.; supervision, L.L. and G.A.; project administration, L.L. and G.A. All authors have read and agreed to the published version of the manuscript.

Funding: This research is supported by an Australian Government Research Training Program Scholarship. This research was funded by RACE for 2030 CRC, Sydney Water Corporation, and the University of Technology Sydney (UTS). Project Code: 21.B5.P.0217.

Data Availability Statement: No new data were created or analysed in this study. Data sharing is not applicable to this article.

Acknowledgments: This research is supported by an Australian Government Research Training Program Scholarship. This project is part of the RACE for 2030 Industry PhD Program, funded by the RACE for 2030 CRC, Sydney Water Corporation, and the University of Technology Sydney.

Conflicts of Interest: Author Gregory Appleby was employed by the company Sydney Water Corporation. The remaining authors declare that the research was conducted in the absence of any commercial or financial relationships that could be construed as a potential conflict of interest.

References

1. Maktabifard, M.; Zaborowska, E.; Makinia, J. Energy neutrality versus carbon footprint minimization in municipal wastewater treatment plants. *Bioresour. Technol.* **2020**, *300*, 122647. [CrossRef] [PubMed]
2. BECA. Opportunities for Renewable Energy in the Australian Water Sector. ARENA. 2015. Available online: https://arena.gov.au/assets/2016/01/Opportunities-for-renewable-energy-in-the-Australian-water-sector.pdf (accessed on 13 September 2022).
3. Government, N. Energy Efficiency Opportunities in Wastewater Treatment Facilities. 2019. Available online: https://www.environment.nsw.gov.au/resources/business/wastewater-treatment-facilities-energy-efficiency-opportunities-190114.pdf (accessed on 10 February 2022).
4. Topare, N.S.; Attar, S.; Manfe, M.M. Sewage/wastewater treatment technologies: A review. *Sci. Revs. Chem. Commun.* **2011**, *1*, 18–24.
5. Gandiglio, M.; Lanzini, A.; Soto, A.; Leone, P.; Santarelli, M. Enhancing the energy efficiency of wastewater treatment plants through co-digestion and fuel cell systems. *Front. Environ. Sci.* **2017**, *5*, 70. [CrossRef]
6. Kirchem, D.; Lynch, M.; Bertsch, V.; Casey, E. Market effects of industrial demand response and flexibility potential from wastewater treatment facilities. In Proceedings of the 2018 15th International Conference on the European Energy Market (EEM), Lodz, Poland, 27–29 June 2018; pp. 1–6.
7. Lima, D.; Appleby, G.; Li, L. A Scoping Review of Options for Increasing Biogas Production from Sewage Sludge: Challenges and Opportunities for Enhancing Energy Self-Sufficiency in Wastewater Treatment Plants. *Energies* **2023**, *16*, 2369. [CrossRef]
8. Saadabadi, S.A.; Thattai, A.T.; Fan, L.; Lindeboom, R.E.; Spanjers, H.; Aravind, P. Solid Oxide Fuel Cells fuelled with biogas: Potential and constraints. *Renew. Energy* **2019**, *134*, 194–214. [CrossRef]
9. Riley, D.M.; Tian, J.; Güngör-Demirci, G.; Phelan, P.; Villalobos, J.R.; Milcarek, R.J. Techno-economic assessment of chp systems in wastewater treatment plants. *Environments* **2020**, *7*, 74. [CrossRef]
10. Ibrahim, A.; Akilli, H. Supercritical water gasification of wastewater sludge for hydrogen production. *Int. J. Hydrogen Energy* **2019**, *44*, 10328–10349. [CrossRef]
11. Fan, Y.; Hornung, U.; Dahmen, N. Hydrothermal liquefaction of sewage sludge for biofuel application: A review on fundamentals, current challenges and strategies. *Biomass Bioenergy* **2022**, *165*, 106570. [CrossRef]
12. Gao, N.; Kamran, K.; Quan, C.; Williams, P.T. Thermochemical conversion of sewage sludge: A critical review. *Prog. Energy Combust. Sci.* **2020**, *79*, 100843. [CrossRef]
13. Ghoneim, W.; Helal, A.; Wahab, M.A. Renewable energy resources and recovery opportunities in wastewater treatment plants. In Proceedings of the 2016 3rd International Conference on Renewable Energies for Developing Countries (REDEC), Zouk Mosbeh, Lebanon, 13–15 July 2016; pp. 1–8.
14. Islam, A.K.; Dunlop, P.S.; Hewitt, N.J.; Lenihan, R.; Brandoni, C. Bio-hydrogen production from wastewater: A comparative study of low energy intensive production processes. *Clean Technol.* **2021**, *3*, 156–182. [CrossRef]

15. Escapa, A.; Mateos, R.; Martínez, E.; Blanes, J. Microbial electrolysis cells: An emerging technology for wastewater treatment and energy recovery. From laboratory to pilot plant and beyond. *Renew. Sustain. Energy Rev.* **2016**, *55*, 942–956. [CrossRef]
16. AlSayed, A.; Soliman, M.; Eldyasti, A. Microbial fuel cells for municipal wastewater treatment: From technology fundamentals to full-scale development. *Renew. Sustain. Energy Rev.* **2020**, *134*, 110367. [CrossRef]
17. Liu, X.; Zhu, F.; Zhang, R.; Zhao, L.; Qi, J. Recent progress on biodiesel production from municipal sewage sludge. *Renew. Sustain. Energy Rev.* **2021**, *135*, 110260. [CrossRef]
18. Do, P.; Chow, C.W.; Rameezdeen, R.; Gorjian, N. Understanding the impact of spot market electricity price on wastewater asset management strategy. *Water Conserv. Sci. Eng.* **2022**, *7*, 101–117. [CrossRef]
19. Riffat, R.; Husnain, T. *Fundamentals of Wastewater Treatment and Engineering*; Crc Press: Boca Raton, FL, USA, 2013.
20. Von Sperling, M. *Wastewater Characteristics, Treatment and Disposal*; IWA Publishing: London, UK, 2007.
21. Kacprzak, M.; Neczaj, E.; Fijałkowski, K.; Grobelak, A.; Grosser, A.; Worwag, M.; Rorat, A.; Brattebo, H.; Almås, Å.; Singh, B.R. Sewage sludge disposal strategies for sustainable development. *Environ. Res.* **2017**, *156*, 39–46. [CrossRef]
22. Kamyab, H.; Yuzir, A.; Ashokkumar, V.; Hosseini, S.E.; Balasubramanian, B.; Kirpichnikova, I. Review of the application of gasification and combustion technology and waste-to-energy technologies in sewage sludge treatment. *Fuel* **2022**, *316*, 123199.
23. Strazzabosco, A.; Kenway, S.J.; Lant, P.A. Solar PV adoption in wastewater treatment plants: A review of practice in California. *J. Environ. Manag.* **2019**, *248*, 109337. [CrossRef]
24. Patrick Dube, S.C.; Collins, D.; Deslauriers, S.; Fillmore, L.; McFadden, L.; Moss, L.; Serfass, P.; Stone, L.; Turgeon, J. *Accelerating Resource Recovery: Biosolids Innovations and Opportunities*; Water Environment Federation: Alexandria, VA, USA, 2017.
25. Shen, Y.; Linville, J.L.; Urgun-Demirtas, M.; Mintz, M.M.; Snyder, S.W. An overview of biogas production and utilization at full-scale wastewater treatment plants (WWTPs) in the United States: Challenges and opportunities towards energy-neutral WWTPs. *Renew. Sustain. Energy Rev.* **2015**, *50*, 346–362. [CrossRef]
26. Nguyen, L.N.; Kumar, J.; Vu, M.T.; Mohammed, J.A.; Pathak, N.; Commault, A.S.; Sutherland, D.; Zdarta, J.; Tyagi, V.K.; Nghiem, L.D. Biomethane production from anaerobic co-digestion at wastewater treatment plants: A critical review on development and innovations in biogas upgrading techniques. *Sci. Total Environ.* **2021**, *765*, 142753. [CrossRef]
27. Kuo-Dahab, W.C.; Amirhor, P.; Zona, M.; Duest, D.; Park, C. Investigating anaerobic co-digestion of sewage sludge and food waste using a bench-scale pilot study. In Proceedings of the WEFTEC 2014, New Orleans, LA, USA, 28 September–1 October 2014; Water Environment Federation: Alexandria, VA, USA, 2014.
28. Azarmanesh, R.; Zonoozi, M.H.; Ghiasinejad, H. Characterization of food waste and sewage sludge mesophilic anaerobic co-digestion under different mixing ratios of primary sludge, secondary sludge and food waste. *Biomass Bioenergy* **2020**, *139*, 105610. [CrossRef]
29. Prabhu, M.S.; Mutnuri, S. Anaerobic co-digestion of sewage sludge and food waste. *Waste Manag. Res.* **2016**, *34*, 307–315. [CrossRef] [PubMed]
30. Pastor, L.; Ruiz, L.; Pascual, A.; Ruiz, B. Co-digestion of used oils and urban landfill leachates with sewage sludge and the effect on the biogas production. *Appl. Energy* **2013**, *107*, 438–445. [CrossRef]
31. Di Maria, F.; Sordi, A.; Cirulli, G.; Micale, C. Amount of energy recoverable from an existing sludge digester with the co-digestion with fruit and vegetable waste at reduced retention time. *Appl. Energy* **2015**, *150*, 9–14. [CrossRef]
32. Grosser, A.; Neczaj, E.; Singh, B.; Almås, Å.; Brattebø, H.; Kacprzak, M. Anaerobic digestion of sewage sludge with grease trap sludge and municipal solid waste as co-substrates. *Environ. Res.* **2017**, *155*, 249–260. [CrossRef]
33. Abudi, Z.N.; Hu, Z.; Xiao, B.; Abood, A.R.; Rajaa, N.; Laghari, M. Effects of pretreatments on thickened waste activated sludge and rice straw co-digestion: Experimental and modeling study. *J. Environ. Manag.* **2016**, *177*, 213–222. [CrossRef]
34. Chu, X.; Wu, G.; Wang, J.; Hu, Z.-H. Dry co-digestion of sewage sludge and rice straw under mesophilic and thermophilic anaerobic conditions. *Environ. Sci. Pollut. Res.* **2015**, *22*, 20143–20153. [CrossRef]
35. Elsayed, M.; Andres, Y.; Blel, W.; Gad, A. Methane production by anaerobic co-digestion of sewage sludge and wheat straw under mesophilic conditions. *Int. J. Sci. Res. Eng. Technol. (IJSET)* **2015**, *4*, 1–6.
36. Potdukhe, R.M.; Sahu, N.; Kapley, A.; Kumar, R. Co-digestion of waste activated sludge and agricultural straw waste for enhanced biogas production. *Bioresour. Technol. Rep.* **2021**, *15*, 100769. [CrossRef]
37. Yalcinkaya, S.; Malina Jr, J.F. Model development and evaluation of methane potential from anaerobic co-digestion of municipal wastewater sludge and un-dewatered grease trap waste. *Waste Manag.* **2015**, *40*, 53–62. [CrossRef]
38. Yalcinkaya, S.; Malina, J.F. Anaerobic co-digestion of municipal wastewater sludge and un-dewatered grease trap waste for assessing direct feed of grease trap waste in municipal digesters. *Int. Biodeterior. Biodegrad.* **2015**, *104*, 490–497. [CrossRef]
39. Alqaralleh, R.M.; Kennedy, K.; Delatolla, R.; Sartaj, M. Thermophilic and hyper-thermophilic co-digestion of waste activated sludge and fat, oil and grease: Evaluating and modeling methane production. *J. Environ. Manag.* **2016**, *183*, 551–561. [CrossRef] [PubMed]
40. Tandukar, M.; Pavlostathis, S.G. Anaerobic co-digestion of municipal sludge with fat-oil-grease (FOG) enhances the destruction of sludge solids. *Chemosphere* **2022**, *292*, 133530. [CrossRef] [PubMed]
41. Alves, I.R.; Mahler, C.F.; Oliveira, L.B.; Reis, M.M.; Bassin, J.P. Investigating the effect of crude glycerol from biodiesel industry on the anaerobic co-digestion of sewage sludge and food waste in ternary mixtures. *Energy* **2022**, *241*, 122818. [CrossRef]
42. Silvestre, G.; Fernández, B.; Bonmatí, A. Addition of crude glycerine as strategy to balance the C/N ratio on sewage sludge thermophilic and mesophilic anaerobic co-digestion. *Bioresour. Technol.* **2015**, *193*, 377–385. [CrossRef]

43. Bai, X.; Chen, Y.-C. Synergistic effect and supernatant nitrogen reduction from anaerobic co-digestion of sewage sludge and pig manure. *Bioresour. Technol. Rep.* **2020**, *10*, 100424. [CrossRef]
44. Borowski, S.; Domański, J.; Weatherley, L. Anaerobic co-digestion of swine and poultry manure with municipal sewage sludge. *Waste Manag.* **2014**, *34*, 513–521. [CrossRef]
45. Sillero, L.; Solera, R.; Perez, M. Improvement of the anaerobic digestion of sewage sludge by co-digestion with wine vinasse and poultry manure: Effect of different hydraulic retention times. *Fuel* **2022**, *321*, 124104. [CrossRef]
46. Alkhrissat, T.; Kassab, G.; Abdel-Jaber, M.t. Impact of Iron Oxide Nanoparticles on Anaerobic Co-Digestion of Cow Manure and Sewage Sludge. *Energies* **2023**, *16*, 5844. [CrossRef]
47. Avila, R.; Justo, Á.; Carrero, E.; Crivillés, E.; Vicent, T.; Blánquez, P. Water resource recovery coupling microalgae wastewater treatment and sludge co-digestion for bio-wastes valorisation at industrial pilot-scale. *Bioresour. Technol.* **2022**, *343*, 126080. [CrossRef]
48. Beltrán, C.; Jeison, D.; Fermoso, F.G.; Borja, R. Batch anaerobic co-digestion of waste activated sludge and microalgae (*Chlorella sorokiniana*) at mesophilic temperature. *J. Environ. Sci. Health Part A* **2016**, *51*, 847–850. [CrossRef]
49. Ortega-Martinez, E.; Sapkaite, I.; Fdz-Polanco, F.; Donoso-Bravo, A. From pre-treatment toward inter-treatment. Getting some clues from sewage sludge biomethanation. *Bioresour. Technol.* **2016**, *212*, 227–235. [CrossRef] [PubMed]
50. Choi, J.-M.; Han, S.-K.; Lee, C.-Y. Enhancement of methane production in anaerobic digestion of sewage sludge by thermal hydrolysis pretreatment. *Bioresour. Technol.* **2018**, *259*, 207–213. [CrossRef] [PubMed]
51. Liu, X.; Xu, Q.; Wang, D.; Zhao, J.; Wu, Y.; Liu, Y.; Ni, B.-J.; Wang, Q.; Zeng, G.; Li, X. Improved methane production from waste activated sludge by combining free ammonia with heat pretreatment: Performance, mechanisms and applications. *Bioresour. Technol.* **2018**, *268*, 230–236. [CrossRef] [PubMed]
52. Trzcinski, A.P.; Tian, X.; Wang, C.; Lin, L.L.; Ng, W.J. Combined ultrasonication and thermal pre-treatment of sewage sludge for increasing methane production. *J. Environ. Sci. Health Part A* **2015**, *50*, 213–223. [CrossRef]
53. Serrano, A.; Siles, J.Á.; Gutiérrez, M.d.C.; Martín, M.d.l.Á. Comparison of Pre-treatment Technologies to Improve Sewage Sludge Biomethanization. *Appl. Biochem. Biotechnol.* **2021**, *193*, 777–790. [CrossRef]
54. Martínez, E.; Gil, M.; Rosas, J.; Moreno, R.; Mateos, R.; Morán, A.; Gómez, X. Application of thermal analysis for evaluating the digestion of microwave pre-treated sewage sludge. *J. Therm. Anal. Calorim.* **2017**, *127*, 1209–1219. [CrossRef]
55. Serrano, A.; Siles, J.; Martín, M.; Chica, A.; Estévez-Pastor, F.; Toro-Baptista, E. Improvement of anaerobic digestion of sewage sludge through microwave pre-treatment. *J. Environ. Manag.* **2016**, *177*, 231–239. [CrossRef]
56. Liu, J.; Yang, M.; Zhang, J.; Zheng, J.; Xu, H.; Wang, Y.; Wei, Y. A comprehensive insight into the effects of microwave-H2O2 pretreatment on concentrated sewage sludge anaerobic digestion based on semi-continuous operation. *Bioresour. Technol.* **2018**, *256*, 118–127. [CrossRef]
57. Lizama, A.C.; Figueiras, C.C.; Herrera, R.R.; Pedreguera, A.Z.; Espinoza, J.E.R. Effects of ultrasonic pretreatment on the solubilization and kinetic study of biogas production from anaerobic digestion of waste activated sludge. *Int. Biodeterior. Biodegrad.* **2017**, *123*, 1–9. [CrossRef]
58. Neumann, P.; González, Z.; Vidal, G. Sequential ultrasound and low-temperature thermal pretreatment: Process optimization and influence on sewage sludge solubilization, enzyme activity and anaerobic digestion. *Bioresour. Technol.* **2017**, *234*, 178–187. [CrossRef]
59. Houtmeyers, S.; Degrève, J.; Willems, K.; Dewil, R.; Appels, L. Comparing the influence of low power ultrasonic and microwave pre-treatments on the solubilisation and semi-continuous anaerobic digestion of waste activated sludge. *Bioresour. Technol.* **2014**, *171*, 44–49. [CrossRef] [PubMed]
60. Tian, X.; Wang, C.; Trzcinski, A.P.; Lin, L.; Ng, W.J. Interpreting the synergistic effect in combined ultrasonication–ozonation sewage sludge pre-treatment. *Chemosphere* **2015**, *140*, 63–71. [CrossRef] [PubMed]
61. BioCNG. Fact Sheet: City of Janesville Wastewater Treatment Plant. 2015. Available online: https://biocng.us/wp-content/uploads/2015/06/Janesville-Fact-Sheet-2015.pdf (accessed on 11 June 2024).
62. Socalgas. Point Loma Wastewater Treatment Plant. Available online: https://www.sandiego.gov/public-utilities/water-quality/water-wastewater-facilities/point-loma (accessed on 15 June 2024).
63. Amaresco, I. Case Study—San Antionio Water System, TX. 2017. Available online: https://www.ameresco.com/wp-content/uploads/2017/11/san-antonio-water-system-tx.pdf (accessed on 17 July 2024).
64. EPA. An Overview of Renewable Natural Gas from Biogas. 2020. Available online: https://www.epa.gov/sites/default/files/2020-07/documents/lmop_rng_document.pdf (accessed on 11 August 2024).
65. Hauptmeier, K.; Penkuhn, M.; Tsatsaronis, G. Economic assessment of a solid oxide fuel cell system for biogas utilization in sewage plants. *Energy* **2016**, *117*, 361–368. [CrossRef]
66. Grasham, O.; Dupont, V.; Camargo-Valero, M.A.; García-Gutiérrez, P.; Cockerill, T. Combined ammonia recovery and solid oxide fuel cell use at wastewater treatment plants for energy and greenhouse gas emission improvements. *Appl. Energy* **2019**, *240*, 698–708. [CrossRef]
67. Sanaye, S.; Imeni, M.; Yazdani, M. Clean production of power and heat for waste water treatment plant by integrating sewage sludge anaerobic digester and solid oxide fuel cell. *Energy Convers. Manag.* **2023**, *288*, 117136. [CrossRef]

68. Basrawi, F.; Ibrahim, T.K.; Habib, K.; Yamada, T.; Idris, D.M.N.D. Techno-economic performance of biogas-fueled micro gas turbine cogeneration systems in sewage treatment plants: Effect of prime mover generation capacity. *Energy* **2017**, *124*, 238–248. [CrossRef]
69. Movahed, P.; Avami, A. Techno-economic optimization of biogas-fueled micro gas turbine cogeneration systems in sewage treatment plant. *Energy Convers. Manag.* **2020**, *218*, 112965. [CrossRef]
70. Chang, C.-C.; Do, M.V.; Hsu, W.-L.; Liu, B.-L.; Chang, C.-Y.; Chen, Y.-H.; Yuan, M.-H.; Lin, C.-F.; Yu, C.-P.; Chen, Y.-H. A case study on the electricity generation using a micro gas turbine fuelled by biogas from a sewage treatment plant. *Energies* **2019**, *12*, 2424. [CrossRef]
71. Gholamian, E.; Mehr, A.; Yari, M.; Carton, J. Dynamic simulation and techno-economic assessment of hydrogen utilization in dual fuel (Hydrogen/biogas) micro gas turbine systems for a wastewater treatment plant. *Process Saf. Environ. Prot.* **2023**, *169*, 220–237. [CrossRef]
72. Di Fraia, S.; Massarotti, N.; Vanoli, L.; Costa, M. Thermo-economic analysis of a novel cogeneration system for sewage sludge treatment. *Energy* **2016**, *115*, 1560–1571. [CrossRef]
73. Silvestre, G.; Fernández, B.; Bonmatí, A. Significance of anaerobic digestion as a source of clean energy in wastewater treatment plants. *Energy Convers. Manag.* **2015**, *101*, 255–262. [CrossRef]
74. Felca, A.T.A.; Barros, R.M.; Tiago Filho, G.L.; dos Santos, I.F.S.; Ribeiro, E.M. Analysis of biogas produced by the anaerobic digestion of sludge generated at wastewater treatment plants in the South of Minas Gerais, Brazil as a potential energy source. *Sustain. Cities Soc.* **2018**, *41*, 139–153. [CrossRef]
75. MosayebNezhad, M.; Mehr, A.; Gandiglio, M.; Lanzini, A.; Santarelli, M. Techno-economic assessment of biogas-fed CHP hybrid systems in a real wastewater treatment plant. *Appl. Therm. Eng.* **2018**, *129*, 1263–1280. [CrossRef]
76. Farahbakhsh, M.T.; Chahartaghi, M. Performance analysis and economic assessment of a combined cooling heating and power (CCHP) system in wastewater treatment plants (WWTPs). *Energy Convers. Manag.* **2020**, *224*, 113351. [CrossRef]
77. De Arespacochaga, N.; Valderrama, C.; Peregrina, C.; Hornero, A.; Bouchy, L.; Cortina, J. On-site cogeneration with sewage biogas via high-temperature fuel cells: Benchmarking against other options based on industrial-scale data. *Fuel Process. Technol.* **2015**, *138*, 654–662. [CrossRef]
78. Hantoko, D.; Yan, M.; Kanchanatip, E.; Adnan, M.A.; Mubeen, I.; Hamid, F.S. Supercritical water gasification of sewage sludge and combined cycle for H2 and power production–a thermodynamic study. *Int. J. Hydrogen Energy* **2019**, *44*, 24459–24470. [CrossRef]
79. Hu, M.; Hu, H.; Ye, Z.; Tan, S.; Yin, K.; Chen, Z.; Guo, D.; Rong, H.; Wang, J.; Pan, Z. A review on turning sewage sludge to value-added energy and materials via thermochemical conversion towards carbon neutrality. *J. Clean. Prod.* **2022**, *379*, 134657. [CrossRef]
80. Jiang, G.; Xu, D.; Hao, B.; Liu, L.; Wang, S.; Wu, Z. Thermochemical methods for the treatment of municipal sludge. *J. Clean. Prod.* **2021**, *311*, 127811. [CrossRef]
81. Adar, E.; Karatop, B.; İnce, M.; Bilgili, M.S. Comparison of methods for sustainable energy management with sewage sludge in Turkey based on SWOT-FAHP analysis. *Renew. Sustain. Energy Rev.* **2016**, *62*, 429–440. [CrossRef]
82. Swann, L.; Downs, D.; Waye, M. Waste to energy solution–The sludge treatment facility in Tuen Mun, Hong Kong. *Energy Procedia* **2017**, *143*, 500–505. [CrossRef]
83. Enebe, N.L.; Chigor, C.B.; Obileke, K.; Lawal, M.S.; Enebe, M.C. Biogas and syngas production from sewage sludge: A sustainable source of energy generation. *Methane* **2023**, *2*, 192–217. [CrossRef]
84. Haghighat, M.; Majidian, N.; Hallajisani, A. Production of bio-oil from sewage sludge: A review on the thermal and catalytic conversion by pyrolysis. *Sustain. Energy Technol. Assess.* **2020**, *42*, 100870. [CrossRef]
85. Al-Rumaihi, A.; Shahbaz, M.; Mckay, G.; Mackey, H.; Al-Ansari, T. A review of pyrolysis technologies and feedstock: A blending approach for plastic and biomass towards optimum biochar yield. *Renew. Sustain. Energy Rev.* **2022**, *167*, 112715. [CrossRef]
86. Wang, S.; Xu, D.; Guo, Y.; Tang, X.; Wang, Y.; Zhang, J.; Ma, H.; Qian, L.; Li, Y. *Supercritical Water Processing Technologies for Environment, Energy and Nanomaterial Applications*; Springer: Berlin/Heidelberg, Germany, 2020.
87. Adar, E.; İnce, M.; Bilgili, M.S. Evaluation of development in supercritical water oxidation technology. *Desalin. Water Treat.* **2019**, *161*, 243–253. [CrossRef]
88. Campoy, M.; Gómez-Barea, A.; Ollero, P.; Nilsson, S. Gasification of wastes in a pilot fluidized bed gasifier. *Fuel Process. Technol.* **2014**, *121*, 63–69. [CrossRef]
89. Thomsen, T.P.; Sárossy, Z.; Gøbel, B.; Stoholm, P.; Ahrenfeldt, J.; Frandsen, F.J.; Henriksen, U.B. Low temperature circulating fluidized bed gasification and co-gasification of municipal sewage sludge. Part 1: Process performance and gas product characterization. *Waste Manag.* **2017**, *66*, 123–133. [CrossRef]
90. Chiang, K.-Y.; Lu, C.-H.; Liao, C.-K.; Ger, R.H.-R. Characteristics of hydrogen energy yield by co-gasified of sewage sludge and paper-mill sludge in a commercial scale plant. *Int. J. Hydrogen Energy* **2016**, *41*, 21641–21648. [CrossRef]
91. Schmid, M.; Beirow, M.; Schweitzer, D.; Waizmann, G.; Spörl, R.; Scheffknecht, G. Product gas composition for steam-oxygen fluidized bed gasification of dried sewage sludge, straw pellets and wood pellets and the influence of limestone as bed material. *Biomass Bioenergy* **2018**, *117*, 71–77. [CrossRef]
92. Ayol, A.; Yurdakos, O.T.; Gurgen, A. Investigation of municipal sludge gasification potential: Gasification characteristics of dried sludge in a pilot-scale downdraft fixed bed gasifier. *Int. J. Hydrogen Energy* **2019**, *44*, 17397–17410. [CrossRef]

93. Schweitzer, D.; Gredinger, A.; Schmid, M.; Waizmann, G.; Beirow, M.; Spörl, R.; Scheffknecht, G. Steam gasification of wood pellets, sewage sludge and manure: Gasification performance and concentration of impurities. *Biomass Bioenergy* **2018**, *111*, 308–319. [CrossRef]
94. Ma, J.; Chen, M.; Yang, T.; Liu, Z.; Jiao, W.; Li, D.; Gai, C. Gasification performance of the hydrochar derived from co-hydrothermal carbonization of sewage sludge and sawdust. *Energy* **2019**, *173*, 732–739. [CrossRef]
95. Freda, C.; Cornacchia, G.; Romanelli, A.; Valerio, V.; Grieco, M. Sewage sludge gasification in a bench scale rotary kiln. *Fuel* **2018**, *212*, 88–94. [CrossRef]
96. Choi, Y.-K.; Ko, J.-H.; Kim, J.-S. Gasification of dried sewage sludge using an innovative three-stage gasifier: Clean and H2-rich gas production using condensers as the only secondary tar removal apparatus. *Fuel* **2018**, *216*, 810–817. [CrossRef]
97. Roche, E.; De Andrés, J.M.; Narros, A.; Rodríguez, M.E. Air and air-steam gasification of sewage sludge. The influence of dolomite and throughput in tar production and composition. *Fuel* **2014**, *115*, 54–61. [CrossRef]
98. Choi, Y.-K.; Cho, M.-H.; Kim, J.-S. Steam/oxygen gasification of dried sewage sludge in a two-stage gasifier: Effects of the steam to fuel ratio and ash of the activated carbon on the production of hydrogen and tar removal. *Energy* **2015**, *91*, 160–167. [CrossRef]
99. Jeong, Y.-S.; Choi, Y.-K.; Park, K.-B.; Kim, J.-S. Air co-gasification of coal and dried sewage sludge in a two-stage gasifier: Effect of blending ratio on the producer gas composition and tar removal. *Energy* **2019**, *185*, 708–716. [CrossRef]
100. Mun, T.-Y.; Cho, M.-H.; Kim, J.-S. Air gasification of dried sewage sludge in a two-stage gasifier. Part 3: Application of olivine as a bed material and nickel coated distributor for the production of a clean hydrogen-rich producer gas. *Int. J. Hydrogen Energy* **2014**, *39*, 5634–5643. [CrossRef]
101. Choi, Y.-K.; Cho, M.-H.; Kim, J.-S. Air gasification of dried sewage sludge in a two-stage gasifier. Part 4: Application of additives including Ni-impregnated activated carbon for the production of a tar-free and H2-rich producer gas with a low NH3 content. *Int. J. Hydrogen Energy* **2016**, *41*, 1460–1467. [CrossRef]
102. Jaramillo-Arango, A.; Fonts, I.; Chejne, F.; Arauzo, J. Product compositions from sewage sludge pyrolysis in a fluidized bed and correlations with temperature. *J. Anal. Appl. Pyrolysis* **2016**, *121*, 287–296. [CrossRef]
103. Huang, F.; Yu, Y.; Huang, H. Temperature influence and distribution of bio-oil from pyrolysis of granular sewage sludge. *J. Anal. Appl. Pyrolysis* **2018**, *130*, 36–42. [CrossRef]
104. Arazo, R.O.; Genuino, D.A.D.; de Luna, M.D.G.; Capareda, S.C. Bio-oil production from dry sewage sludge by fast pyrolysis in an electrically-heated fluidized bed reactor. *Sustain. Environ. Res.* **2017**, *27*, 7–14. [CrossRef]
105. Alvarez, J.; Amutio, M.; Lopez, G.; Barbarias, I.; Bilbao, J.; Olazar, M. Sewage sludge valorization by flash pyrolysis in a conical spouted bed reactor. *Chem. Eng. J.* **2015**, *273*, 173–183. [CrossRef]
106. Alvarez, J.; Amutio, M.; Lopez, G.; Bilbao, J.; Olazar, M. Fast co-pyrolysis of sewage sludge and lignocellulosic biomass in a conical spouted bed reactor. *Fuel* **2015**, *159*, 810–818. [CrossRef]
107. Xie, Q.; Peng, P.; Liu, S.; Min, M.; Cheng, Y.; Wan, Y.; Li, Y.; Lin, X.; Liu, Y.; Chen, P. Fast microwave-assisted catalytic pyrolysis of sewage sludge for bio-oil production. *Bioresour. Technol.* **2014**, *172*, 162–168. [CrossRef]
108. Deng, S.; Tan, H.; Wang, X.; Yang, F.; Cao, R.; Wang, Z.; Ruan, R. Investigation on the fast co-pyrolysis of sewage sludge with biomass and the combustion reactivity of residual char. *Bioresour. Technol.* **2017**, *239*, 302–310. [CrossRef]
109. Ruiz-Gómez, N.; Quispe, V.; Ábrego, J.; Atienza-Martínez, M.; Murillo, M.B.; Gea, G. Co-pyrolysis of sewage sludge and manure. *Waste Manag.* **2017**, *59*, 211–221. [CrossRef]
110. Zhou, Y.; Liu, Y.; Jiang, W.; Shao, L.; Zhang, L.; Feng, L. Effects of pyrolysis temperature and addition proportions of corncob on the distribution of products and potential energy recovery during the preparation of sludge activated carbon. *Chemosphere* **2019**, *221*, 175–183. [CrossRef]
111. Zhu, J.; Yang, Y.; Yang, L.; Zhu, Y. High quality syngas produced from the co-pyrolysis of wet sewage sludge with sawdust. *Int. J. Hydrogen Energy* **2018**, *43*, 5463–5472. [CrossRef]
112. Patel, S.; Kundu, S.; Paz-Ferreiro, J.; Surapaneni, A.; Fouche, L.; Halder, P.; Setiawan, A.; Shah, K. Transformation of biosolids to biochar: A case study. *Environ. Prog. Sustain. Energy* **2019**, *38*, 13113. [CrossRef]
113. He, C.; Wang, K.; Giannis, A.; Yang, Y.; Wang, J.-Y. Products evolution during hydrothermal conversion of dewatered sewage sludge in sub-and near-critical water: Effects of reaction conditions and calcium oxide additive. *Int. J. Hydrogen Energy* **2015**, *40*, 5776–5787. [CrossRef]
114. Su, Y.; Liu, D.; Gong, M.; Zhu, W.; Yu, Y.; Gu, H. Investigation on the decomposition of chemical compositions during hydrothermal conversion of dewatered sewage sludge. *Int. J. Hydrogen Energy* **2019**, *44*, 26933–26942. [CrossRef]
115. Kim, D.; Lee, K.; Park, K.Y. Hydrothermal carbonization of anaerobically digested sludge for solid fuel production and energy recovery. *Fuel* **2014**, *130*, 120–125. [CrossRef]
116. Zhao, P.; Shen, Y.; Ge, S.; Yoshikawa, K. Energy recycling from sewage sludge by producing solid biofuel with hydrothermal carbonization. *Energy Convers. Manag.* **2014**, *78*, 815–821. [CrossRef]
117. Peng, C.; Zhai, Y.; Zhu, Y.; Xu, B.; Wang, T.; Li, C.; Zeng, G. Production of char from sewage sludge employing hydrothermal carbonization: Char properties, combustion behavior and thermal characteristics. *Fuel* **2016**, *176*, 110–118. [CrossRef]
118. Silva, R.D.V.K.; Lei, Z.; Shimizu, K.; Zhang, Z. Hydrothermal treatment of sewage sludge to produce solid biofuel: Focus on fuel characteristics. *Bioresour. Technol. Rep.* **2020**, *11*, 100453. [CrossRef]

119. Gaur, R.Z.; Khoury, O.; Zohar, M.; Poverenov, E.; Darzi, R.; Laor, Y.; Posmanik, R. Hydrothermal carbonization of sewage sludge coupled with anaerobic digestion: Integrated approach for sludge management and energy recycling. *Energy Convers. Manag.* **2020**, *224*, 113353. [CrossRef]
120. Zhao, J.; Liu, C.; Hou, T.; Lei, Z.; Yuan, T.; Shimizu, K.; Zhang, Z. Conversion of biomass waste to solid fuel via hydrothermal co-carbonization of distillers grains and sewage sludge. *Bioresour. Technol.* **2022**, *345*, 126545. [CrossRef]
121. Lee, J.; Sohn, D.; Lee, K.; Park, K.Y. Solid fuel production through hydrothermal carbonization of sewage sludge and microalgae Chlorella sp. from wastewater treatment plant. *Chemosphere* **2019**, *230*, 157–163. [CrossRef]
122. Nazari, L.; Yuan, Z.; Ray, M.B.; Xu, C.C. Co-conversion of waste activated sludge and sawdust through hydrothermal liquefaction: Optimization of reaction parameters using response surface methodology. *Appl. Energy* **2017**, *203*, 1–10. [CrossRef]
123. Anastasakis, K.; Biller, P.; Madsen, R.B.; Glasius, M.; Johannsen, I. Continuous hydrothermal liquefaction of biomass in a novel pilot plant with heat recovery and hydraulic oscillation. *Energies* **2018**, *11*, 2695. [CrossRef]
124. Biller, P.; Johannsen, I.; Dos Passos, J.S.; Ottosen, L.D.M. Primary sewage sludge filtration using biomass filter aids and subsequent hydrothermal co-liquefaction. *Water Res.* **2018**, *130*, 58–68. [CrossRef] [PubMed]
125. Malins, K.; Kampars, V.; Brinks, J.; Neibolte, I.; Murnieks, R.; Kampare, R. Bio-oil from thermo-chemical hydro-liquefaction of wet sewage sludge. *Bioresour. Technol.* **2015**, *187*, 23–29. [CrossRef] [PubMed]
126. Shah, A.A.; Toor, S.S.; Conti, F.; Nielsen, A.H.; Rosendahl, L.A. Hydrothermal liquefaction of high ash containing sewage sludge at sub and supercritical conditions. *Biomass Bioenergy* **2020**, *135*, 105504. [CrossRef]
127. Xu, D.; Lin, G.; Liu, L.; Wang, Y.; Jing, Z.; Wang, S. Comprehensive evaluation on product characteristics of fast hydrothermal liquefaction of sewage sludge at different temperatures. *Energy* **2018**, *159*, 686–695. [CrossRef]
128. Qian, L.; Wang, S.; Savage, P.E. Hydrothermal liquefaction of sewage sludge under isothermal and fast conditions. *Bioresour. Technol.* **2017**, *232*, 27–34. [CrossRef]
129. Wang, W.; Yu, Q.; Meng, H.; Han, W.; Li, J.; Zhang, J. Catalytic liquefaction of municipal sewage sludge over transition metal catalysts in ethanol-water co-solvent. *Bioresour. Technol.* **2018**, *249*, 361–367. [CrossRef]
130. Zhai, Y.; Chen, H.; Xu, B.; Xiang, B.; Chen, Z.; Li, C.; Zeng, G. Influence of sewage sludge-based activated carbon and temperature on the liquefaction of sewage sludge: Yield and composition of bio-oil, immobilization and risk assessment of heavy metals. *Bioresour. Technol.* **2014**, *159*, 72–79. [CrossRef]
131. Das, P.; Khan, S.; AbdulQuadir, M.; Thaher, M.; Waqas, M.; Easa, A.; Attia, E.S.M.; Al-Jabri, H. Energy recovery and nutrients recycling from municipal sewage sludge. *Sci. Total Environ.* **2020**, *715*, 136775. [CrossRef]
132. Rahman, T.; Jahromi, H.; Roy, P.; Adhikari, S.; Hassani, E.; Oh, T.-S. Hydrothermal liquefaction of municipal sewage sludge: Effect of red mud catalyst in ethylene and inert ambiences. *Energy Convers. Manag.* **2021**, *245*, 114615. [CrossRef]
133. Yang, T.; Liu, X.; Li, R.; Li, B.; Kai, X. Hydrothermal liquefaction of sewage sludge to produce bio-oil: Effect of co-pretreatment with subcritical water and mixed surfactants. *J. Supercrit. Fluids* **2019**, *144*, 28–38. [CrossRef]
134. Liu, R.; Tian, W.; Kong, S.; Meng, Y.; Wang, H.; Zhang, J. Effects of inorganic and organic acid pretreatments on the hydrothermal liquefaction of municipal secondary sludge. *Energy Convers. Manag.* **2018**, *174*, 661–667. [CrossRef]
135. Qian, L.; Wang, S.; Xu, D.; Guo, Y.; Tang, X.; Wang, L. Treatment of municipal sewage sludge in supercritical water: A review. *Water Res.* **2016**, *89*, 118–131. [CrossRef] [PubMed]
136. Bharathi, P.; Pennarasi, M. Production of biodiesel from municipal sewage sludge by transesterification process. In *Biomass Valorization Bioenergy*; Springer: Singapore, 2020; pp. 97–111.
137. Capodaglio, A.G.; Callegari, A.; Dondi, D. Microwave-induced pyrolysis for production of sustainable biodiesel from waste sludges. *Waste Biomass Valorization* **2016**, *7*, 703–709. [CrossRef]
138. Gong, M.; Zhu, W.; Fan, Y.; Zhang, H.; Su, Y. Influence of the reactant carbon–hydrogen–oxygen composition on the key products of the direct gasification of dewatered sewage sludge in supercritical water. *Bioresour. Technol.* **2016**, *208*, 81–86. [CrossRef] [PubMed]
139. Chen, Y.; Yi, L.; Wei, W.; Jin, H.; Guo, L. Hydrogen production by sewage sludge gasification in supercritical water with high heating rate batch reactor. *Energy* **2022**, *238*, 121740. [CrossRef]
140. Weijin, G.; Zizheng, Z.; Yue, L.; Qingyu, W.; Lina, G. Hydrogen production and phosphorus recovery via supercritical water gasification of sewage sludge in a batch reactor. *Waste Manag.* **2019**, *96*, 198–205. [CrossRef]
141. Gong, M.; Zhu, W.; Zhang, H.; Ma, Q.; Su, Y.; Fan, Y. Influence of NaOH and Ni catalysts on hydrogen production from the supercritical water gasification of dewatered sewage sludge. *Int. J. Hydrogen Energy* **2014**, *39*, 19947–19954. [CrossRef]
142. Wang, C.; Zhu, W.; Gong, M.; Su, Y.; Fan, Y. Influence of H2O2 and Ni catalysts on hydrogen production and PAHs inhibition from the supercritical water gasification of dewatered sewage sludge. *J. Supercrit. Fluids* **2017**, *130*, 183–188. [CrossRef]
143. Fan, Y.; Zhu, W.; Gong, M.; Su, Y.; Zhang, H.; Zeng, J. Catalytic gasification of dewatered sewage sludge in supercritical water: Influences of formic acid on hydrogen production. *Int. J. Hydrogen Energy* **2016**, *41*, 4366–4373. [CrossRef]
144. Amrullah, A.; Matsumura, Y. Supercritical water gasification of sewage sludge in continuous reactor. *Bioresour. Technol.* **2018**, *249*, 276–283. [CrossRef]
145. Sawai, O.; Nunoura, T.; Yamamoto, K. Supercritical water gasification of sewage sludge using bench-scale batch reactor: Advantages and drawbacks. *J. Mater. Cycles Waste Manag.* **2014**, *16*, 82–92. [CrossRef]
146. Wang, C.; Zhu, W.; Chen, C.; Zhang, H.; Lin, N.; Su, Y. Influence of reaction conditions on the catalytic activity of a nickel during the supercritical water gasification of dewatered sewage sludge. *J. Supercrit. Fluids* **2018**, *140*, 356–363. [CrossRef]

147. Hantoko, D.; Kanchanatip, E.; Yan, M.; Weng, Z.; Gao, Z.; Zhong, Y. Assessment of sewage sludge gasification in supercritical water for H2-rich syngas production. *Process Saf. Environ. Prot.* **2019**, *131*, 63–72. [CrossRef]
148. Adar, E.; Ince, M.; Bilgili, M.S. Supercritical water gasification of sewage sludge by continuous flow tubular reactor: A pilot scale study. *Chem. Eng. J.* **2020**, *391*, 123499. [CrossRef]
149. Qian, L.; Wang, S.; Xu, D.; Guo, Y.; Tang, X.; Wang, L. Treatment of sewage sludge in supercritical water and evaluation of the combined process of supercritical water gasification and oxidation. *Bioresour. Technol.* **2015**, *176*, 218–224. [CrossRef]
150. Qi, J.; Zhu, F.; Wei, X.; Zhao, L.; Xiong, Y.; Wu, X.; Yan, F. Comparison of biodiesel production from sewage sludge obtained from the A2/O and MBR processes by in situ transesterification. *Waste Manag.* **2016**, *49*, 212–220. [CrossRef]
151. Melero, J.; Sánchez-Vázquez, R.; Vasiliadou, I.; Castillejo, F.M.; Bautista, L.; Iglesias, J.; Morales, G.; Molina, R. Municipal sewage sludge to biodiesel by simultaneous extraction and conversion of lipids. *Energy Convers. Manag.* **2015**, *103*, 111–118. [CrossRef]
152. Choi, O.; Song, J.; Cha, D.; Lee, J. Biodiesel production from wet municipal sludge: Evaluation of in situ transesterification using xylene as a cosolvent. *Bioresour. Technol.* **2014**, *166*, 51–56. [CrossRef]
153. Patiño, Y.; Mantecón, L.G.; Polo, S.; Faba, L.; Díaz, E.; Ordóñez, S. Effect of sludge features and extraction-esterification technology on the synthesis of biodiesel from secondary wastewater treatment sludges. *Bioresour. Technol.* **2018**, *247*, 209–216. [CrossRef]
154. Patiño, Y.; Faba, L.; Díaz, E.; Ordóñez, S. Effect of pretreatments and catalytic route in the quality and productivity of biodiesel obtained from secondary sludge. *Biomass Bioenergy* **2021**, *152*, 106195. [CrossRef]
155. Supaporn, P.; Yeom, S.H. Optimization of a two-step biodiesel production process comprised of lipid extraction from blended sewage sludge and subsequent lipid transesterification. *Biotechnol. Bioprocess Eng.* **2016**, *21*, 551–560. [CrossRef]
156. Wu, X.; Zhu, F.; Qi, J.; Zhao, L.; Yan, F.; Li, C. Challenge of biodiesel production from sewage sludge catalyzed by KOH, KOH/activated carbon, and KOH/CaO. *Front. Environ. Sci. Eng.* **2017**, *11*, 1–11. [CrossRef]
157. Zhang, R.; Zhu, F.; Dong, Y.; Wu, X.; Sun, Y.; Zhang, D.; Zhang, T.; Han, M. Function promotion of $SO_4^{2-}/Al_2O_3–SnO_2$ catalyst for biodiesel production from sewage sludge. *Renew. Energy* **2020**, *147*, 275–283. [CrossRef]
158. Arazo, R.O.; de Luna, M.D.G.; Capareda, S.C. Assessing biodiesel production from sewage sludge-derived bio-oil. *Biocatal. Agric. Biotechnol.* **2017**, *10*, 189–196. [CrossRef]
159. Jung, S.; Kim, M.; Kim, Y.-H.; Lin, K.-Y.A.; Chen, W.-H.; Tsang, Y.F.; Kwon, E.E. Use of sewage sludge biochar as a catalyst in production of biodiesel through thermally induced transesterification. *Biochar* **2022**, *4*, 67. [CrossRef]
160. Katuri, K.P.; Ali, M.; Saikaly, P.E. The role of microbial electrolysis cell in urban wastewater treatment: Integration options, challenges, and prospects. *Curr. Opin. Biotechnol.* **2019**, *57*, 101–110. [CrossRef]
161. Ge, Z.; He, Z. Long-term performance of a 200 liter modularized microbial fuel cell system treating municipal wastewater: Treatment, energy, and cost. *Environ. Sci. Water Res. Technol.* **2016**, *2*, 274–281. [CrossRef]
162. Hiegemann, H.; Herzer, D.; Nettmann, E.; Lübken, M.; Schulte, P.; Schmelz, K.-G.; Gredigk-Hoffmann, S.; Wichern, M. An integrated 45 L pilot microbial fuel cell system at a full-scale wastewater treatment plant. *Bioresour. Technol.* **2016**, *218*, 115–122. [CrossRef]
163. Heidrich, E.S.; Edwards, S.R.; Dolfing, J.; Cotterill, S.E.; Curtis, T.P. Performance of a pilot scale microbial electrolysis cell fed on domestic wastewater at ambient temperatures for a 12 month period. *Bioresour. Technol.* **2014**, *173*, 87–95. [CrossRef]
164. Baeza, J.A.; Martínez-Miró, À.; Guerrero, J.; Ruiz, Y.; Guisasola, A. Bioelectrochemical hydrogen production from urban wastewater on a pilot scale. *J. Power Sources* **2017**, *356*, 500–509. [CrossRef]
165. Cotterill, S.; Dolfing, J.; Jones, C.; Curtis, T.; Heidrich, E. Low temperature domestic wastewater treatment in a microbial electrolysis cell with 1 m2 anodes: Towards system scale-up. *Fuel Cells* **2017**, *17*, 584–592. [CrossRef]
166. Zakaria, B.S.; Lin, L.; Dhar, B.R. Shift of biofilm and suspended bacterial communities with changes in anode potential in a microbial electrolysis cell treating primary sludge. *Sci. Total Environ.* **2019**, *689*, 691–699. [CrossRef] [PubMed]
167. Ahn, Y.; Im, S.; Chung, J.-W. Optimizing the operating temperature for microbial electrolysis cell treating sewage sludge. *Int. J. Hydrogen Energy* **2017**, *42*, 27784–27791. [CrossRef]
168. Moreno, R.; San-Martín, M.; Escapa, A.; Morán, A. Domestic wastewater treatment in parallel with methane production in a microbial electrolysis cell. *Renew. Energy* **2016**, *93*, 442–448. [CrossRef]
169. Liang, P.; Duan, R.; Jiang, Y.; Zhang, X.; Qiu, Y.; Huang, X. One-year operation of 1000-L modularized microbial fuel cell for municipal wastewater treatment. *Water Res.* **2018**, *141*, 1–8. [CrossRef]
170. Rossi, R.; Jones, D.; Myung, J.; Zikmund, E.; Yang, W.; Gallego, Y.A.; Pant, D.; Evans, P.J.; Page, M.A.; Cropek, D.M. Evaluating a multi-panel air cathode through electrochemical and biotic tests. *Water Res.* **2019**, *148*, 51–59. [CrossRef]
171. He, W.; Zhang, X.; Liu, J.; Zhu, X.; Feng, Y.; Logan, B.E. Microbial fuel cells with an integrated spacer and separate anode and cathode modules. *Environ. Sci. Water Res. Technol.* **2016**, *2*, 186–195. [CrossRef]
172. Park, Y.; Park, S.; Yu, J.; Torres, C.I.; Rittmann, B.E.; Lee, T. Complete nitrogen removal by simultaneous nitrification and denitrification in flat-panel air-cathode microbial fuel cells treating domestic wastewater. *Chem. Eng. J.* **2017**, *316*, 673–679. [CrossRef]
173. Blatter, M.; Delabays, L.; Furrer, C.; Huguenin, G.; Cachelin, C.P.; Fischer, F. Stretched 1000-L microbial fuel cell. *J. Power Sources* **2021**, *483*, 229130. [CrossRef]
174. Sonawane, J.M.; Marsili, E.; Ghosh, P.C. Treatment of domestic and distillery wastewater in high surface microbial fuel cells. *Int. J. Hydrogen Energy* **2014**, *39*, 21819–21827. [CrossRef]

175. Ali, A.E.-H.; Gomaa, O.M.; Fathey, R.; Abd El Kareem, H.; Abou Zaid, M. Optimization of double chamber microbial fuel cell for domestic wastewater treatment and electricity production. *J. Fuel Chem. Technol.* **2015**, *43*, 1092–1099. [CrossRef]
176. He, W.; Dong, Y.; Li, C.; Han, X.; Liu, G.; Liu, J.; Feng, Y. Field tests of cubic-meter scale microbial electrochemical system in a municipal wastewater treatment plant. *Water Res.* **2019**, *155*, 372–380. [CrossRef] [PubMed]
177. Abbas, A.I.; Qandil, M.D.; Al-Haddad, M.R.; Saravani, M.S.; Amano, R.S. Utilization of Hydroturbines in Wastewater Treatment Plants. *J. Energy Resour. Technol.* **2019**, *141*, 062011. [CrossRef]
178. Llácer-Iglesias, R.M.; López-Jiménez, P.A.; Pérez-Sánchez, M. Hydropower Technology for Sustainable Energy Generation in Wastewater Systems: Learning from the Experience. *Water* **2021**, *13*, 3259. [CrossRef]
179. Simoes, S.G.; Catarino, J.; Picado, A.; Lopes, T.F.; Di Berardino, S.; Amorim, F.; Girio, F.; Rangel, C.; de Leao, T.P. Water availability and water usage solutions for electrolysis in hydrogen production. *J. Clean. Prod.* **2021**, *315*, 128124. [CrossRef]
180. Zawadzki, P.; Kończak, B.; Smoliński, A. Municipal wastewater reclamation: Reclaimed water for hydrogen production by electrolysis–A case study. *Measurement* **2023**, *216*, 112928. [CrossRef]
181. Loots, I.; Van Dijk, M.; Barta, B.; Van Vuuren, S.; Bhagwan, J. A review of low head hydropower technologies and applications in a South African context. *Renew. Sustain. Energy Rev.* **2015**, *50*, 1254–1268. [CrossRef]
182. Power, C.; McNabola, A.; Coughlan, P. Development of an evaluation method for hydropower energy recovery in wastewater treatment plants: Case studies in Ireland and the UK. *Sustain. Energy Technol. Assess.* **2014**, *7*, 166–177. [CrossRef]
183. Tomczyk, P.; Mastalerek, K.; Wiatkowski, M.; Kuriqi, A.; Jurasz, J. Assessment of a Francis Micro Hydro Turbine Performance Installed in a Wastewater Treatment Plant. *Energies* **2023**, *16*, 7214. [CrossRef]
184. Power, C.; Coughlan, P.; McNabola, A. Microhydropower energy recovery at wastewater-treatment plants: Turbine selection and optimization. *J. Energy Eng.* **2017**, *143*, 04016036. [CrossRef]
185. Chae, K.-J.; Kim, I.-S.; Ren, X.; Cheon, K.-H. Reliable energy recovery in an existing municipal wastewater treatment plant with a flow-variable micro-hydropower system. *Energy Convers. Manag.* **2015**, *101*, 681–688. [CrossRef]
186. Durrani, A.M.; Mujahid, O.; Uzair, M. Micro hydro power plant using sewage water of Hayatabad Peshawar. In Proceedings of the 2019 15th International Conference on Emerging Technologies (ICET), Peshawar, Pakistan, 2–3 December 2019; pp. 1–5.
187. Hasan, A.; Salem, A.R.; Hadi, A.A.; Qandil, M.; Amano, R.S.; Alkhalidi, A. The power reclamation of utilizing micro-hydro turbines in the aeration basins of wastewater treatment plants. *J. Energy Resour. Technol.* **2021**, *143*, 081301. [CrossRef]
188. Ak, M.; Kentel, E.; Kucukali, S. A fuzzy logic tool to evaluate low-head hydropower technologies at the outlet of wastewater treatment plants. *Renew. Sustain. Energy Rev.* **2017**, *68*, 727–737. [CrossRef]
189. Petran, V.; Fernandes, W.; Topiwala, H.; McKeown, N.; Nardi, J.; Phillis, M.; Richardson, T.; Raza, S.; O'Sullivan, D. The Power of Wastewater-Micro Hydro Turbine at the Clarkson Wastewater Treatment Plant. *Proc. Water Environ. Fed.* **2015**, *2015*, 4271–4285. [CrossRef]
190. Bousquet, C.; Samora, I.; Manso, P.; Rossi, L.; Heller, P.; Schleiss, A.J. Assessment of hydropower potential in wastewater systems and application to Switzerland. *Renew. Energy* **2017**, *113*, 64–73. [CrossRef]
191. Gretzschel, O.; Schäfer, M.; Steinmetz, H.; Pick, E.; Kanitz, K.; Krieger, S. Advanced wastewater treatment to eliminate organic micropollutants in wastewater treatment plants in combination with energy-efficient electrolysis at WWTP Mainz. *Energies* **2020**, *13*, 3599. [CrossRef]
192. Chauhan, D.; Ahn, Y.-H. Alkaline electrolysis of wastewater and low-quality water. *J. Clean. Prod.* **2023**, *397*, 136613. [CrossRef]
193. Guo, Z.; Sun, Y.; Pan, S.-Y.; Chiang, P.-C. Integration of green energy and advanced energy-efficient technologies for municipal wastewater treatment plants. *Int. J. Environ. Res. Public Health* **2019**, *16*, 1282. [CrossRef]
194. Musabandesu, E.; Loge, F. Load shifting at wastewater treatment plants: A case study for participating as an energy demand resource. *J. Clean. Prod.* **2021**, *282*, 124454. [CrossRef]
195. EMWD. Energy Efficiency Programs. *EMWD—Eastern Municipal Water District.* Available online: https://www.emwd.org/post/energy-efficiency-programs (accessed on 4 June 2022).
196. Ali, I.; Abdelkader, L.; El Houssayne, B.; Mohamed, K.; El Khadir, L. Solar convective drying in thin layers and modeling of municipal waste at three temperatures. *Appl. Therm. Eng.* **2016**, *108*, 41–47. [CrossRef]
197. Khanlari, A.; Tuncer, A.D.; Sözen, A.; Şirin, C.; Gungor, A. Energetic, environmental and economic analysis of drying municipal sewage sludge with a modified sustainable solar drying system. *Sol. Energy* **2020**, *208*, 787–799. [CrossRef]
198. Wang, P.; Mohammed, D.; Zhou, P.; Lou, Z.; Qian, P.; Zhou, Q. Roof solar drying processes for sewage sludge within sandwich-like chamber bed. *Renew. Energy* **2019**, *136*, 1071–1081. [CrossRef]
199. Huang, Y.; Chen, M.; Jia, L. Assessment on thermal behavior of municipal sewage sludge thin-layer during hot air forced convective drying. *Appl. Therm. Eng.* **2016**, *96*, 209–216. [CrossRef]
200. Khanlari, A.; Sözen, A.; Afshari, F.; Şirin, C.; Tuncer, A.D.; Gungor, A. Drying municipal sewage sludge with v-groove triple-pass and quadruple-pass solar air heaters along with testing of a solar absorber drying chamber. *Sci. Total Environ.* **2020**, *709*, 136198. [CrossRef] [PubMed]
201. Boncescu, C.; Robescu, L.D. The Impact of Photovoltaic Power System on Carbon Footprint in the Wastewater Treatment Plant. In Proceedings of the 2021 10th International Conference on ENERGY and ENVIRONMENT (CIEM), Bucharest, Romania, 14–15 October 2021; pp. 1–5.
202. Piergrossi, V.; De Sanctis, M.; Chimienti, S.; Di Iaconi, C. Energy recovery capacity evaluation within innovative biological wastewater treatment process. *Energy Convers. Manag.* **2018**, *172*, 529–539. [CrossRef]

203. Quintero Pulido, D.F.; Barreto, C.M.; Ten Kortenaar, M.V.; Balda, R.R.; Hurink, J.L.; Smit, G.J. Simulation of sizing of energy storage for off-grid decentralized wastewater treatment units: A case study in the Netherlands. *Water Pract. Technol.* **2018**, *13*, 771–779. [CrossRef]
204. Mbarga, A.H.A.; Rainwater, K.; Song, L.; Cleveland, T.; Williams, W.R. Economic Analyses of the Seadrift Wind-Aided Wastewater Treatment Plant Operations. *Tex. Water J.* **2021**, *12*, 42–57. [CrossRef]
205. Duarte, P.T.; Duarte, E.A.; Murta-Pina, J. Increasing self-sufficiency of a wastewater treatment plant with integrated implementation of anaerobic co-digestion and photovoltaics. In Proceedings of the 2018 International Young Engineers Forum (YEF-ECE), Costa da Caparica, Portugal, 4 May 2018; pp. 103–108.
206. del Moral, A.; Petrakopoulou, F. Evaluation of the coupling of a hybrid power plant with a water generation system. *Appl. Sci.* **2019**, *9*, 4989. [CrossRef]
207. Andrei, H.; Badea, C.A.; Andrei, P.; Spertino, F. Energetic-environmental-economic feasibility and impact assessment of grid-connected photovoltaic system in wastewater treatment plant: Case study. *Energies* **2020**, *14*, 100. [CrossRef]
208. Jacob, R.; Short, M.; Belusko, M.; Bruno, F. Maximising renewable gas export opportunities at wastewater treatment plants through the integration of alternate energy generation and storage options. *Sci. Total Environ.* **2020**, *742*, 140580. [CrossRef]
209. Islam, A.K. Hydropower coupled with hydrogen production from wastewater: Integration of micro-hydropower plant (MHP) and microbial electrolysis cell (MEC). *Int. J. Hydrogen Energy* **2024**, *49*, 1–14. [CrossRef]
210. Mehr, A.; Gandiglio, M.; MosayebNezhad, M.; Lanzini, A.; Mahmoudi, S.; Yari, M.; Santarelli, M. Solar-assisted integrated biogas solid oxide fuel cell (SOFC) installation in wastewater treatment plant: Energy and economic analysis. *Appl. Energy* **2017**, *191*, 620–638. [CrossRef]
211. Gandiglio, M.; Mehr, A.S.; MosayebNezhad, M.; Lanzini, A.; Santarelli, M. Solutions for improving the energy efficiency in wastewater treatment plants based on solid oxide fuel cell technology. *J. Clean. Prod.* **2020**, *247*, 119080. [CrossRef]
212. ACUA. Jersey-Atlantic Wind Farm—Fact Sheet. ACUA—Atlantic City Utilities Authority. Available online: https://travel.sygic.com/en/poi/acua-wastewater-treatment-facility-poi:12844 (accessed on 26 May 2024).
213. Day, D. Wind Turbines From Goldwind USA Help The Field's Point Treatment Plant Save $1 Million A Year. Available online: https://www.tpomag.com/editorial/2015/02/wind_turbines_from_goldwind_usa_help_the_fields_point_treatment_plant_save (accessed on 28 May 2024).
214. City of Pueblo (Colo.). Recognized for Using Solar Power at Wastewater Treatment Facility. US EPA—Environmental Protection Agency. Available online: https://www.environmental-expert.com/news/city-of-pueblo-colo-recognized-for-using-solar-power-at-wastewater-treatment-facility-258583 (accessed on 25 May 2024).
215. Greer, D. Directed Biogas To Power Fuel Cells. BioCycle Energy. Available online: https://www.biocycle.net/directed-biogas-to-power-fuel-cells/ (accessed on 27 April 2022).
216. OYJ, O. Outotec to Build the Largest and Most Advanced Sewage Sludge Thermal Treatment Plant in Switzerland. Available online: https://www.metso.com/corporate/media/news/2012/11/outotec-to-build-the-largest-and-most-advanced-sewage-sludge-thermal-treatment-plant-in-switzerland/ (accessed on 10 September 2024).
217. Massachusetts Department of Energy Resources. Renewable Energy Case Study: Deer Island Wastewater Treatment Plant, Mass. Water Resources Authority. 2017. Available online: https://www.mass.gov/doc/renewable-energy-installations-at-deer-island/download (accessed on 17 July 2024).
218. Enel-x. Eastern Municipal Water District Works with Enel X to Earn Payments in California Demand Response. Available online: https://www.enelx.com/n-a/en/resources/case-studies/eastern-municipal-water-district (accessed on 7 June 2024).
219. Enbala. Revenue Generating Smart Grid Technology—A Pennsylvania American Water Case Study. Available online: https://www.paawwa.org/wp-content/uploads/2013/06/23Hufton.pdf (accessed on 8 June 2024).
220. Generac. *Puget Sound Energy: Using Real-Time Demand Response Technology to Defer Investments in Fossil Fuel*; Generac: Waukesha, WI, USA, 2021.
221. Abdin, I.; Zio, E. Optimal planning of electric power systems. In *Optimization in Large Scale Problems*; Springer: Berlin/Heidelberg, Germany, 2019; pp. 53–65.
222. Hunter Water Corporation. Renewable Energy Project. Available online: https://www.hunterwater.com.au/community/major-projects-in-your-area/renewables (accessed on 15 June 2024).
223. Carbon and Energy. Icon Water. Available online: https://www.iconwater.com.au/water-education/sustainability-and-environment/sustainability-and-environment-programs/carbon-and-energy (accessed on 15 June 2024).
224. ARENA. Sewage Treatment Plants Turn Sludge into Liquid Fuel. Available online: https://arena.gov.au/blog/sewage-treatment-plants-turn-sludge-into-liquid-fuel/ (accessed on 5 June 2024).
225. ARENA. Sewage Treatment Plant Smells Success in Synthetic Gas Trial. Available online: https://arena.gov.au/blog/logan-gasification-sewage-treatment-plant/ (accessed on 11 June 2024).
226. Melbourne Water Annual Report. Melbourne Water. 2023. Available online: http://www.melbournewater.com.au (accessed on 10 August 2024).
227. Energy Management and Climate. SA Water. Available online: https://www.sawater.com.au/water-and-the-environment/recycling-and-the-environment/energy-management-and-climate (accessed on 15 June 2022).
228. Energy Management and Climate Change. Sydney Water. Available online: https://www.sydneywater.com.au/water-the-environment/what-we-are-doing/energy-management-climate-change.html (accessed on 15 June 2022).

229. Annual Report 2021. Water Corporation. 2021. Available online: https://pw-cdn.watercorporation.com.au/-/media/WaterCorp/Documents/About-us/Our-performance/Annual-Reports/2021-Annual-Report/Water_Corporation_Annual_Report_2021.pdf?rev=06dffd8387764bf6ad20708d7abab6ed&hash=319404485B22EBA7C62A3E80ABDBE5F1 (accessed on 11 September 2024).
230. Schäfer, M.; Gretzschel, O.; Schmitt, T.G.; Knerr, H. Wastewater treatment plants as system service provider for renewable energy storage and control energy in virtual power plants–a potential analysis. *Energy Procedia* **2015**, *73*, 87–93. [CrossRef]
231. Brok, N.B.; Munk-Nielsen, T.; Madsen, H.; Stentoft, P.A. Unlocking energy flexibility of municipal wastewater aeration using predictive control to exploit price differences in power markets. *Appl. Energy* **2020**, *280*, 115965. [CrossRef]
232. Nguyen, H.T.; Safder, U.; Nguyen, X.N.; Yoo, C. Multi-objective decision-making and optimal sizing of a hybrid renewable energy system to meet the dynamic energy demands of a wastewater treatment plant. *Energy* **2020**, *191*, 116570. [CrossRef]
233. Kehrein, P.; Van Loosdrecht, M.; Osseweijer, P.; Garfí, M.; Dewulf, J.; Posada, J. A critical review of resource recovery from municipal wastewater treatment plants–market supply potentials, technologies and bottlenecks. *Environ. Sci. Water Res. Technol.* **2020**, *6*, 877–910. [CrossRef]

energies

MDPI

Article

Economic Analysis of the Operation of a Hydropower Plant in a Wastewater Treatment Plant: Toruń, Poland

Paweł Tomczyk [1], Krzysztof Mastalerek [2], Karol Kociszewski [3], Wojciech Orzepowski [4] and Mirosław Wiatkowski [1,*]

[1] Institute of Environmental Engineering, Wroclaw University of Environmental and Life Sciences, pl. Grunwaldzki 24, 50-363 Wrocław, Poland; pawel.tomczyk@upwr.edu.pl
[2] Eneris Alternative Fuels Sp. z o.o., Koszykowa 65, 00-667 Warsaw, Poland; krzysztof.mastalerek@wp.pl
[3] Department of Ecological Economics, Wroclaw University of Economics and Business, Komandorska 118/120, 53-3405 Wrocław, Poland; karol.kociszewski@ue.wroc.pl
[4] Department of Environmental Protection and Development, Wroclaw University of Environmental and Life Sciences, pl. Grunwaldzki 24, 50-363 Wrocław, Poland; wojciech.orzepowski@upwr.edu.pl
* Correspondence: miroslaw.wiatkowski@upwr.edu.pl

Abstract: The energy industry is constantly changing and evolving. One of the visible solutions is the pursuit of low-emission solutions. One such solution is investment in renewable energy sources (RESs), including hydropower, which in 2023 will constitute a dominant share in the production of electricity from RESs (47.02%) in the world. One of the alternative solutions is the use of hydropower in wastewater treatment plants, whose global potential is estimated at about 4350 TWh. An important issue in the operation of these facilities is their profitability. This article presents an economic analysis of a hydropower plant in the wastewater treatment plant in Toruń (northern Poland) in order to assess its profitability in the context of profits, costs, and payback period. The analyses showed the profitability of 9 out of 12 characterized variants, with the payback period estimated for real variants at 10.75–23.74 years and for theoretical variants at 5.06–5.32 years. The most significant factor in reducing the payback period was the level of electricity production between years, while different types of settlements and changes in electricity sales prices played a lesser role. Taking into account all profitable variants, after 25 years of operation, the net income will amount to PLN 1.07 million for the actual variants and PLN 3.18 million for the theoretical variants. The work can contribute to understanding the specifics of the operation of such facilities, which is consistent with, among others, the goals of sustainable development, climate and energy policies, or the circular economy.

Keywords: renewable energy sources; sustainable development; bioeconomy; climate change adaptation; cost-benefit analysis; investment payback period

1. Introduction

Currently, the energy industry is constantly changing and evolving. There is a visible drive towards zero-emission or low-emission solutions [1,2], which are part of the idea of adaptation to climate change and a circular economy [3,4]. As a result, there is a continuous increase in the share of renewable energy sources (RESs) in electricity production and in various sectors of the economy. According to data from 2023 [5], RESs constitute 30.2% of the electricity generation (installed capacity equal to 3.865 GW) in the world. In 2000, it was 18.3%. Globally, the dominant source of RESs is hydropower (47.02% of production from RESs), followed by solar and wind energy (25.83% and 17.88%, respectively). In

2030, the share of renewable energy in electricity production worldwide is expected to be around 30%, with a similar share of solar, wind, and water energy (the importance of the first two is growing due to the decreasing installation and operating costs and subsidies and reliefs for this type of investment [6,7], e.g., the levelized cost of energy for photovoltaic panels dropped by 90% comparing 2023 with 2010 [8]). Lower RES emissions compared to conventional sources are noticeable. For coal and gas, these values are 820 and 490 gCO_2eq/kWh, respectively, while for wind, water (run-of-river power plants), geothermal, and rooftop solar energy, they are 12 (4.9), 24, 38, and 41 gCO_2eq/kWh, respectively (median value) [9,10]. The described RES investments are consistent with, among others, the European Green Deal, which assumes achieving zero net greenhouse gas emissions by 2050 [11,12]. Similarly, the development of the RES sector is consistent with the sustainable development goals (SDGs), especially SDG 7 ("affordable and clean energy") and SDG 13 ("climate action") [13].

As mentioned, hydropower plays an important role in the energy sector, being the dominant source of renewable energy in 2023 on a global scale. This is mainly due to the flexibility of this energy source [14] and, in the case of accompanying reservoirs, also for recreational, navigation, flood protection, or water supply purposes [15,16]. The largest amounts of this energy source for electricity production per capita are used in Iceland, Norway, and Bhutan due to favorable hydrographic and geomorphological conditions (in 2023: 37,890, 24,836, and 11,502 kWh, respectively) [17]. However, according to forecasts, this energy source may be used less and less in the future due to the increasing installation costs and higher levelized energy costs compared to other renewable energy sources, especially solar and wind [8]. Another major obstacle is often the long and complicated administrative and legal procedures, as well as the potential negative impacts related to the construction and operation of such facilities (e.g., impact on hydrological characteristics, conditions for the migration of aquatic organisms, quality of water and bottom sediments, and potential population resettlements) [18,19].

An alternative to standard run-of-river, reservoir, or pumped-storage hydroelectric power plants, which require the construction of appropriate infrastructure, are hydropower plants in wastewater treatment plants (WWTPs) [20]. Their advantages include the use of existing infrastructure, usually using a constant inflow of wastewater from the treatment plant, the possibility of constant monitoring due to the location in an existing facility, and cost reduction (e.g., land lease and construction of damming devices) [21]. Thanks to this solution, the impact on the environment and society is minimized because no damming devices, water reservoirs, or dams are created in the riverbed, so such a solution can be more easily incorporated into the existing infrastructure [22]. Hydropower plants in WWTPs can also be easier to manage due to their greater predictability than in rivers, as well as the higher stability of electricity production from water energy. It can also be a method of energy recovery that will function well, for example, alongside hybrid water supply systems, the use of sewage sludge, or co-fermentation systems [23–25]. This is, therefore, consistent with the idea of a closed-loop economy in which available raw materials are used to the maximum extent possible. The potential for using energy from WWTPs is large. It is estimated that on a global scale, this value may amount to 4347.77 TWh in 2022 [26]; for comparison, the total energy production from hydropower plants in the world in the same year was equal to 4326.76 TWh [27].

An important aspect of hydropower plant operations is their profitability [28]. In this case, profit is derived from the sale of generated electricity, while costs are incurred both in the pre-operational and operational phases [29]. These costs result from the preparation of documentation, construction works related to the construction of the power plant, equipping the hydropower plant, connection to the power grid, land lease, taxes and fees related to the sale of energy and the use of land for hydrotechnical structures, modernization and maintenance of equipment, payments, and costs related to mitigating environmental effects [30]. The distribution and size of costs depend on the type of hydropower plant (reservoir, run-of-river, or pumped-storage), its location (region/country), and technical parameters (e.g., head, power, and types of installed equipment) [31,32]. For example, it is estimated that maintenance and modernization costs in 2020 for 270 hydropower plants in the United States ranged from 16 USD/kW for plants over 500 MW to 213 USD/kW for plants under 10 MW [33]. They typically constitute from 1% to 6% of the total annual installation costs [34] (in 2023, the average installation cost for hydropower plants was 2806 USD/kW [8]). A cross-sectional study by Sovacool et al. (2014) [35] showed that the value of 401 hydropower projects with a capacity of over 325 GW was 820 billion USD, which is approximately 2.52 million USD per MW.

As can be seen, the above considerations mainly concern run-of-river, reservoir, and pumped-storage hydropower plants. However, there are few publications so far related to the economic analysis of hydropower plants in WWTPs (examples: Tatkar WWTP in Ankara, Turkey [36], WWTPs in Great Britain and Ireland [37], comprehensive analysis of 46 facilities in 14 countries [38]). In the case of Poland, Central Europe, i.e., the country on which this publication focuses, no results in this area have been published so far, so it was decided to fill this knowledge gap. Currently (December 2024), there are three hydropower plants in WWTPs, i.e., in Szczecin, Kraków, and Toruń, using Francis or Kaplan turbines with capacities of 140, 130, and 55 kW, respectively [26,39,40]. The distinguishing element of this article from similar economic analyses of WWTPs is the use of variants based on different criteria, i.e., electricity settlement principles (households, competitive market), energy sales prices (2018, 2023), the amount of electricity produced (2018, 2023), and using actual and potential electricity production data. In addition, the actual results of the active power of the turbine, i.e., delivered to the power grid, are used, which increases the quality of the presented analyses. In other studies, estimated values of head, efficiency, and flows were usually used, and only on this basis were the power and electricity production determined, e.g., [36,38,41]. The methods used in similar analyses include sustainable management models (a combination of economic, environmental, and social analyses) [42], determination of energy recovery potential, economic profitability and sensitivity analysis [36], fuzzy logic tool (technical, economic, and environmental aspects) [35], algorithm using GIS data, inflows, local investment costs, and electricity sales tariffs [18].

The aim of the article is to assess the profitability of the operation of a hydropower plant located in a WWTP in Toruń (northern Poland, Kuyavian-Pomeranian Voivodeship) based on the analysis of revenues, costs, and payback period of this type of investment. The work is a contribution to understanding the specifics of the operation of this type of facility, thanks to which it can contribute to the implementation of sustainable development goals related to national and international energy policies, assuming an increase in the use of RES, such as hydropower. The described case study also promotes activities related to rational water and wastewater management, conducted in the spirit of the circular economy, which also takes into account activities related to adaptation to climate change. The work may also be of practical importance, e.g., for investors or institutions wishing to invest in the construction of a hydropower plant in WWTPs and constitute a certain compendium in the assessment of the justification for launching this type of facilities (e.g., as an element of the strengths, weaknesses, opportunities, and threats analysis = SWOT analysis).

2. Materials and Methods

2.1. Research Facility

The hydropower plant operating in the Central WWTP in Toruń (northern Poland, Kuyavian-Pomeranian Voivodeship, Central Europe; 53°00′50″ N 18°32′19″ E) was selected as the research object for analysis. The function of the treatment plant located in an area of 9.5 ha is to treat municipal and industrial wastewater, infiltration water, and rainwater from the city of Toruń and neighboring communes, which are supplied by the wastewater network and transported by wastewater disposal vehicles. The maximum daily capacity of the facility is 90 thousand cubic meters. On average, about 15 million cubic meters of wastewater are treated per year. There are three stages of treatment, i.e., mechanical treatment, biological treatment, and processing of wastewater sludge. The technology of Passawant company is used. The facility has been operating since 1998 [43,44].

Since 2016, a hydropower plant has been operating at the outlet of the treatment plant, where the treated wastewater is directed to the Vistula River. It uses the natural accumulation of water in the treated wastewater chamber, which is located below the secondary settling tanks. The hydropower plant is equipped with a 55 kW Francis turbine. The energy produced is used to compensate for the electricity needed for wastewater treatment. Its location is shown in Figure 1. The exact operating principle of the power plant and construction diagrams are included in the previously published article [26].

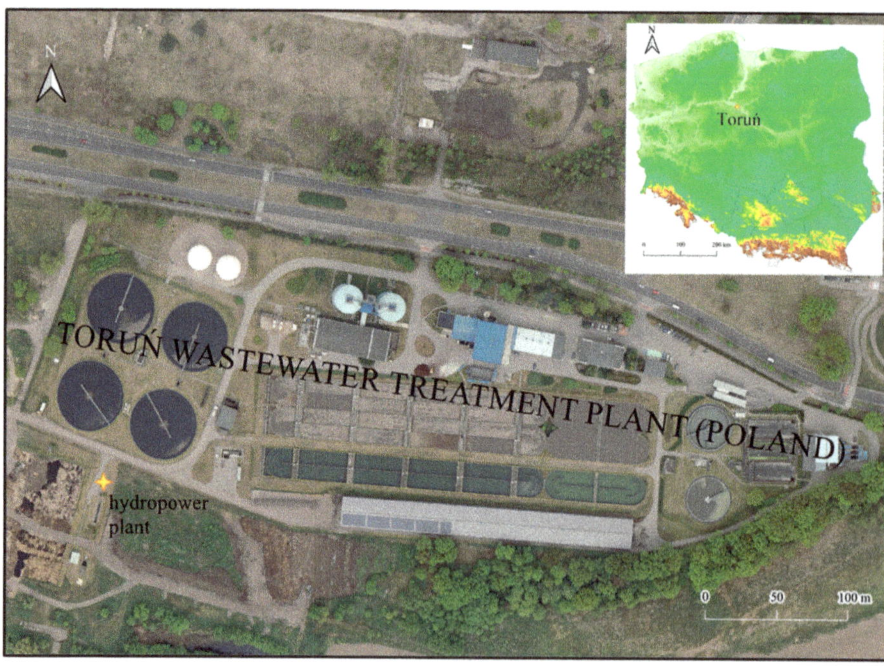

Figure 1. Location of the research facility: Central Wastewater Treatment Plant in Toruń (Poland) (map background: https://mapy.geoportal.gov.pl/imap/Imgp_2.html; accessed on 14 January 2025).

2.2. Calculations of Electricity Production in a Hydropower Plant

The article calculates the daily electricity production from the hydropower plant at the WWTP in Toruń in 2018 and 2023 (these years were selected to capture potential time variability of revenues and costs, as well as potential different operating conditions of the facility). To determine it, the hourly active power of the Francis turbine for the analyzed years, sent by Toruńskie Wodociągi Sp. z o.o. was used. These data were then converted to daily, calculating the average for each day.

The final electricity production was determined based on Formula (1) [45]:

$$EP = P \cdot t, \tag{1}$$

where

EP—daily electricity production (kWh);

P—turbine power (kW);

t—operating time of the hydropower plant in the WWTP (t = 24 h).

As this was actual data, multiplying by 24 h produced appropriate results, taking into account variability in turbine performance.

In the absence of such data, electricity production would be determined using the following Formula (2) [46,47]:

$$EP = H \cdot Q \cdot \eta_t \cdot g \cdot t, \tag{2}$$

where

H—hydraulic head (m);

Q—flow through the turbine (m^3/s);

η_t—turbine set efficiency, i.e., the product of the turbine, transmission, generator, and power output efficiency (-);

g—gravitational acceleration (m/s^2);

t—operating time of the hydropower plant in the WWTP (h).

The work also used information on hourly amounts of wastewater at the outlet of the WWTP, which were converted into daily totals. This data also comes from Toruńskie Wodociągi Sp. z o.o. (Toruń Waterworks). Figure 2 presents the individual steps of the analyses performed in this article, described in Sections 2.2 and 2.3.

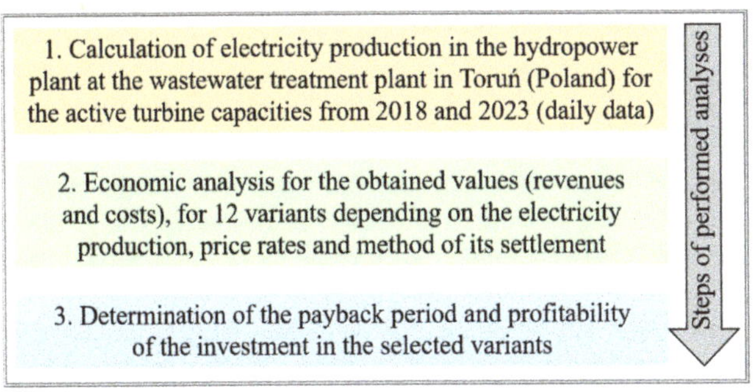

Figure 2. Analyses performed as part of work on the article.

2.3. Assumptions for Economic Analysis

The economic analysis includes the balance of revenues and costs related to the construction and operation of a hydropower plant in the WWTP in Toruń, along with the determination of the estimated payback period and its profitability. Due to limitations resulting from the availability of data and the scope of the research carried out, we did not take into account external costs and benefits. Due to the objectivity of the research results, taxes and public subsidies were also not taken into account, including revenues from green certificates, which in Poland are awarded to renewable energy producers. Thus, the analysis takes into account net revenue, which includes only revenues from the sale of products and services related to the main activity of a given company.

In the case of the selected research facility, the net revenue is the value of electricity production from the turbine, which depends on the factors given in Formula (2). Eight evaluation variants were considered in the economic analysis:

1. Value of electricity production in 2018 at the average selling price of electricity on the competitive market in Poland in 2018 (according to Energy Regulatory Office in Poland = URE [48]: PLN 0.1943/1 kWh; PLN 1 = USD 0.245 = EUR 0.233, as of 3 December 2024);
2. Value of electricity production from 2018 according to the average selling price of electricity for households in Poland in 2018 (according to URE [49]: PLN 0.5055/1 kWh);
3. Value of electricity production from 2023 at the average selling price of electricity on the competitive market in Poland in 2018 (according to URE [48]: PLN 0.1943/1 kWh);
4. The value of electricity production from 2023 at the average selling price of electricity for households in Poland in 2018 (according to URE [49]: PLN 0.5055/1 kWh);
5. Value of electricity production from 2018 at the average selling price of electricity on the competitive market in Poland in 2023 (according to URE [50]: PLN 0.75929/1 kWh);
6. Value of electricity production from 2018 at the average selling price of electricity for households in Poland in 2023 (according to URE [51]: PLN 0.7840/1 kWh);
7. Value of electricity production from 2023 at the average selling price of electricity on the competitive market in Poland in 2023 (according to URE [50]: PLN 0.75929/1 kWh);
8. The value of electricity production from 2023 at the average selling price of electricity for households in Poland in 2023 (according to URE [51]: PLN 0.7840/1 kWh).

The above variants take into account changes in rates between 2018 and 2023, differences between settlements on different principles, and variability in electricity production in the selected research facility in the analyzed years.

The described profit should include the cost of electricity production. For hydropower plants, in 2018, it amounted to PLN 0.1898/kWh [52], and in 2023 it was PLN 0.2301/kWh [8]. Moreover, according to Polish law, the sale of electricity from hydropower plants is not subject to taxation if the producer only holds a concession for electricity production (in other cases, excise duty should be assumed, e.g., if the WWTP was the final purchaser); additionally, it is renewable energy, and its capacity is lower than 1 MW, which also qualifies for exemption from excise duty [53]. The current costs assumed were the costs of repairs and modernization in the amount of 2.5% of the installation costs per year [34]. Moreover, in accordance with the applicable Water Law, an annual fee must be paid for the use of land covered with water for hydropower plant construction facilities, which amounts to PLN 8.90/m^2/year (assuming 360 m^2 of land area, calculated from the terrain map) [54].

Before the hydropower plant was launched, it was also necessary to incur administrative and legal costs related to the conducted proceedings, land lease costs, and the cost of building the hydropower plant (including the purchase of its individual components, e.g., turbines, generators, and gears). The price levels from 2018 and 2023 were assumed, thus taking into account inflation (in accordance with the variants analyzed above) [8,50]. These were the costs of documentation related to the Water, Construction, and Environmental Protection Law: PLN 90,357 and PLN 126,734 (estimation based on [55], original prices from December 2021), lease costs: PLN 41,086 and PLN 56,954 (5 years of lease assumed; calculations based on data from [56] for January 2013, taking into account inflation), and installation costs: PLN 6022.17/kW and PLN 11,325.89/kW (average costs for total hydropower plants at the global level for 2018 and 2023) [8,50].

The last element of the economic analysis is the estimation of the investment payback period, i.e., the moment when the balance of revenues and costs is equal. The payback period was calculated by dividing the net revenue (balance) on day zero (taking into account the costs before the start of operation) by the net revenue (balance) on an annual basis. In addition, the profitability of the project in the cited variants was assessed, citing the literature on the subject.

3. Results and Discussion

3.1. Annual Electricity Production at the Hydropower Plant in Toruń in 2018 and 2023

Figure 3 shows the average daily power of the turbine installed in the water power plant of the WWTP in Toruń in 2018 and 2023, divided into months. The average variability of active power in both years, expressed by the coefficient of variation (CV), was average and small, i.e., 20.38% in 2018 and 18.11% in 2023. In individual months, these values in 2018 ranged from 6.79% in February to 49.24% in June, while in 2023, it was from 0.00% in August, September, and October to 57.83% in November. In the latter case, the high variability resulted from the largest difference between the powers achieved in this month, i.e., from 0 kW (from 1 to 6 November) to 33.40 kW (21 November). It should be noted that in 2023, the turbine did not work for over 3 months (98 days); in 2018, this period was 7 days (late May and early June). The average turbine power was 19.34 and 11.38 kW in 2018 and 2023, respectively (which is 35.17% and 20.69% of the maximum active power). Broken down by month, it was in 2018 from 15.63 kW in June to 25.89 kW in January, and for 2023, it was from 0.00 kW in August, September, and October to 21.23 kW in December. Looking at the time variability of the turf power, there is no correlation between the two analyzed years. Moreover, the full power potential of the turbine was never used (the highest power was 36.75 kW on 4 January 2018).

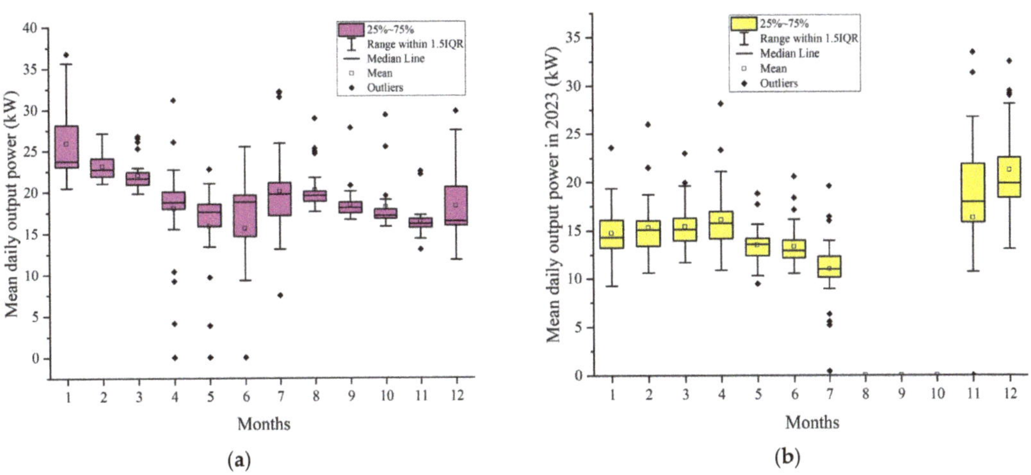

Figure 3. Mean daily output power of the turbine at the hydropower plant of the wastewater treatment plant in Toruń, divided into months: (**a**) in 2018, (**b**) in 2023.

Translating the described turbine capacities into electricity production (Figure 4), the difference between 2018 and 2023 is visible up to the 300th day in 2018. These values were usually higher (daily average for the first 200 days: 2018: 482.18 kWh, 2023: 340.39 kWh), and in the last 60 days, they were higher in 2023 than in 2018 (respectively, 418.30 kWh and 500.28 kWh). Total estimated electricity production based on actual data amounted to 169.43 MWh in 2018 and 99.68 MWh in 2023.

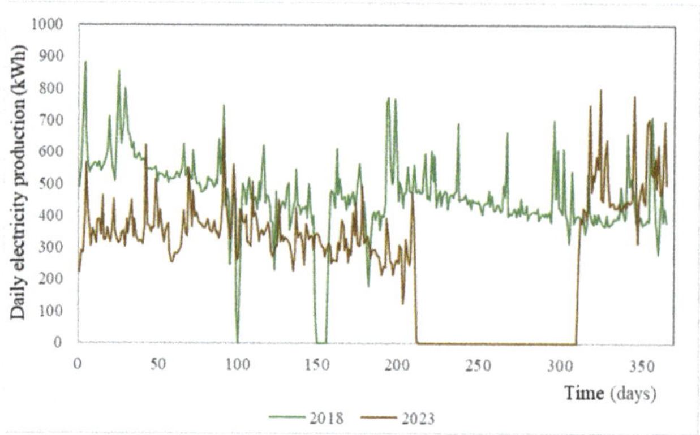

Figure 4. Variability of the calculated daily electricity production in the hydropower plant at the wastewater treatment plant in Toruń in 2018 and 2023.

Downtime periods in both years resulted from modernization and repairs and not from the lack of wastewater flowing into the hydropower plant, as shown in Figure 5. The lowest daily volume of wastewater at the outlet of the treatment plant in 2018 was 33.42 thousand m^3 (1 November), while in 2023, it was 8.69 thousand m^3 (16 September, low value due to the lack of flow meter readings for 19 h; such situations occurred in 0.62% of cases during the year, so they are negligible). In 2018, the average daily volume of outflowing wastewater ranged from 38.52 thousand m^3 in November to 49.71 thousand m^3 in January, while in 2023, it was from 37.42 thousand m^3 in July to 47.65 thousand m^3 in December. On an annual basis, the average daily volume of wastewater in 2018 and 2023 was 42.74 thousand m^3 and 43.04 thousand m^3, and the sum was 15.60 million m^3 and 15.71 million m^3. These values were, therefore, comparable to each other; therefore, the potential for electricity production with continuous turbine operation should also be similar. It is worth noting that the WWTP operated properly in the context of not exceeding the maximum daily wastewater throughput of 90 thousand m^3 (maximum volumes: 2018: 76.07 thousand m^3, 2023: 76.86 thousand m^3).

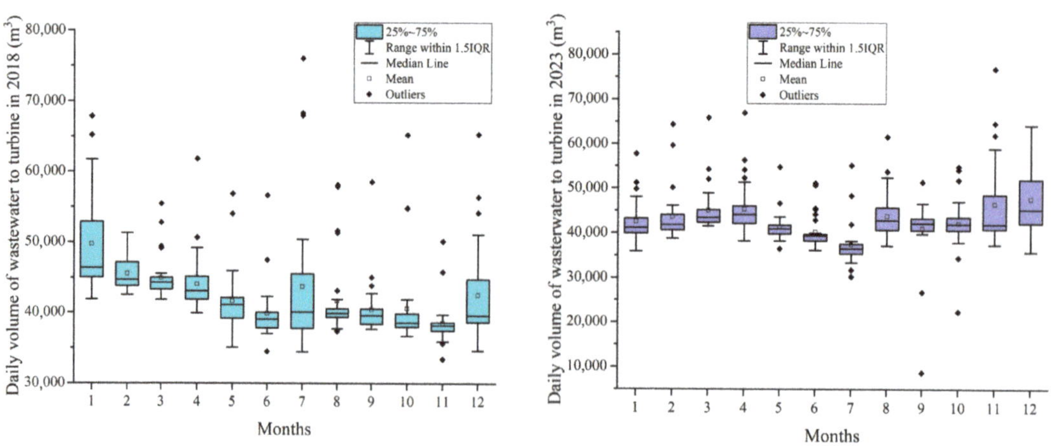

Figure 5. Daily volume of wastewater to the turbine in the hydropower plant at the wastewater treatment plant in Toruń: (**a**) in 2018, (**b**) in 2023.

3.2. Simplified Cost-Benefit Analysis

According to the information shown in Figure 6, the total estimated costs before the launch of the hydropower plant are about PLN 462.66 thousand in 2018 and PLN 806.61 thousand in 2023. This indicates an increase in total costs by 74.34%. The largest share is made up of costs related to installation (construction of the hydropower plant with accompanying infrastructure), i.e., 71.59% and 77.23% for 2018 and 2023. In second place were the costs of documentation (e.g., water law reports, water law permits, environmental impact assessments, building permits), i.e., 19.53% and 15.71%, and in last place were land lease costs (assumed as perpetual usufruct): 8.88% and 7.06%. The price increase is most visible, especially for the first position, i.e., 88.07%, while for documentation and lease costs, it was about 40%. It should be noted that the estimated costs of individual items may differ. The largest variable is the installed capacity of the hydropower plant, the type of hydropower plant (run-of-river, pump-storage, reservoir), the applied construction solutions (turbines, generators, gears, fish ladders, location in the created energy channel, etc.), and the location affecting the size of the costs. According to International Hydropower Association (IHA) data (2022) [57], in small run-of-river hydropower plants in Europe, the installation costs are distributed as follows: civil: 50% (construction works), mechanical: 30% (construction element), and electrical: 20% (related to connection to the power grid). According to the IHA report (2024) [58], the variability of averaged installation costs is also variable for different regions of the world. In the years 2016–2023, the lowest costs for small hydropower plants (<10 MW) were recorded in China (1829 USD/kW) and India (approximately 2000 USD/kW), while the highest were in Africa and Europe (approximately 3900–4000 USD/kW, respectively).

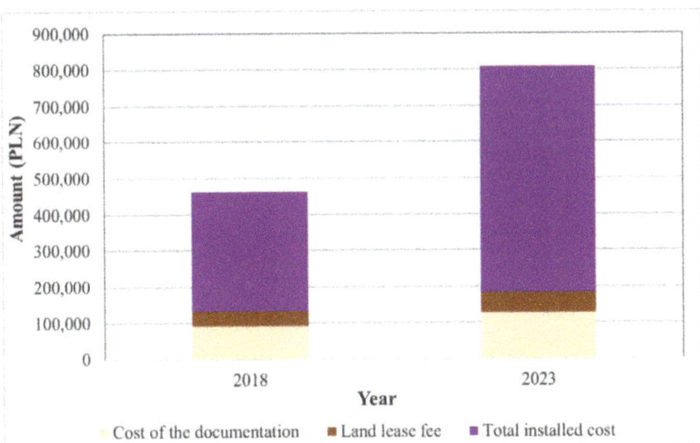

Figure 6. Comparison of estimated costs incurred before the construction of a hydropower plant at the wastewater treatment plant in Toruń in 2018 and 2023.

In the context of other renewable energy sources, currently the most cost-effective in terms of installation costs are photovoltaic panels, the price of which has fallen by 86% compared to 2023 with 2010 (758 USD/kW vs. 5310 USD/kW) [8]. Onshore and offshore wind energy is in second place (2023: 1160 and 2800 USD/kW), the prices of which have fallen by almost 50%. The highest price increases were recorded for hydropower and geothermal energy, 92% and 52%, respectively (costs: 2806 and 4589 USD/kWh). The most expensive installation costs are for concentrated solar energy (2023: 6589 USD/kWh). Bioenergy remains at a similar price level.

The first element of the balance of current revenues and costs is the value of gross electricity sales, assuming variants for different price rates (2018 and 2023), different types of entities (competitive market, households), and different volumes of electricity production in the hydropower plant at the WWTP in Toruń (2018 and 2023). The results of this analysis are presented in Figure 7. The lowest revenue was recorded for variant 3, i.e., electricity production from 2023 and settlement according to the rate for 2018 on the competitive market (PLN 19,367.52), while the highest was for variant 6 (production from 2018, settlement for households according to the rate of 2023), equal to PLN 132,830.57. The difference between these results is almost 7-fold. Analyzing the average production, rate, and settlement method, it can be seen that the greatest changes occurred for the rate. In 2023, it was higher by 120.53% than in 2018; different electricity production is of lesser importance. In 2018, it was 69.97% higher than in 2023, and the smallest different settlement method, i.e., 35.23% higher for households than in the competitive market.

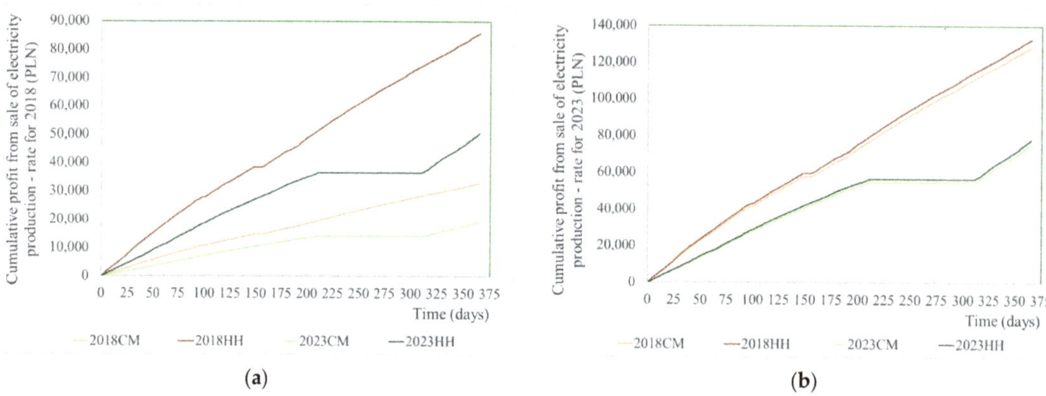

Figure 7. Cumulative profit from the sale of electricity production according to different rates: (**a**) from 2018, (**b**) from 2023 (CM—calculations according to the average price on the competitive market, HH—according to the average price for households, the year determines the amount of electricity produced in the hydropower plant at the wastewater treatment plant in Toruń).

Table 1 summarizes the current revenues and costs during the operation of the hydropower plant. In the context of the costs incurred, the largest share concerns the costs of electricity production, i.e., between 54.59% in variants 7 and 8 to 73.68% in variants 1 and 2. In second place were the costs related to maintenance (18.97%–37.33%), and in third place was the land use fee, which has been the same since 2018 (5.55%–10.54% of current costs). In the overall balance, this gives an annual loss in two variants (approximately PLN 11,000 in variants 1 and 3) and a profit in the remaining 6 (in the range from approximately PLN 20,000 in variant 4 to over PLN 75,000 in variant 6).

Only variants 2, 4, 5, 6, 7, and 8 were taken into account for further analyses because in cases 1 and 3, the estimated current costs exceed the profits, so the investment is unprofitable (average selling price of electricity on the competitive market, production as for 2018 and 2023). The estimated average costs of electricity production apply to hydropower plants worldwide in general, and they may vary depending on the type and size of the hydropower plant (especially the installed capacity). In reality, therefore, the balance could be positive in all variants. For example, in Europe, the cost of electricity production in small hydropower plants (<10 MW) for the period 2016–2023 ranged from 0.05 USD/kWh to 0.28 USD/kWh [8]. In the context of maintenance costs, run-of-river hydropower plants are assumed to be in the range of 1% to 4% of the installation costs, according to the IRENA study from 2018 [52].

Table 1. Balance of revenues and costs during the operation of the hydropower plant in Toruń for various variants, taking into account data on electricity production from 2018 and 2023.

<table>
<tr><th>Variant</th><th>Income from Electricity Production (PLN)</th><th>The Levelized Cost of Electricity-LCOE (PLN)</th><th>Operations and Maintenance Costs (PLN)</th><th>Land Use Fee-Water Law (PLN)</th><th>BALANCE (PLN)</th></tr>
<tr><td>1</td><td>32,919.62</td><td>32,157.20</td><td rowspan="4">8280.48</td><td rowspan="8">3204.00</td><td>−10,722.06</td></tr>
<tr><td>2</td><td>85,645.22</td><td>32,157.20</td><td>42,003.54</td></tr>
<tr><td>3</td><td>19,367.52</td><td>18,918.97</td><td>−11,035.93</td></tr>
<tr><td>4</td><td>50,387.46</td><td>18,918.97</td><td>19,984.01</td></tr>
<tr><td>5</td><td>128,644.03</td><td>38,985.09</td><td rowspan="4">15,573.10</td><td>70,881.84</td></tr>
<tr><td>6</td><td>132,830.57</td><td>38,985.09</td><td>75,068.38</td></tr>
<tr><td>7</td><td>75,684.86</td><td>22,936.01</td><td>33,971.75</td></tr>
<tr><td>8</td><td>78,147.91</td><td>22,936.01</td><td>36,434.8</td></tr>
</table>

According to Lewandowski's study from 2014 [56], who conducted a cost analysis for 89 run-of-river hydropower plants in Poland (which also includes the plant in the wastewater treatment plant in Toruń producing energy from flowing water), in the context of the incurred indirect costs of electricity production, in most cases, they result from taxes on hydrotechnical structures (98.88%), statutory fees for the use of land covered with flowing water (76.40%), fees related to perpetual usufruct of land (67.42%), and the provision of free services and participation in the costs of operation and maintenance of hydrotechnical facilities (57.30%). The first position is no longer valid in light of the law in force in Poland, while the remaining ones were included in the analysis of current costs.

Comparing the costs of electricity production for other renewable energy sources, currently hydropower and geothermal power have recorded the largest increases (change from 2023 to 2010: 33% and 31%, respectively) [8]. In the remaining RES, there were decreases, the highest in solar and wind power (from 63% to 90%). In the context of absolute values, the average costs in 2023 were the highest for concentrated solar power, offshore wind power, bioenergy, and geothermal power (respectively, 0.117, 0.075, 0.072, and 0.071 USD/kWh). The lowest costs were for onshore wind power, photovoltaic panels, and hydropower (respectively, 0.033, 0.044, and 0.057 USD/kWh).

3.3. Payback Period and Profitability of Investment

The analysis of the balance of revenues and costs (Figure 8) shows that the investment will pay off, depending on the variant, from 10.75 years after launch (variant 6) to 23.74 years (variant 7), ignoring price changes. The shortest payback period, therefore, concerns electricity production from 2018 at the average selling price of electricity for households in Poland in 2023; the longest was electricity production from 2023 at the average selling price of electricity on the competitive market in Poland in 2023. A significantly shorter payback period was achieved for 2018 production (10.75–11.38) compared to 2023 (22.13–23.74), so the settlement itself according to rates for different types of entities (competitive market: odd variants, households: even variants) and price increases (2018: variants 1–4, 2023: variants 5–6) are of lesser importance.

Assuming electricity production for the maximum achieved turbine power in 2018 and 2023 (36.75 kW), i.e., 321,930 kWh, this time could be shortened. The cost comparison for this theoretical example is presented in Table 2 (variant 9: electricity sales rate from 2018 on the competitive market in Poland, variant 10: electricity sales rate from 2018 for households, variant 11: sales rate from 2023 on the competitive market, variant 12: sales rate from 2023 for households). As can be seen, variant 9 is unprofitable, while for 10, 11, and 12, the payback period will be checked.

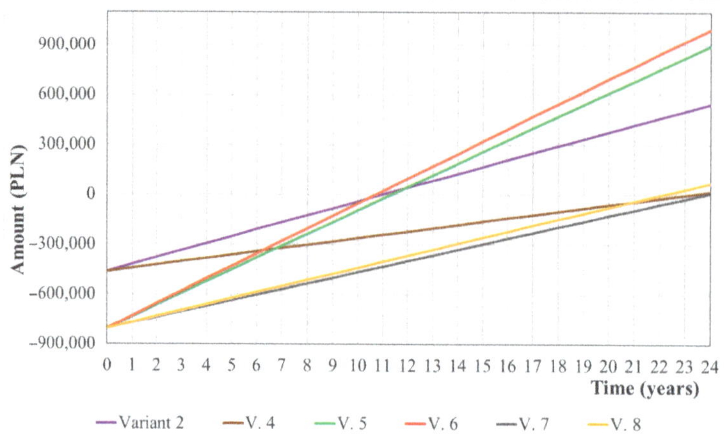

Figure 8. Temporal balance of profits and losses during the operation of the hydropower plant at the Toruń wastewater treatment plant according to different variants for data from 2018 and 2023.

Table 2. The balance of revenues and costs during the operation of the hydropower plant in Toruń for various theoretical variants assuming annual electricity production at the level of 321,930 kWh (constant turbine power of 36.75 kW throughout the year).

Variant	Income from Electricity Production (PLN)	The Levelized Cost of Electricity-LCOE (PLN)	Operations and Maintenance Costs (PLN)	Land Use Fee-Water Law (PLN)	BALANCE (PLN)
9	62,551.00	61,102.31	8280.48	3204.00	−10,035.79
10	162,735.62	61,102.31			90,148.83
11	244,438.23	74,076.09	15,573.10		151,585.04
12	252,393.12	74,076.09			159,539.93

With the new assumption, the investment payback period would be between 5.06 (variant 12) and 5.32 years (variant 11; Figure 9). With a higher wastewater inflow, it would be possible to operate with such power. Since the active power of the turbine is 55 kW, and the assumed value is 66.81% of this value, the potential electricity production and the investment payback period could be even more favorable. For Francis turbines, the efficiency is usually assumed to be between 80% and 95% [59], so the example given may occur in real conditions. Taking into account the remaining elements of the turbine set (gears, generators, and block transformers), this efficiency may vary between 73.7% and 92.6% [46]. On a global scale, for small hydropower plants (<10 MW), however, the average net capacity factor is lower, i.e., 52% [60], and varies between 30% and 75% [8], so the value adopted above is consistent with current knowledge. For 49 hydropower systems in WWTPs assessed for facilities in Switzerland, Japan, Spain, and Germany, among others, the capacity factor in the context of energy generation ranged from 15.5% to 52.7% [38]. Taking into account all analyzed variants, after 25 years of operation, the net revenue (balance) will amount to PLN 1.07 million for actual variants and PLN 3.18 million for theoretical variants (average PLN 0.468 million and PLN 2.65 million, respectively).

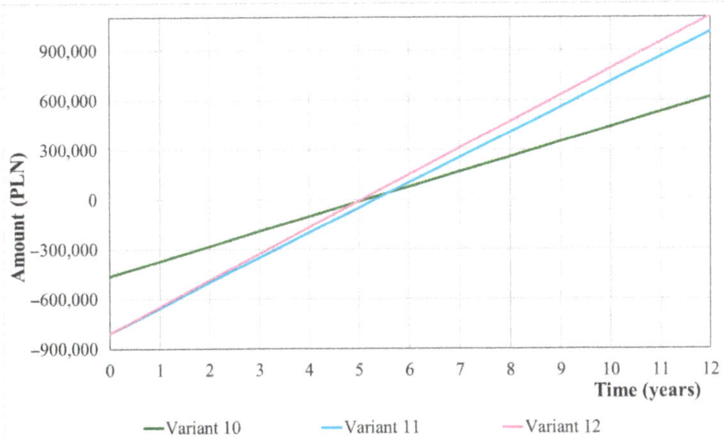

Figure 9. Temporal balance of profits and losses during the operation of the hydropower plant at the Toruń wastewater treatment plant according to different variants for the theoretical assumption of a constant turbine power of 36.75 kW.

According to the study by Capua et al., 2014 [61], in the context of research on turbines in hydropower plants in WWTPs, their profitability can be assessed by taking into account the average daily wastewater flow and the hydraulic drop corresponding to this flow. Using previously published data related to the analysis of the relationship between turbine power and hydraulic drop [26], the missing data for 2018 were determined, assuming the literature efficiency of the turbine equal to 80% and a flow occurring throughout the year (Q = 0.387 m^3/s); a hydraulic drop of 7.84 m was obtained, which corresponds to the average turbine power of 19.34 kW. In addition, the quoted average wastewater outflow per day for 2018, i.e., 42.742 m^3/d, was taken into account. According to Figure 10, developed based on the above study, the estimated payback period for the data, flow = 11.29 MGD (42,742 m^3/d) and head = 25.72 feet (7.84 m), is between 2 and 3 years, so according to the classification adopted therein, it is a profitable investment (payback less than 5 years). A payback period $\geq$ 8 years was assumed as unprofitable and 6–7 as potentially profitable; so, in the context of the payback periods calculated above, the profitable variants are 10, 11, and 12. Of course, the assumed payback period for the investment defined as profitable may be different. For example, Power et al. (2014) [36] specify a time of 10 years, and for the studied 25 hydropower plants in WWTPs in Ireland and Great Britain, the estimated payback period ranged from 5.1 to 19.7 years and from 3.1 to 23.6 years, respectively. In other studies, the payback period depended on the above-mentioned parameters (hydraulic head and flow) but also on the generated power and investment costs, and was, e.g., 7.4 years (head = 4.6 m, flow = 0.35 m^3/s, generated power = 15 kW, investment cost = 208,000 USD) [62] and 40 years (respectively, 10 m, 0.55 m^3/s, 13.5 kW, 60,000 USD) [63].

Looking at the current trends in the energy sector, the profitability of investments can be increased by using highly efficient, environmentally friendly solutions (e.g., Archimedes turbines) adapted to local conditions [64]. Unconventional solutions will also become increasingly important, e.g., hydrokinetic hydropower plants [65], combining different types of renewable energy sources (e.g., hybrid systems of hydropower plants with photovoltaic panels, wind-solar-hydro coupling) [66,67], storing surpluses of produced energy [68], solar-hydro cable pooling [69]. The advantage of using different types of energy sources within one installation may be their complementarity, i.e., mutual supplementation of the produced energy in different periods (e.g., in low flow conditions, but in sunny weather, hydropower plants will produce little electricity, and photovoltaic panels will be able to compensate for these deficiencies at the same time) [70]. Hydropower systems themselves (alone or in combination with other solutions) are considered a good way to recover energy in WWTPs, which is the subject of many studies, e.g., [71–73].

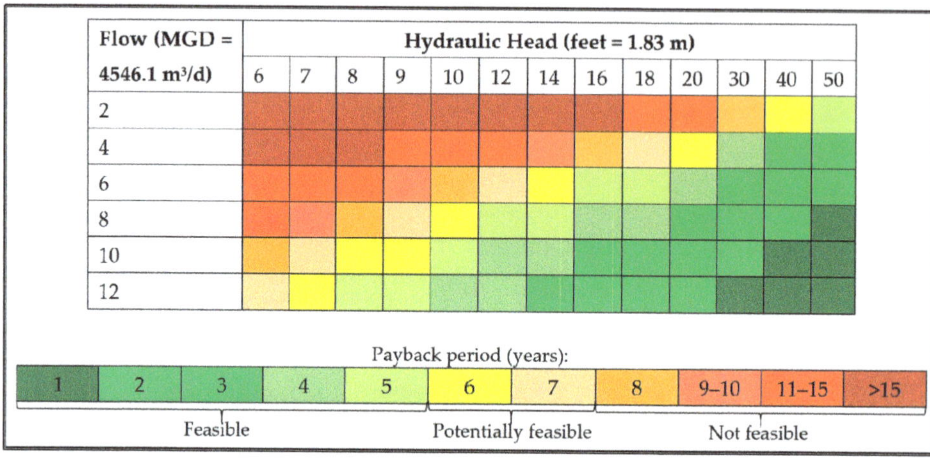

Figure 10. Profitability matrix of a hydropower plant depending on the average daily hydraulic drop and the average daily flow per turbine according to [59]; data for the hydropower plant in Central Wastewater Treatment Plant in Toruń: flow = 11.29 MGD, head = 25.72 feet.

4. Conclusions

The analyses carried out have shown that hydropower plants in wastewater treatment plants (WWTPs) can be a potential alternative solution in the context of increasing the share of renewable energy sources (RES) in electricity production, in line with the assumptions of the circular economy by using the locally available resource, which is treated wastewater flowing out of the WWTP, as exemplified by the WWTP in Toruń (Poland).

Twelve variants were examined, assuming different settlement of energy sales rates (households, competitive market), different sales prices (levels from 2018 and 2023), and different levels of electricity production (calculated for the Francis turbine power from 2018 and 2023, as well as for a theoretical example of the maximum recorded power for both analyzed years). The profit was assumed to be the sale of electricity and the costs of preparation of documentation, land lease, installation costs (pre-launch costs), electricity production costs, land use fee, operation, and modernization (current costs).

A total of 25% of the variants proved unprofitable due to costs exceeding revenues. For the remaining variants, the payback periods for the actual variants ranged from 10.75 to 23.74 years, while for the theoretical variants, it was between 5.06 and 5.32 years. The most significant factor in reducing the payback period was the level of electricity production between years. Taking into account all the profitable variants, after 25 years of operation, the net revenue will amount to PLN 1.07 million for the actual variants and PLN 3.18 million for the theoretical variants.

The profitability of the investment was also checked in terms of the average daily flow through the turbine and the corresponding hydraulic gradient, according to the literature. It was shown that in this approach, the payback period is about 2.5 years, and the operation of the hydropower plant in accordance with the proposed approach is profitable ($\leq$5 years).

This variant corresponds to 35.16% of the installed capacity, so the payback period could potentially be shorter.

As indicated, the profitability of the investment could be increased by using, for example, other types of technological solutions (e.g., other types of turbines: Kaplan, Archimedes, Pelton, propeller, etc.), connecting the existing hydropower plant with another RES (such as solar and wind energy), storing energy in generators and batteries or optimizing hydropower management. This will be the subject of future research.

The presented studies have their limitations because the actual conditions in the research facility could be different in the context of the electricity produced, the applied rates for billing and selling electricity, or the amount of costs incurred. For example, the cost of electricity production depends on many factors and may vary in real conditions. Similarly, it is difficult to predict price changes or variability in the amount of wastewater, which is why the potential economic balance may differ from the one presented over the years. Another limitation may be the fact that, in this case, Polish legislation was taken into account, so potential taxes or fees may differ for other countries. From a sustainable development perspective, in addition to economic issues, it would be worth considering environmental and social issues as well so that the conclusions are more cross-sectional.

In summary, the presented analysis can be one of the case studies in the context of the profitability of alternative solutions in hydropower, such as hydropower systems in WWTPs. It can also be the basis for developing recommendations and guidelines for the management of these facilities. The work is also important from the perspective of implemented energy and water policies related to sustainable development goals, adaptation to climate change, and the circular economy.

Future studies may focus on a more detailed analysis of profits and losses in other such water and wastewater management facilities in Poland, which operate in Kraków and Szczecin and have greater power. In the case of Kraków, a different type of turbine (Kaplan instead of Francis), and in Szczecin, a different arrangement of two turbines instead of one, which uses water from the water treatment plant. It would also be worth considering other analytical methods, using mathematical and statistical models or indexes used in economics or SWOT analyses. In addition, it would be possible to assume greater variability in the context of prediction, using different forecasting scenarios based on the literature on the subject.

Author Contributions: Conceptualization, K.M., P.T. and M.W.; methodology, K.M., P.T., K.K., W.O. and M.W.; software, K.M. and P.T.; validation, K.M., P.T. and M.W; formal analysis, K.M., P.T., K.K., W.O. and M.W.; investigation, K.M., P.T. and M.W.; resources, K.M., P.T. and M.W.; data curation, K.M. and P.T.; writing—original draft preparation, P.T.; writing—review and editing, K.K., W.O. and M.W.; visualization, K.M. and P.T.; supervision, M.W. All authors have read and agreed to the published version of the manuscript.

Funding: This research received no external funding.

Data Availability Statement: The data presented in this study are available on request from the corresponding author.

Acknowledgments: We would like to thank Toruńskie Wodociągi Sp. z o.o. for providing data on technologies used in the WWTP in Toruń (including those related to the hydropower plant), wastewater volume and turbine output power in 2018 and 2023. Paweł Tomczyk is grateful for the Foundation for Polish Science's (FNP) support through funding from the START scholarship.

Conflicts of Interest: Author Krzysztof Mastalerek was employed by the company ENERIS, Koszykowa 65, 00-667 Warsaw, Poland. The remaining authors declare that the research was conducted in the absence of any commercial or financial relationships that could be construed as a potential conflict of interest.

Abbreviations

CV	Coefficient of Variation
EUR	Euro
IHA	International Hydropower Association
IRENA	International Renewable Energy Agency
PLN	Polish Zloty
RES	Renewable Energy Source
SDG	Sustainable Development Goal
URE	Energy Regulatory Office in Poland (Urząd Regulacji Energetyki)
USD	The United States Dollar
WWTP	Wastewater treatment plant

References

1. Aziz, S.; Ahmed, I.; Khan, K.; Khalid, M. Emerging Trends and Approaches for Designing Net-Zero Low-Carbon Integrated Energy Networks: A Review of Current Practices. *Arab. J. Sci. Eng.* **2024**, *49*, 6163–6185. [CrossRef]
2. Ortega, A.; Gkoumas, K.; Tsakalidis, A.; Pekár, F. Low-Emission Alternative Energy for Transport in the EU: State of Play of Research and Innovation. *Energies* **2021**, *14*, 7764. [CrossRef]
3. Yang, M.; Chen, L.; Wang, J.; Msigwa, G.; Osman, A.I.; Fawzy, S.; Rooney, D.W.; Yap, P.S. Circular economy strategies for combating climate change and other environmental issues. *Environ. Chem. Lett.* **2022**, *21*, 55–80. [CrossRef]
4. Suroso, D.S.A.; Setiawan, B.; Pradono, P.; Iskandar, Z.S.; Hastari, M.A. Revisiting the role of international climate finance (ICF) towards achieving the nationally determined contribution (NDC) target: A case study of the Indonesian energy sector. *Environ. Sci. Policy* **2022**, *131*, 188–195. [CrossRef]
5. IEA—International Energy Agency. *Renewables 2024. Analysis and Forecast to 2030*; IEA Publications: Paris, France, 2024.
6. Iskandarova, M.; Dembek, A.; Fraaije, M.; Matthews, W.; Stasik, A.; Wittmayer, J.M.; Sovacool, B.K. Who finances renewable energy in Europe? Examining temporality, authority and contestation in solar and wind subsidies in Poland, the Netherlands and the United Kingdom. *Energy Strategy Rev.* **2021**, *38*, 100730. [CrossRef]
7. Hassan, Q.; Algburi, S.; Sameen, A.Z.; Salman, H.M.; Jaszczur, M. A review of hybrid renewable energy systems: Solar and wind-powered solutions: Challenges, opportunities, and policy implications. *Results Eng.* **2023**, *20*, 101621. [CrossRef]
8. IRENA—International Renewable Energy Agency. *Renewable Power Generation Costs in 2023*; International Renewable Energy Agency: Abu Dhabi, United Arab Emirates, 2024.
9. Schlömer, S.; Bruckner, T.; Fulton, L.; Hertwich, E.; McKinnon, A.; Perczyk, D.; Roy, J.; Schaeffer, R.; Sims, R.; Smith, P.; et al. 2014: Annex III: Technology-specific cost and performance parameters. In *Climate Change 2014: Mitigation of Climate Change. Contribution of Working Group III to the Fifth Assessment Report of the Intergovernmental Panel on Climate Change*; Edenhofer, O., Pichs-Madruga, R., Sokona, Y., Farahani, E., Kadner, S., Seyboth, K., Adler, A., Baum, I., Brunner, S., Eickemeier, P., et al., Eds.; Cambridge University Press: Cambridge, UK; New York, NY, USA, 2015.
10. Bäckstrand, K. Towards a Climate-Neutral Union by 2050? The European Green Deal, Climate Law, and Green Recovery. In *Routes to a Resilient European Union: Interdisciplinary European Studies*; Palgrave Macmillan: Cham, Switzerland, 2022; pp. 39–61. [CrossRef]
11. Levasseur, A.; Mercier-Blais, S.; Prairie, Y.T.; Tremblay, A.; Turpin, C. Improving the accuracy of electricity carbon footprint: Estimation of hydroelectric reservoir greenhouse gas emissions. *Renew. Sustain. Energy Rev.* **2021**, *136*, 110433. [CrossRef]
12. Fetting, C. *The European Green Deal*; ESDN Report; ESDN Office: Vienna, Austria, 2020.
13. Büyüközkan, G.; Karabulut, Y.; Mukul, E. A novel renewable energy selection model for United Nations' sustainable development goals. *Energy* **2018**, *165*, 290–302. [CrossRef]
14. Hirth, L. The benefits of flexibility: The value of wind energy with hydropower. *Appl. Energy* **2016**, *181*, 210–223. [CrossRef]
15. Kumar, S.; Vardhan, A.S.S.; Vardhan, A.S.S.; Saket, R.K.; Kothari, D.P.; Eslamian, S. Hydropower and Floods. In *Flood Handbook*; CRC Press: Boca Raton, FL, USA, 2022; pp. 111–142. [CrossRef]
16. Tomczyk, P.; Wiatkowski, M. Shaping changes in the ecological status of watercourses within barrages with hydropower schemes—Literature review. *Arch. Environ. Prot.* **2020**, *46*, 78–94. [CrossRef]
17. Ember. Per Capita Electricity Generation from Hydropower. 2023. Available online: https://ourworldindata.org/grapher/hydro-electricity-per-capita (accessed on 11 December 2024).
18. Bousquet, C.; Samora, I.; Manso, P.; Rossi, L.; Heller, P.; Schleiss, A.J. Assessment of hydropower potential in wastewater systems and application to Switzerland. *Renew. Energy* **2017**, *113*, 64–73. [CrossRef]

19. Hauer, C.; Wagner, B.; Aigner, J.; Holzapfel, P.; Flödl, P.; Liedermann, M.; Tritthart, M.; Sindelar, C.; Pulg, U.; Klösch, M.; et al. State of the art, shortcomings and future challenges for a sustainable sediment management in hydropower: A review. *Renew. Sustain. Energy Rev.* **2018**, *98*, 40–55. [CrossRef]
20. Kuriqi, A.; Pinheiro, A.N.; Sordo-Ward, A.; Bejarano, M.D.; Garrote, L. Ecological impacts of run-of-river hydropower plants—Current status and future prospects on the brink of energy transition. *Renew. Sustain. Energy Rev.* **2021**, *142*, 110833. [CrossRef]
21. García, A.M.; Díaz JA, R.; Morillo, J.G.; McNabola, A. Energy Recovery Potential in Industrial and Municipal Wastewater Networks Using Micro-Hydropower in Spain. *Water* **2021**, *13*, 691. [CrossRef]
22. Yüksel, I. Hydropower for sustainable water and energy development. *Renew. Sustain. Energy Rev.* **2010**, *14*, 462–469. [CrossRef]
23. Jucherski, A.; Walczowski, A.; Bugajski, P.; Operacz, A. Reliability of organic and biogenic pollutant removal in selected technologies used in domestic wastewater treatment plants: A comparative analysis. *J. Environ. Manag.* **2024**, *354*, 120381. [CrossRef] [PubMed]
24. Wdowczyk, A.; Szymańska-Pulikowska, A. Micro- and Macroelements Content of Plants Used for Landfill Leachate Treatment Based on *Phragmites australis* and *Ceratophyllum demersum*. *Int. J. Environ. Res. Public Health* **2022**, *19*, 6035. [CrossRef]
25. Wdowczyk, A.; Szymańska-Pulikowska, A. Effect of substrates on the potential of *Phragmites australis* to accumulate and translocate selected contaminants from landfill leachate. *Water Resour. Ind.* **2023**, *29*, 100203. [CrossRef]
26. Tomczyk, P.; Mastalerek, K.; Wiatkowski, M.; Kuriqi, A.; Jurasz, J. Assessment of a Francis Micro Hydro Turbine Performance Installed in a Wastewater Treatment Plant. *Energies* **2023**, *16*, 7214. [CrossRef]
27. IHA–International Hydropower Association. *2022 Hydropower Status Report: Sector Trends and Insights*; International Hydropower Association: London, UK, 2022.
28. Daneshgar, S.; Zahedi, R. Investigating the hydropower plants production and profitability using system dynamics approach. *J. Energy Storage* **2022**, *46*, 103919. [CrossRef]
29. Tajziehchi, S.; Karbassi, A.; Nabi, G.; Yoo, C.; Ifaei, P. A Cost-Benefit Analysis of Bakhtiari Hydropower Dam Considering the Nexus between Energy and Water. *Energies* **2022**, *15*, 871. [CrossRef]
30. Kałuża, T.; Hämmerling, M.; Zawadzki, P.; Czekała, W.; Kasperek, R.; Sojka, M.; Mokwa, M.; Ptak, M.; Szkudlarek, A.; Czechlowski, M.; et al. The hydropower sector in Poland: Barriers and the outlook for the future. *Renew. Sustain. Energy Rev.* **2022**, *163*, 112500. [CrossRef]
31. Kougias, I.; Aggidis, G.; Avellan, F.; Deniz, S.; Lundin, U.; Moro, A.; Muntean, S.; Novara, D.; Pérez-Díaz, J.I.; Quaranta, E.; et al. Analysis of emerging technologies in the hydropower sector. *Renew. Sustain. Energy Rev.* **2019**, *113*, 109257. [CrossRef]
32. Sasthav, C.; Oladosu, G. Environmental design of low-head run-of-river hydropower in the United States: A review of facility design models. *Renew. Sustain. Energy Rev.* **2022**, *160*, 112312. [CrossRef]
33. Singh, V.K.; Singal, S.K. Operation of hydro power plants-a review. *Renew. Sustain. Energy Rev.* **2017**, *69*, 610–619. [CrossRef]
34. IRENA—International Renewable Energy Agency. *Renewable Energy Technologies: Cost Analysis Series: Hydropower*; International Renewable Energy Agency: Bonn, Germany, 2012; Volume 1.
35. Ak, M.; Kentel, E.; Kucukali, S. A fuzzy logic tool to evaluate low-head hydropower technologies at the outlet of wastewater treatment plants. *Renew. Sustain. Energy Rev.* **2017**, *68*, 727–737. [CrossRef]
36. Power, C.; McNabola, A.; Coughlan, P. Development of an evaluation method for hydropower energy recovery in wastewater treatment plants: Case studies in Ireland and the UK. *Sustain. Energy Technol. Assess.* **2014**, *7*, 166–177. [CrossRef]
37. Sovacool, B.K.; Gilbert, A.; Nugent, D. An international comparative assessment of construction cost overruns for electricity infrastructure. *Energy Res. Soc. Sci.* **2014**, *3*, 152–160. [CrossRef]
38. Llácer-Iglesias, R.M.; López-Jiménez, P.A.; Pérez-Sánchez, M. Hydropower Technology for Sustainable Energy Generation in Wastewater Systems: Learning from the Experience. *Water* **2021**, *13*, 3259. [CrossRef]
39. ZWiK Szczecin. Przepis na Czystą Energię z Biogazu, Słońca i Wody. 2020. Available online: https://zwik.szczecin.pl/firma/wydarzenia/przepis-na-czyst-energi-z-biogazu-soca-i-wody?fbclid=IwAR36fvSutDCp9Enio6RzvV0uVr7Y2avZ_UZ_DRMo9pfTeimMJCyxMKf2Z8c (accessed on 11 December 2024).
40. Biedrzycka, A. Mikroelektrownia Wodna w Oczyszczalni Płaszów. *Nowoczesne Budownictwo Inżynieryjne*, March–April, 18–21. Available online: https://nbi.com.pl/content/uploads/assets/NBI-pdf/2017/2_71_2017/pdf/3_Mikroelektrownia_wodna_w_oczyszczalni_Plaszow.pdf (accessed on 11 December 2024).
41. Bekker, A.; van Dijk, M.; Niebuhr, C.M. A review of low head hydropower at wastewater treatment works and development of an evaluation framework for South Africa. *Renew. Sustain. Energy Rev.* **2022**, *159*, 112216. [CrossRef]
42. Llácer-Iglesias, R.M.; López-Jiménez, P.A.; Pérez-Sánchez, M. Exploring options for energy recovery from wastewater: Evaluation of hydropower potential in a sustainability framework. *Sustain. Cities Soc.* **2023**, *95*, 104576. [CrossRef]
43. Breza-Boruta, B.; Paluszak, Z. Influence of Water Treatment Plant on Microbiological Composition of Air Bioaerosol. *Pol. J. Environ. Stud.* **2007**, *16*, 663–670.

44. Paluszak, Z.; Ligocka, A.; Breza-Boruta, B. Effectiveness of Sewage Treatment Based on Selected Faecal Bacteria Elimination in Municipal Wastewater Treatment Plant in Toruń. *Pol. J. Environ. Stud.* **2003**, *12*, 345–349.
45. Khan, M.A.; Badshah, S. Research Article Design and Analysis of Cross Flow Turbine for Micro Hydro Power Application using Sewerage Water. *Res. J. Appl. Sci. Eng. Technol.* **2014**, *8*, 821–828. [CrossRef]
46. Adhikari, R.; Wood, D. The Design of High Efficiency Crossflow Hydro Turbines: A Review and Extension. *Energies* **2018**, *11*, 267. [CrossRef]
47. Wiatkowski, M.; Gruss, Ł.; Tomczyk, P. Hydrological and environmental analysis of the location of a small hydropower plant on the Budkowiczanka River in the village of Krzywa Góra. *Ecol. Eng.* **2018**, *19*, 103–113. [CrossRef]
48. URE—Energy Regulatory Office in Poland. *Informacja Prezesa Urzędu Regulacji Energetyki (nr 20/2019) w Sprawie Średniej ceny Sprzedaży Energii Elektrycznej na Rynku Konkurencyjnym za rok 2018*; URE: Warsaw, Poland, 2019.
49. URE—Energy Regulatory Office in Poland. *Informacja Prezesa URE (nr 23/2019) w spr. Średniej ceny Energii Elektrycznej dla Odbiorcy w Gospodarstwie Domowym Uwzględniającej Opłatę za Świadczenie Usługi Dystrybucji Energii Elektrycznej, Obliczonej na Podst. Cen Zawartych w Umowach Kompleksowych w 2018 r*; URE: Warsaw, Poland, 2019.
50. URE—Energy Regulatory Office in Poland. *Informacja Prezesa URE nr 19/2024 w Sprawie Średniej ceny Sprzedaży Energii Elektrycznej na Rynku Konkurencyjnym za rok 2023*; URE: Warsaw, Poland, 2024.
51. URE—Energy Regulatory Office in Poland. *Informacja Prezesa URE nr 20/2024 w Sprawie Średniej ceny Energii Elektrycznej dla Odbiorcy w Gospodarstwie Domowym Uwzględniającej Opłatę za Świadczenie Usługi Dystrybucji Energii Elektrycznej, Obliczonej na Podstawie cen Zawartych w Umowach Kompleksowych w roku 2023*; URE: Warsaw, Poland, 2024.
52. IRENA—International Renewable Energy Agency. *Renewable Generation Costs in 2018*; International Renewable Energy Agency: Abu Dhabi, United Arab Emirates, 2019.
53. Mazurek-Czarnecka, A.; Rosiek, K.; Salamaga, M.; Wąsowicz, K.; Żaba-Nieroda, R. Study on Support Mechanisms for Renewable Energy Sources in Poland. *Energies* **2022**, *15*, 4196. [CrossRef]
54. RM—Council of Ministers in Poland. Rozporządzenie Rady Ministrów z dnia 18 grudnia 2017 r. w sprawie wysokości stawek opłaty rocznej za użytkowanie gruntów pokrytych wodami. *J. Laws Repub. Pol.* **2017**. Position 2496 (Dz.U. 2017 poz. 2496). Available online: https://isap.sejm.gov.pl/isap.nsf/download.xsp/WDU20170002496/O/D20172496.pdf (accessed on 11 December 2024).
55. UNGC—UN Global Compact Network Poland. *Małe Elektrownie Wodne w Polsce*; UNGC: Warsaw, Poland, 2022.
56. Lewandowski, S. *Analiza Kosztów Związanych Pośrednią z Produkcją Energii Elektrycznej w Elektrowniach Wodnych*; Towarzystwo Elektrowni Wodnych: Reda, Poland, 2014.
57. IHA—International Hydropower Association. Sector trend and insights. In *2022 Hydropower Status Report*; IHA: London, UK, 2023.
58. IHA—International Hydropower Association. Opportunities to advance net zero. In *2024 World Hydropower Outlook*; IHA: London, UK, 2024.
59. Tiwari, G.; Kumar, J.; Prasad, V.; Patel, V.K. Derivation of cavitation characteristics of a 3MW prototype Francis turbine through numerical hydrodynamic analysis. *Mater. Today Proc.* **2020**, *26*, 1439–1448. [CrossRef]
60. Turner, S.W.D.; Ghimire, G.R.; Hansen, C.; Singh, D.; Kao, S.C. Hydropower capacity factors trending down in the United States. *Nat. Commun.* **2024**, *15*, 5445. [CrossRef] [PubMed]
61. Capua, M.R.; Dzwonkoski, J.; Harris, C.D. *Reclamation of Power in Wastewater Treatment Facilities*; Worcester Polytechnic Institute: Millbury, MA, USA, 2014.
62. Berger, V.; Niemann, A.; Frehmann, T.; Brockmann, H. Advanced energy recovery strategies for wastewater treatment plants and sewer systems using small hydropower. *Water Util. J.* **2013**, *5*, 15–24.
63. Islam, A.K. Hydropower coupled with hydrogen production from wastewater: Integration of micro-hydropower plant (MHP) and microbial electrolysis cell (MEC). *Int. J. Hydrogen Energy* **2024**, *49*, 1–14. [CrossRef]
64. Lavrič, H.; Rihar, A.; Fišer, R. Influence of equipment size and installation height on electricity production in an Archimedes screw-based ultra-low head small hydropower plant and its economic feasibility. *Renew. Energy* **2019**, *142*, 468–477. [CrossRef]
65. Holanda, P.d.S.; Blanco, C.J.C.; Mesquita, A.L.A.; Brasil Junior, A.C.P.; de Figueiredo, N.M.; Macêdo, E.N.; Secretan, Y. Assessment of hydrokinetic energy resources downstream of hydropower plants. *Renew. Energy* **2017**, *101*, 1203–1214. [CrossRef]
66. Chen, C.; Liu, H.; Xiao, Y.; Zhu, F.; Ding, L.; Yang, F. Power Generation Scheduling for a Hydro-Wind-Solar Hybrid System: A Systematic Survey and Prospect. *Energies* **2022**, *15*, 8747. [CrossRef]
67. Zhao, C.; Wang, F.; Xu, J.; Tan, C.; Østergaard, P.A. Optimal planning and operation for a grid-connected solar–wind–hydro energy system in wastewater treatment plants. *Renew. Energy* **2024**, *230*, 120791. [CrossRef]
68. Javed, M.S.; Zhong, D.; Ma, T.; Song, A.; Ahmed, S. Hybrid pumped hydro and battery storage for renewable energy based power supply system. *Appl. Energy* **2020**, *257*, 114026. [CrossRef]
69. Jurasz, J.; Tomczyk, P.; Bochenek, B.; Kuriqi, A.; Kasiulis, E.; Chen, D.; Ming, B. Solar-hydro cable pooling—Utilizing the untapped potential of existing grid infrastructure. *Energy Convers. Manag.* **2024**, *306*, 118307. [CrossRef]

70. François, B.; Borga, M.; Creutin, J.D.; Hingray, B.; Raynaud, D.; Sauterleute, J.F. Complementarity between solar and hydro power: Sensitivity study to climate characteristics in Northern-Italy. *Renew. Energy* **2016**, *86*, 543–553. [CrossRef]
71. Sinagra, M.; Picone, C.; Picone, P.; Aricò, C.; Tucciarelli, T.; Ramos, H.M. Low-Head Hydropower for Energy Recovery in Wastewater Systems. *Water* **2022**, *14*, 1649. [CrossRef]
72. Punys, P.; Jurevičius, L. Assessment of Hydropower Potential in Wastewater Systems and Application in a Lowland Country, Lithuania. *Energies* **2022**, *15*, 5173. [CrossRef]
73. Gallagher, J.; Harris, I.M.; Packwood, A.J.; McNabola, A.; Williams, A.P. A strategic assessment of micro-hydropower in the UK and Irish water industry: Identifying technical and economic constraints. *Renew. Energy* **2015**, *81*, 808–815. [CrossRef]

MDPI AG
Grosspeteranlage 5
4052 Basel
Switzerland
Tel.: +41 61 683 77 34

Energies Editorial Office
E-mail: energies@mdpi.com
www.mdpi.com/journal/energies

www.ingramcontent.com/pod-product-compliance
Lightning Source LLC
Chambersburg PA
CBHW061841210926
51850CB00006B/941

* 9 7 8 3 7 2 5 8 6 4 5 6 0 *